THE ECONOMETRICS OF MA

GW01465426

G

Other Advanced Texts in Econometrics

ARCH: Selected Readings
Edited by Robert F. Engle

Asymptotic Theory for Integrated Processes
By H. Peter Boswijk

Bayesian Inference in Dynamic Econometric Models
By Luc Bauwens, Michel Lubrano, and Jean-François Richard

Co-integration, Error Correction, and the Econometric Analysis of Non-Stationary Data
By Anindya Banerjee, Juan J. Dolado, John W. Galbraith, and David Hendry

Dynamic Econometrics
By David F. Hendry

Finite Sample Econometrics
By Aman Ullah

Generalized Method of Moments
By Alastair Hall

Likelihood-Based Inference in Cointegrated Vector Autoregressive Models
By Søren Johansen

Long-Run Econometric Relationships: Readings in Cointegration
Edited by R. F. Engle and C. W. J. Granger

Micro-Econometrics for Policy, Program, and Treatment Effect
By Myoung-jae Lee

Modelling Economic Series: Readings in Econometric Methodology
Edited by C. W. J. Granger

Modelling Non-Linear Economic Relationships
By Clive W. J. Granger and Timo Teräsvirta

Modelling Seasonality
Edited by S. Hylleberg

Non-Stationary Times Series Analysis and Cointegration
Edited by Colin P. Hargeaves

Outlier Robust Analysis of Economic Time Series
By André Lucas, Philip Hans Franses, and Dick van Dijk

Panel Data Econometrics
By Manuel Arellano

Periodicity and Stochastic Trends in Economic Time Series
By Philip Hans Franses

Progressive Modelling: Non-nested Testing and Encompassing
Edited by Massimiliano Marcellino and Grayham E. Mizon

Readings in Unobserved Components
Edited by Andrew Harvey and Tommaso Proietti

Stochastic Limit Theory: An Introduction for Econometricians
By James Davidson

Stochastic Volatility
Edited by Neil Shephard

Testing Exogeneity
Edited by Neil R. Ericsson and John S. Irons

The Econometrics of Macroeconomic Modelling
By Gunnar Bårdsen, Øyvind Eitrheim, Eilev S. Jansen, and Ragnar Nymoen

Time Series with Long Memory
Edited by Peter M. Robinson

Time-Series-Based Econometrics: Unit Roots and Co-integrations
By Michio Hatanaka

Workbook on Cointegration
By Peter Reinhard Hansen and Søren Johansen

The Econometrics of Macroeconomic Modelling

GUNNAR BÅRDSEN
ØYVIND EITRHEIM
EILEV S. JANSEN
AND
RAGNAR NYMOEN

OXFORD
UNIVERSITY PRESS

OXFORD
UNIVERSITY PRESS

Great Clarendon Street, Oxford OX2 6DP

Oxford University Press is a department of the University of Oxford.
It furthers the University's objective of excellence in research, scholarship,
and education by publishing worldwide in

Oxford New York

Auckland Cape Town Dar es Salaam Hong Kong Karachi
Kuala Lumpur Madrid Melbourne Mexico City Nairobi
New Delhi Shanghai Taipei Toronto

With offices in

Argentina Austria Brazil Chile Czech Republic France Greece
Guatemala Hungary Italy Japan Poland Portugal Singapore
South Korea Switzerland Thailand Turkey Ukraine Vietnam

Oxford is a registered trade mark of Oxford University Press
in the UK and in certain other countries

Published in the United States
by Oxford University Press Inc., New York

British Library Cataloguing in Publication Data
Data available

Library of Congress Cataloging in Publication Data
Data available

Typeset by Newgen Imaging Systems (P) Ltd., Chennai, India
Printed in Great Britain
on acid-free paper by
Biddles Ltd., King's Lynn

ISBN 0-19-924649-1   978-0-19-9246496
ISBN 0-19-924650-5   978-0-19-9246502

1 3 5 7 9 10 8 6 4 2

Preface

At the European Meeting of the Econometric Society in Santiago de Compostela in September 1999, Clive Granger asked if we would like to write a book for the *Advanced Texts in Econometrics* series about the approach to macroeconometric modelling we had adopted at the Research Department of Norges Bank over the past 15 years. It has taken us 5 years to comply with his request, and the result is found within these covers.

This book is about building models by testing hypotheses of macroeconomic theories–rather than by imposing theories untested. This is quite a crucial distinction in macroeconometric model building. For an empirical model to be useful, be it as a basis for economic policy decisions or for forecasting, it needs to describe the relevant aspects of reality. Simplification is the main virtue of theoretical model building. In empirical modelling it might easily become a vice. A theoretical model is often reduced to just those equations that are required to make it work for the problem at hand. A good empirical model should also be able to explain problems that might occur. Einstein's advice that 'everything should be as simple as possible...but no simpler' is as relevant as ever. If a model does not describe the data, it may just be too simple to be used as a tool for macroeconomic decision making.

The main target group for the book is researchers and practitioners of macroeconomic model building in academia, private agencies and governmental services. As a textbook it can be used in graduate courses on applied macroeconometrics in general and—more specifically—in courses focusing on wage and price formation in the open economy. In that context it is obvious that a companion text on econometric methods and practice will be useful, and we recommend *Dynamic Econometrics* by David F. Hendry (Hendry 1995a) and *Empirical Modeling of Economic Time Series* by Neil R. Ericsson (Ericsson 2005) for this purpose.

The work on the book has formed a joint research agenda for the authors since its conception. Hence, we draw extensively on our published papers, many of which was written with the demands of this book in mind: Section 1.4 and Chapter 2 are based on Jansen (2002); Sections 5.6 and 6.7.2 on Bårdsen *et al.* (1998); Sections 6.1–6.3 on Kolsrud and Nymoen (1998) and Bårdsen and Nymoen (2003); Section 6.8 on Holden and Nymoen (2002) and Nymoen and Rødseth (2003); Chapter 7 on Bårdsen *et al.* (2004), Section 8.4 on Eitrheim

(1998); Chapter 9 on Bårdsen *et al.* (2003); Section 11.2 on Eitrheim *et al.* (1999, 2002*a*) and Section 11.3 on Bårdsen *et al.* (2002*a*).

Also, we have used material from unpublished joint work with other authors. In particular we would like to thank Q. Farooq Akram, Neil R. Ericsson and Neva A. Kerbeshian for their permission to do so: Akram *et al.* (2003) underlies Chapter 10 and we draw on Ericsson et al. (1997) in Section 4.4.

The views are those of the authors and should not be interpreted to reflect those of their respective institutions. Throughout the book our main econometric tools have been the programs developed by Jurgen A. Doornik, David F. Hendry and Hans-Martin Krolzig, i.e., the *Oxmetrics* package (provided by Timberlake Consultants), in particular *PcGive*, *PcFIML* and *PcGets*. In Chapter 7 and Sections 9.5 and 10.3 we have used *Eviews* (provided by Quantitative Micro Software) and the simulations in Section 11.2.2 are carried out with *TROLL* (provided by Intex Solutions).

Data documentation, data series, programs and detailed information about the software used are available from a homepage for the book:

http://www.svt.ntnu.no/iso/macroectrics.

We are indebted to many colleagues and friends for comments, discussions and critisism to the various parts of the book. The editors of the series—Clive W. J. Granger and Grayham E. Mizon—have given us advice and constant encouragement. David F. Hendry and Bjørn E. Naug have read the entire manuscript and given us extensive, constructive and very helpful comments. In addition to those already acknowledged, grateful thanks goes to: Q. Farooq Akram, Olav Bjerkholt, Neil R. Ericsson, Paul G. Fisher, Roger Hammersland, Steinar Holden, Tore Anders Husebø, Kåre Johansen, Søren Johansen, Adrian Pagan, Asbjørn Rødseth, Timo Teräsvirta, Anders Vredin, Kenneth F. Wallis, and Fredrik Wulfsberg. Last, but not least, we are indebted to Jurgen A. Doornik for his generosity with both time, patience, and effort throughout the project.

While working on the book Gunnar Bårdsen has visited the School of Economics and Finance, Queensland University of Technology (November 2000–January 2001) and Department of Economics, University of California San Diego (March 2003), and Eilev S. Jansen has been a visitor at Department of Economics, University of Oslo (August 2001–January 2003), DG Research, European Central Bank, Frankfurt (February 2003–June 2003) and Department of Economics, University of California San Diego (August 2003–July 2004). The hospitality and excellent working conditions offered at those institutions are gratefully acknowledged.

Finally, we are grateful to our respective employers—Norges Bank, Norwegian University of Science and Technology, and University of Oslo—for allocating resources and time for this project. That said, the time spent on the book has often gone beyond normal hours, which is but one reason why this book is dedicated to our wonderful and wise wives.

Trondheim/Oslo, November 2004
Gunnar Bårdsen, Øyvind Eitrheim, Eilev S. Jansen and Ragnar Nymoen

Contents

List of Figures

List of Tables

List of Abbreviations

2SLS	two-stage least squares
AR	autoregressive process
ARCH	autoregressive conditional heteroscedasticity
ARIMA	autoregressive integrated moving-average process
ARMA	autoregressive moving-average process
AWM	Area Wide Model
AWSU	average wage-share rate of unemployment
B&N	Brodin and Nymoen (1992)
CF	consumption function
CIRU	constant rate of inflation rate of unemployment
CPI	consumer price index
DGP	data generating process
DSGE	dynamic stochastic general equilibrium
dVAR	vector autoregressive model in differences
EE	Euler equation
EqCM	equilibrium-correction model
FIML	full information maximum likelihood
GDP	gross domestic product
GG	Galí and Gertler (1999)
GGL	Galí, Gertler, and López-Salido (2001)
GMM	generalised method of moments
GUM	general unrestricted model
HP	Hodrick-Prescott (filter)
ICM	Incomplete Competition Model
LIML	limited information maximum liklihood
MMSFE	minimum mean squared forecast error
MSFE	mean squared forecast error
NAIRU	non-accelerating inflation rate of unemployment
NAWRU	non-accelerating wage rate of unemployment,

NPC	New Keynesian Phillips curve
NPCM	New Keynesian Phillips curve model
OLS	ordinary least squares
PCM	Phillips curve model
pGUM	parsimonious general unrestricted model
PPP	purchasing power parity
QNA	quarterly national accounts
RMSFE	root mean squared forecast error
RMSTE	root mean squared target error
sdev	standard deviaton
SEM	simultaneous equation model
VAR	vector autoregressive model
VEqCM	vector equilibrium-correction model

1

Introduction

Macroeconometric modelling is one of the 'big' projects in economics, dating back to Tinbergen and Frisch. This introductory chapter first discusses the state of the project. We advocate the view that, despite some noteworthy setbacks, the development towards more widespread use of econometric models, is going to continue. However, models change as research progresses, as the economy develops, and as the demand and needs of model users change. We point to evidence of this kind of adaptive changes going on in current day macroeconometric models. We then discuss the aspects of the macroeconometric modelling project that we have contributed to in our own research, and where in the book the different dimensions and issues are presented.

1.1   The case for macroeconometric models

Macroeconometric models, in many ways the flagships of the economics profession in the 1960s, came under increasing attack from both theoretical economics and practitioners in the late 1970s. The onslaught came on a wide front: lack of microeconomic theoretical foundations, ad hoc modelling of expectations, lack of identification, neglect of dynamics and non-stationarity, and poor forecasting properties. As a result, by the start of the 1990s, the status of macroeconometric models had declined markedly, and had fallen completely out of (and with!) academic economics. Specifically, it has become increasingly rare that university programmes in economics give courses in large-scale empirical macroeconomic modelling.

Nevertheless, unlike the dinosaurs which they often have been likened to, macroeconometric models never completely disappeared from the scene. Moreover, if we use the term econometric model in a broad sense, it is fair to say that such models continue to play a role in economic policy. Model building and maintenance, and model based economic analyses, continue to be an important part of many economists' working week, either as a producer (e.g. member of modelling staff) or as a consumer (e.g. chief economists and consultants). Thus, the discipline of macroeconometric modelling has been able to adapt to changing demands, both with regards to what kind of problems users expect that models can help them answer, and with regard to quality and reliability.

Consider, for example, the evolution of Norwegian macroeconometric models (parallel developments no doubt have taken place in other countries): the models of the 1960s were designed to meet the demands of governments which attempted to run the economy through regulated markets. Today's models have adapted to a situation with liberalised financial and credit markets. In fact, the process of deregulation has resulted in an increased demand for econometric analysis and forecasting.

The recent change in monetary policy towards inflation targeting provides an example of how political and institutional changes might affect econometric modelling. The origins of inflation targeting seem to be found in the practical and operational issues which the governments of small open economies found themselves with after installing floating exchange rate regimes. As an alternative to the targeting of monetary aggregates, several countries (New Zealand, Canada, United Kingdom, and Sweden were first) opted for inflation targeting, using the interest rate as the policy instrument. In the literature which followed in the wake of the change in central bank practice (see, for example, Svensson 2000), it was made clear that under inflation targeting, the central bank's conditional inflation forecast becomes the operational target of monetary policy. At the back of the whole idea of inflation targeting is therefore the assumption that the inflation forecast is significantly affected by adjustment of the interest rate 'today'. It follows that the monetary authority's inflation forecasts have to be rooted in a model (explicit or not) of the transmission mechanism between the interest rate and inflation.

This characterisation of inflation targeting leads to a set of interesting questions, around which a lively debate evolves. For example: how should the size and structure of the model be decided, and its parameters quantified, that is, by theoretical design, by estimation using historical data or by some method of calibration—or perhaps by emulating the views of the 'monetary policy committee' (since at the end of the day the beliefs of the policy makers matter). A second set of issues follows from having the *forecasted* rate of inflation (rather than the current or historical rate) as the target. As emphasised by, for example, Clements and Hendry (1995*b*), modelling and forecasting are distinct processes (see also Chapter 11). In particular non-stationarities which are not removed by differencing or cointegration impinge on macroeconomic data. One consequence is that even well-specified policy models produce intermittent forecast failure, by which we in this book mean a significant deterioration in forecast quality relative to within sample tracking performance (see Clements and Hendry 1999*b*: ch. 2). Both theory and practical experience tell us that the source of forecast failure is usually to be found in shifts in the means of equilibrium relationships and in the growth rates of exogenous variables. Neither of these factors affect a model's usefulness in policy analysis, yet either of them can destroy the model's forecasts, unless the model user is able to correct them (e.g. by intercept corrections).

The integration of modelling, policy analysis, and forecasting in the mandate given to an inflation targeting central bank raises some important issues. For example, it must be decided to what extent the policy model should affect the forecasts, and how forecasts are best robustified in order to reduce the hazards of forecast-based interest rate setting.

Inflation targeting has already spurred a debate about the role of econometric specification and evaluation of models—that is, not only as an aid in the preparation of inflation forecasts, but also as a way of testing, quantifying, and elucidating the importance of transmission mechanisms in the inflationary process. In this way, inflation targeting actually moves the discussion about the quality and usefulness of econometric methodology and practice into the limelight of the economic policy debate (see Bårdsen *et al.* 2003).

However, even though a continued and even increasing demand for macro-econometric analysis is encouraging for the activity of macroeconometric modelling, it cannot survive as a discipline within economics unless the models reflect the developments in academic research and teaching. But, also in this respect macroeconometric modelling has fared much better than many of its critics seem to acknowledge. Already by the end of the 1980s, European macroeconometric models had a much better representation of price- and wage-setting (i.e. the supply-side) than before. There was also marked improvement in the modelling of the transmission mechanism between the real and financial sectors of the economy (see, for example, Wallis 1989). In the course of the last 20 years of the last century macroeconometric models also took advantage of the methodological and conceptual advances within time-series economet-rics. Use of dynamic behavioural equations are now the rule rather than the exception. Extensive testing of mis-specification is usually performed. The dangers of spurious regressions (see Granger and Newbold 1974) have been reduced as a consequence of the adoption of new inference procedures for integrated variables. No doubt, an important factor behind these advances has been the development of (often research based) software packages for estimation, model evaluation, and simulation.

In an insightful paper about the trends and problems facing econometric models, the Norwegian economist Leif Johansen stated that the trendlike development in the direction of more widespread use of econometric models will hardly be reversed completely (see Johansen 1982). But Johansen also noted that both the models' own conspicuous failures from time to time, and certain political developments, will inflict breaks or temporary setbacks in the trend. However, we think that we are in line with Johansen's views when we suggest that a close interchange between academic economics, theoretical econometrics, and software development are key elements that are necessary to sustain macroeconomic modelling. The present volume is meant as a contribution to macroeconomic modelling along these lines.

Four themes in particular are emphasised in this book:

(1) methodological issues of macroeconometric models;
(2) the supply-side of macroeconometric models;
(3) the transmission mechanism;
(4) the forecasting properties of macroeconometric models.

In the following, we review the main issues connected to these themes, and explain where they are covered in the book.

1.2   Methodological issues (Chapter 2)

The specification of a macroeconomic model rests in both economic theory and the econometric analysis of historical data. Different model builders place different weight on these two inputs to model specification, which is one reason why models differ and controversies remain, cf. the report on macroeconomic modelling and forecasting at the Bank of England (Pagan 2003).

The balance between theoretical consistency and empirical relevance is also of interest for model users, model owners, and research funding institutions. In the case where the model is used in a policy context, model-users may have a tendency to put relatively more weight on 'closeness to theory', on the grounds that theory consistency ensures model properties (e.g. impulse responses of dynamic multipliers) which are easy to understand and to communicate to the general public. While a high degree of theory consistency is desirable in our discipline, it does not by itself imply unique models. This is basically because, in macroeconomics, no universally accepted theory exists. Thus, there is little reason to renounce the requirement that empirical modelling and confrontation of theories with the data are essential ingredients in the process of specifying a serious macro model. In particular, care must be taken to avoid that theory consistency is used rhetorically to impute specific and controversial properties on the models that influence policy-making.

Recently, Pagan (2003) claimed that 'state of the art modelling' in economics would entail a dynamic stochastic general equilibrium (DSGE) model, since that would continue the trend taken by macroeconomic modelling in academia into the realm of policy-oriented modelling. However, despite its theory underpinnings, it is unclear if DSGE models have structural properties in the sense of being invariant over time, across regimes and with respect to additional information (e.g. the information embedded in existing studies, see Chapter 7).

A failure on any of these three requirements means that the model is non-structural according to the wider understanding of 'structure' adopted in this book: a structural representation of an economy embodies not only theory content, but explanatory power, stability, and robustness to regime shifts (see Hendry (1995*a*) and Section 2.3.2 for an example). Since structural representation is a many-faceted model feature, it cannot be uniquely identified with closeness to theory. Instead, theory-driven models are prone to well-known econometric problems, which may signal mis-specification with damaging implications for policy recommendations (see Nymoen 2002).

The approach advocated in this book is therefore a more balanced view. Although theory is a necessary ingredient in the modelling process, empirical determination is always needed to specify the 'final model'. Moreover, as noted, since there are many different theoretical approaches already available in macroeconomics, DSGE representing only one, there is always the question about which theory to use. In our view, economists have been too ready to accept theoretical elegance and rigour as a basis for macroeconomic relationships, even though the underlying assumptions are unrealistic and the representative agent a dubious construct at the macro level. Our approach is instead to favour models that are based on realistic assumptions which are at least consistent with such well-documented phenomena as, for example, involuntary unemployment, a non-unique 'natural rate', and the role of fairness in wage-setting. Such theories belong to behavioural macroeconomics as defined by Akerlof (2002). In Chapters 3–7 of this book, one recurrent theme is to gauge the credibility and usefulness of rival theories of wage- and price-setting from that perspective.

Many macroeconometric models are rather large systems of equations constructed piece-by-piece, for example, equation-by-equation, or, at best, sector-by-sector (the consumption expenditure system, the module for labour demand, and investment, etc.). Thus, there is no way around the implication that the models' overall properties only can be known when the construction is complete. The literature on macroeconometric modelling has produced methods of evaluation of the system of equations as a whole (see, for example, Klein *et al.* 1999).

Nevertheless, the piecewise construction of macroeconometric models is the source of much of the criticism levied against them. First, the specification process may become inefficient, as a seemingly valid single equation or module may either lead to unexpected or unwanted model properties. This point is related to the critique of structural econometric models in Sims (1980), where the author argues that such models can only be identified if one imposes 'incredible' identifying restrictions to the system of equations (see Section 2.2.2). Second, the statistical assumptions underlying single equation analysis may be invalidated when the equation is grafted into the full model. The most common examples are probably that the single equation estimation method is seen to become inconsistent with the statistical model implied by the full set of equations, or that the equation is too simple in the light of the whole model (e.g. omits a variable). These concerns are real, but they may also be seen as unavoidable costs of formulating models that go beyond a handful of equations, and which must therefore be balanced against the benefits of a more detailed modelling of the functional relationships of the macro economy. Chapter 2 discusses operational strategies that promise to reduce the cost of piece-by-piece model specification.

In Section 1.4, we briefly outline the transmission mechanism as represented in the medium scale macroeconometric model RIMINI (an acronym for a model for the Real economy and Income accounts—a MINI version—see Section 1.4),[1] which illustrates the complexity and interdependencies in a realistic macroeconometric model and also why one has to make sense out of bits and pieces rather than handling a complete model. The modelling of subsystems implies making simplifications of the joint distribution of all observable variables in the model through sequential conditioning and marginalisations, as discussed in Section 2.3.

The methodological approach of sequential subsector modelling is highlighted by means of two case studies. First, the strategy of sequential simplification is illustrated for the household sector in RIMINI, see Section 2.4. The empirical consumption function we derive has been stable for more than a decade. Thus, it is of particular interest to compare it with rival models in the literature, as we do in Section 2.4.2. Second, in Chapter 9 we describe a stepwise procedure for modelling wages and prices. This is an exercise that includes all ingredients regarded as important for establishing an econometrically relevant submodel. In this case we are in fact entertaining two models: one core model for wage and price determination, where we condition on a number of explanatory variables and a second model, which is a small aggregated econometric model for the entire economy. Although different, the embedding model shares many properties of the full RIMINI model.

The credentials of the core model within the embedding aggregated model can be seen as indirect evidence for the validity of the assumptions underlying the use of the core model as part of the larger model, that is, RIMINI. The small econometric model is, however, a model of interest in its own right. First, it is small enough to be estimated as a simultaneous system of equations, and the size makes it suitable for model developments and experiments that are cumbersome, time-consuming, and in some cases impossible to carry out with the full-blown RIMINI model. When we analyse the transmission mechanism in the context of econometric inflation targeting in Chapter 9 and evaluate different monetary policy rules in Chapter 10, this is done by means of the small econometric model, cf. Section 9.5.

---

[1] RIMINI has been used by the Central Bank of Norway for more than a decade to make forecasts for the Norwegian economy 4–8 quarters ahead as part of the Inflation report of the Bank; see Olsen and Wulfsberg (2001).

1.3 The supply-side and wage- and price-setting (Chapters 3–8)

In the course of the 1980s and 1990s the supply-side of macroeconometric models received increased attention, correcting the earlier overemphasis on the demand-side of the economy. Although there are many facets of the supply-side, for example, price-setting, labour demand, and investment in fixed capital and R&D, the main theoretical and methodological developments and controversies have focused on wage- and price-setting.

Arguably, the most important conceptual development in this area has been the Phillips curve—the relationship between the rate of change in money wages and the rate of unemployment (Phillips 1958)—and the 'natural rate of unemployment' hypothesis (Phelps 1967 and Friedman 1968). Heuristically, the natural rate hypothesis says that there is only one unemployment rate that can be reconciled with nominal stability of the economy (constant rates of wage and price inflation). Moreover, the natural rate equilibrium is asymptotically stable. Thus the natural rate hypothesis contradicted the demand-driven macroeconometric models of its day, which implied that the rate of unemployment could be kept at any (low) level by means of fiscal policy. A step towards reconciliation of the conflicting views was made with the discovery that a constant ('structural') natural rate is not necessarily inconsistent with a demand driven ('Keynesian') model. The trick was to introduce an 'expectations augmented' Phillips curve relationship into an IS-LM type model. The modified model left considerable scope for fiscal policy in the short run, but due to the Phillips curve, a long-term natural rate property was implied (see, for example, Calmfors 1977).

However, a weak point of the synthesis between the natural rate and the Keynesian models was that the supply-side equilibrating mechanisms were left unspecified and open to interpretation. Thus, new questions came to the forefront, like: How constant is the natural rate? Is the concept inextricably linked to the assumption of perfect competition, or is it robust to more realistic assumptions about market forms and firm behaviour, such as monopolistic competition? And what is the impact of bargaining between trade unions and confederations over wages and work conditions, which in some countries has given rise to a high degree of centralisation and coordination in wage-setting? Consequently, academic economists have discussed the theoretical foundations and investigated the logical, theoretical, and empirical status of the natural rate hypothesis, as for example in the contributions of Layard *et al.* (1991, 1994), Cross (1988, 1995), Staiger *et al.* (1997), and Fair (2000).

In the current literature, the term 'Non-Accelerating Inflation Rate of Unemployment', or NAIRU, is used as a synonym to the 'natural rate of unemployment'. Historically, the need for a new term, that is, NAIRU, arose because the macroeconomic rhetoric of the natural rate suggested inevitability, which is

something of a straitjacket since the long-run rate of unemployment is almost
certainly conditioned by socioeconomic factors, policy and institutions (see
for example, Layard *et al.* 1991 ch. 1.3).[2] The acronym NAIRU itself is something of a misnomer since, taken literally, it implies $\dddot{p} \leq 0$ where p is the log of
the price level and $\dddot{p}$ is the third derivative with respect to time. However, as
a synonym for the natural rate it implies $\ddot{p} = 0$, which would be constant rate
of inflation rate of unemployment (CIRU). We follow established practice and
use the natural rate—NAIRU—terminology in the following.

There is little doubt that the natural rate counts as one of the most influential conceptual developments in the history of macroeconomics. Governments
and international organisations customarily refer to NAIRU calculations in
their discussions of employment and inflation prospects,[3] and the existence of
a NAIRU consistent with a vertical long-run Phillips curve is a main element
in the rhetoric of modern monetary policy (see for example, King 1998).

The 1980s saw a marked change in the consensus view on the model suitable for deriving NAIRU measures. There was a shift away from a Phillips
curve framework that allowed estimation of a natural rate NAIRU from a single
equation for the rate of change of wages (or prices). The modern approach combined a negative relationship between the level of the real wage and the rate
of unemployment, dubbed the wage curve by Blanchflower and Oswald (1994),
with an equation representing firms' price-setting. The wage curve, originally
pioneered by Sargan (1964), is consistent with a wide range of economic theories (see Blanchard and Katz 1997), but its original impact among European
economists was due to the explicit treatment of union behaviour and imper-
fectly competitive product markets, pioneered by Layard and Nickell (1986).
In the same decade, time-series econometrics constituted itself as a separate
branch of econometrics, with its own methodological issues, controversies and
solutions, as explained in Chapter 2.

It is interesting to note how early new econometric methodologies were
applied to wage–price modelling, for example, equilibrium-correction modelling,
the Lucas critique, cointegration, and dynamic modelling. Thus, wage formation became an area where economic theory and econometric methodology
intermingled fruitfully. In this chapter, we draw on these developments when we
discuss how the different theoretical models of wage formation and price-setting
can be estimated and evaluated empirically.

The move from the Phillips curve to a wage curve in the 1980s was, however,
mainly a European phenomenon. The Phillips curve held its ground well in the
United States (see Fuhrer 1995, Gordon 1997, and Blanchard and Katz 1999).
But also in Europe the case has been reopened. For example, Manning (1993)

---

[2] Cross (1995, p. 184) notes that an immutable and unchangeable natural rate was not
implied by Friedman (1968).

[3] Elmeskov and MacFarland (1993), Scarpetta (1996), and OECD (1997*b*: ch. 1) contain
examples.

showed that a Phillips curve specification was consistent with union wage-setting, and that the Layard–Nickell wage equation was not identifiable. In academia, the Phillips curve has been revived and plays a prolific role in New Keynesian macroeconomics and in the modern theory of monetary policy (see Svensson 2000). The defining characteristics of the New Keynesian Phillips curve (NPC) are strict microeconomic foundations together with rational expectations of 'forward' variables (see Clarida *et al.* 1999, Galí and Gertler 1999, and Galí *et al.* 2001).

There is a long list of issues connected to the idea of a supply-side determined NAIRU, for example, the existence and estimation of such an entity, and its eventual correspondence to a steady-state solution of a larger system explaining wages, prices as well as real output and labour demand and supply. However, at an operational level, the NAIRU concept is model dependent. Thus, the NAIRU issues cannot be seen as separated from the wider question of choosing a framework for modelling wage, price, and unemployment dynamics in open economies. In the following chapters we therefore give an appraisal of what we see as the most important macroeconomic models in this area. We cover more than 40 years of theoretical development, starting with the Norwegian (*aka* Scandinavian) model of inflation of the early 1960s, followed by the Phillips curve models of the 1970s and ending up with the modern incomplete competition model and the NPC.

In reviewing the sequence of models, we find examples of newer theories that generalise on the older models that they supplant, as one would hope in any field of knowledge. However, just as often new theories seem to arise and become fashionable because they, by way of specialisation, provide a clear answer on issues that older theories were vague on. The underlying process at work here may be that as society evolves, new issues enter the agenda of politicians and their economic advisers. For example, the Norwegian model of inflation, though rich in insight about how the rate of inflation can be stabilised (i.e. $\ddot{p} = 0$), does not count the adjustment of the rate of unemployment to its natural rate as even a necessary requirement for $\ddot{p} = 0$. Clearly, this view is conditioned by a socioeconomic situation in which 'full employment' with moderate inflation was seen as attainable and almost a 'natural' situation. In comparison, both the Phelps/Friedman Phillips curve model of the natural rate, and the Layard–Nickell NAIRU model specialise their answers to the same question, and take for granted that it is necessary for $\ddot{p} = 0$ that unemployment equals a natural rate or NAIRU which is entirely determined by long-run supply factors.

Just as the Scandinavian model's vagueness about the equilibrating role of unemployment must be understood in a historical context, it is quite possible that the natural rate thesis is a product of socioeconomic developments. However, while relativism is an interesting way of understanding the origin and scope of macroeconomic theories, we do not share Dasgupta's (1985) extreme relativistic stance, that is, that successive theories belong to different epochs, each defined by their answers to a new set of issues, and that one cannot speak of progress in economics. On the contrary, our position is that the older models of wage–price inflation and unemployment often represent insights that remain of interest today.

Chapter 3 starts with a reconstruction of the Norwegian model of inflation, in terms of modern econometric concepts of cointegration and causality. Today this model, which stems back to the 1960s, is little known outside Norway. Yet, in its reconstructed forms, it is almost a time traveller, and in many respects resembles the modern theory of wage formation with unions and price-setting firms. In its time, the Norwegian model of inflation was viewed as a contender to the Phillips curve, and in retrospect it is easy to see that the Phillips curve won. However, the Phillips curve and the Norwegian model are in fact not mutually exclusive. A conventional open economy version of the Phillips curve can be incorporated into the Norwegian model, and in Chapter 4 we approach the Phillips curve from that perspective. However, the bulk of the chapter concerns issues which are quite independent of the connection between the Phillips curve and the Norwegian model of inflation. As perhaps the ultimate example of a consensus model in economics, the Phillips curve also became a focal point for developments in both economic theory and in econometrics. In particular we focus on the development of the natural rate doctrine, and on econometric advances and controversies related to the stability of the Phillips curve (the origin of the Lucas critique).

In Chapter 6 we present a unifying framework for all of the three main models, the Norwegian model, the Phillips curve and the Layard–Nickell wage curve model. In that chapter, we also discuss at some length the NAIRU doctrine: is it a straitjacket for macroeconomic modelling, or an essential ingredient? Is it a truism, or can it be tested? What can be put in its place if it is rejected? We give answers to all these questions, and the thrust of the argument represents an intellectual rationale for macroeconometric modelling of larger systems of equations.

An important underlying assumption of Chapters 3–6 is that inflation and unemployment follow causal or future-independent processes (see Brockwell and Davies 1991 ch. 3), meaning that the roots of the characteristic polynomials of the difference equations are inside the unit circle. This means that all the different economic models can be represented within the framework of linear difference equations with constant parameters. Thus the econometric framework is the vector autoregressive model (VAR), and identifies systems of equations that encompass the VAR (see Hendry and Mizon 1993, Bårdsen and Fisher 1999). Non-stationarity is assumed to be of a kind that can be modelled away by differencing, by establishing cointegrating relationships, or by inclusion of deterministic dummy variables in the non-homogeneous part of the difference equations.

In Chapter 7, we discuss the NPC of Galí and Gertler (1999), where the stationary solution for the rate of inflation involves leads (rather than lags) of the non-modelled variables. However, non-causal stationary solutions could also exist for the 'older' price–wage models in Chapters 3–6 if they are specified with 'forward looking' variables (see Wren-Lewis and Moghadam 1994). Thus, the discussion of testing issues related to forward vs. backward looking models in Chapter 7 is relevant for a wider class of forward-looking models, not just the NPC.

The role of money in the inflation process is an old issue in macroeconomics, yet money plays no essential part in the models appearing up to and including Chapter 7. This reflects how all models, despite the very notable differences existing between them, conform to the same overall view of inflation: namely that inflation is best understood as a complex socioeconomic phenomenon reflecting imbalances in product and labour markets, and generally the level of conflict in society. This is inconsistent with, for example, a simple quantity theory of inflation, but arguably not with having excess demand for money as a source of inflation pressure. Chapter 8 uses that perspective to investigate the relationship between money demand and supply, and inflation.

Econometric analysis of wage, price, and unemployment data serve to substantiate the discussion in this part of the book. An annual data set for Norway is used throughout Chapters 4–6 to illustrate the application of three main models (Phillips curve, wage curve, and wage price dynamics) to a common data set. But frequently we also present analysis of data from the other Nordic countries, as well as of quarterly data from the United Kingdom, the Euro area, and Norway.

1.4   The transmission mechanism (Chapters 9 and 10)

All macroeconometric models contain a quantitative picture of how changes in nominal variables bring about real effects, the so-called transmission mechanism. Sometimes representations of the transmission mechanism are the main objective of the whole modelling exercise, as when central banks seek to understand (and to convey to the public) how changes in the nominal interest rate affect real variables like the GDP growth rate and the rate of unemployment, and through them, the rate of inflation. Clearly, the wage and price submodel is one key element in the model of the transmission mechanism.

In modern economies, the transmission mechanism can be seen as a complex system where different groups of agents interact through markets which are often strongly interlinked, and an attractive feature of a macroeconomic model is that it represents the different linkages in a consistent framework. As an example, we take a closer look at the transmission mechanism of the medium term macroeconomic model, RIMINI.

By Norwegian standards, RIMINI is an aggregated macroeconometric model.[4] The core consists of some 30 important stochastic equations, and there are about 100 exogenous variables which must be projected by the forecaster. Such projections involve judgements, and they are best made manually based

---

[4] See Bjerkholt (1998) for an account of the Norwegian modelling tradition.

on information from a wide set of sources. The model should be run repeatedly to check for consistency between the exogenous assumptions and the results before one arrives at a baseline forecast. In this way the model serves as a tool taking account of international business cycle development, government policy, and market information, for example, forward market interest rates.

The RIMINI is a fairly closed model in the sense that the most important variables for the Norwegian economy are determined by the model, while the model conditions upon 'outside' variables like foreign prices and output and domestic policy variables like interest rates and tax rates. The model distinguishes between five production sectors. The oil and shipping sectors are not modelled econometrically, nor is the sector for agriculture, forestry, and fishing. The two main sectors for which there exist complete submodels are manufacturing and construction (traded goods) and services and retail trade (non-traded goods). There are reasons to expect important differences in, for instance, the responses to changes in interest rates and exchange rates between traded and non-traded goods.

In RIMINI there are two main channels through which monetary policy instruments affect employment, output, and prices—the interest rate channel and the exchange rate channel. For the first channel—the effect of the interest rate—Figure 1.1 shows the roles of households and enterprises in RIMINI and also the main interaction between the demand-side (upper shaded box) and the supply-side (lower shaded box). The main point here is to illustrate the complexity and interdependencies that are typical of macroeconometric systems.

Assuming fixed exchange rates, an increase in the central bank interest rate for loans to the banks (the signal rate) immediately affects the money market interest rate. The money market rate in turn feeds into the deposit and lending rates of commercial and savings banks with a lag. Aggregate demand is affected through several mechanisms: there is a negative effect on housing prices (for a given stock of housing capital), which causes real household wealth to decline, thus suppressing total consumer expenditure. Also, there are negative direct and indirect effects on real investment in the traded and non-traded sectors and on housing investment.

CPI inflation is reduced after a lag, mainly through the effects from changes in aggregate demand on aggregate output and employment, but also from changes in unit labour costs. Notably, productivity first decreases and then increases—due to temporary labour hoarding—to create a cyclical pattern in the effects of the change in the interest rate.

An appreciation of the Krone has a more direct effect on CPI inflation compared to the interest rate. As illustrated by the upper left box in Figure 1.2, it mainly works through reduced import prices with a lagged response which entails a complete pass-through to import and export prices after about 2 years. The model specification is consistent with a constant markup on unit labour costs in the long run. A currency appreciation has a negative effect on the

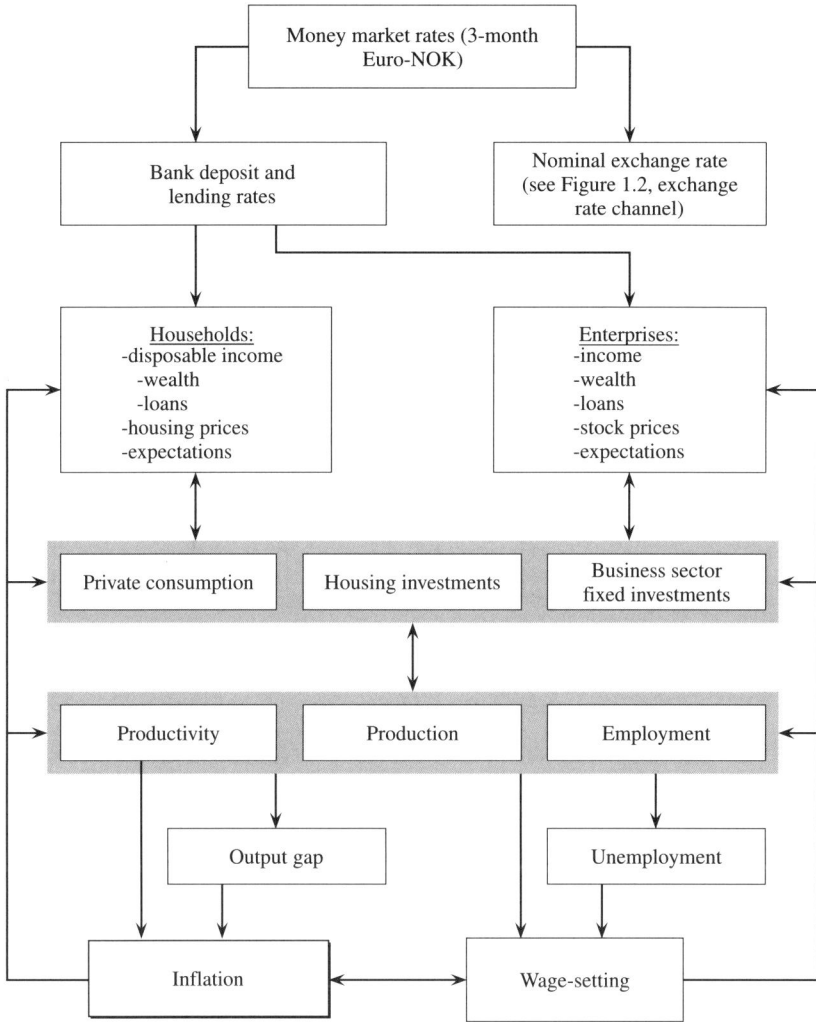

Figure 1.1. Interest rate channels in RIMINI. Given constant exchange rates

demand for traded goods. The direct effects are not of a large magnitude, because there are small relative price elasticities in the export equations and secondly because export prices (in local currency) adjust with a lag and tend to restore the relative prices. However, there are also important feedback mechanisms as the decrease in the price level caused by the appreciation feeds back into aggregate demand from domestic sectors.

If we abandon the assumption of a fixed exchange rate, an increase of interest rates affects the money market rate and this induces an appreciation of the Krone. Hence, we obtain the combined effect of an interest rate increase

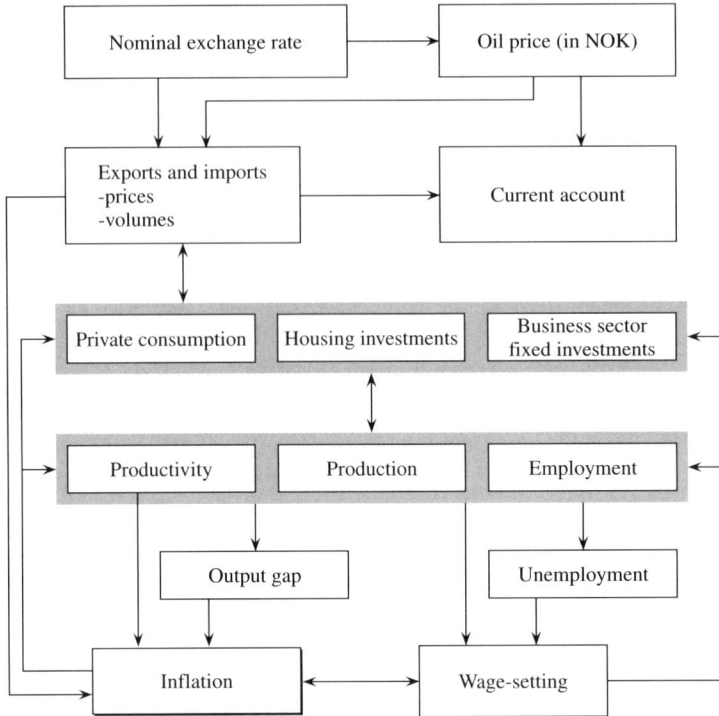

Figure 1.2. Exchange rate channels in RIMINI. Given constant interest rates

through both channels and the exchange rate channel strengthens the effect of interest rate changes on the rate of inflation. This will be analysed further in Section 9.5 in the context of the small macroeconometric model for Norway, which, as we alluded to in Section 1.2, shares many properties of the full RIMINI model.

This brief presentation of the transmission mechanism of an operational model also serves to demonstrate the complexity and interdependencies of an operational macroeconometric model. Again, it is evident that such a model is too big and complex to be formulated in one step, or to be estimated simultaneously. Thus, there is a need to deal with subsectors of the economy—that is, we try to make sense out of bits and pieces rather than handling a complete model. The modelling of subsystems implies making simplifications of the joint distribution of all observable variables in the model through sequential conditioning and marginalisations, as discussed in Section 2.3.

The estimated model in Chapter 9 is based on the assumption that the short-run interest rate is an exogenous policy variable, and the chapter highlights estimation results and model properties along with a discussion about the model's potential to address monetary policy issues which are at the forefront of inflation targeting central banks. Inflation targeting means that the policy instrument (the interest rate) is set with the aim of controlling the conditional forecast of inflation 2–3 years ahead. In practice, this means that central bank economists will need to form a clear opinion about how the inflation forecasts are affected by different future interest rate paths, which in turn amounts to quantitative knowledge of the transmission mechanism in the new regime. The main monetary policy channels in the small macroeconometric model are discussed on the basis of an analysis of dynamic multipliers.

In Chapter 10, we relax the assumption that the short-run interest rate is exogenous. We evaluate the performance of different types of reaction functions or Taylor-type interest rate rules. We perform counterfactual simulations over the period from 1995q1 to 2000q4. In addition to analysing the outcome from employing standard Taylor-type rules, including rules with interest rate smoothing, we also employ *inter alia* interest rate rules dubbed 'real time' rules since they are based on variables less prone to measurement errors, and 'open economy' rules which allow for interest rate responses to exchange rate misalignments. The performance of the employed rules is evaluated by standard efficiency measures and by deriving the mean deviations from targets, which may be of interest for policy makers, especially over short time horizons. We also introduce the root mean squared target error (RMSTE), an analogue to the well-known root mean squared forecast error. Finally we conduct simulation experiments where we vary the weights in the interest rate rules as well as the weights of the variables in the policy maker's loss function. The results are summarised by estimating response surfaces on the basis of the range of weights considered in the simulations. We will assume that monetary policy rules aim at stabilising inflation around the inflation target, and that the monetary authorities potentially put some weight also on the stabilisation of unemployment, output, and interest rates. The performance of different monetary policy rules can then be evaluated on the basis of the monetary authorities' loss function.

1.5 Forecast properties (Chapter 11)

When studies of macroeconometric models' forecast performance started to appear in the 1960s and 1970s, it was considered a surprise that they were found to be outperformed by very simple forecasting mechanisms. As pointed out by Granger and Newbold (1986), many theory-driven macro models largely ignored dynamics and temporal properties of the data, so that it should not come as a surprise why they produced suboptimal forecasts. Forecasting is a time-oriented activity, and a procedure that pays only rudimentary attention to temporal aspects is likely to lose out to rival procedures that put dynamics in the foreground. Such competing procedures were developed and gained ground in the 1970s in the form of Box–Jenkins time-series analysis and ARIMA models.

As we alluded to in Section 1.1, macroeconometric modelling has progressed in the last two decades through the adoption of new techniques and insights from time-series econometrics, with more emphasis on dynamic specification and testing against mis-specification. The dangers of spurious regressions have been reduced as a consequence of the adoption of new inference procedures for integrated variables. As a result, modern macroeconometric forecasting models are less exposed to Granger and Newbold's diagnosis.

In particular, one might reasonably expect that equilibrium-correcting models (EqCMs) will forecast better than models that only use differenced data, so-called differenced vector autoregressions (dVARs) or other member of pure time-series models. In fact, the typical case will be that the dVAR is mis-specified relative to an econometrically specified EqCM, and dVAR forecasts will therefore be suboptimal.

However, as shown by the work of Michael Clements and David Hendry in several books and papers, the expected favourable forecasting properties of econometric models rest on the assumption of constant parameters in the forecast period. This is of course an untenable basis for the theory and practice of economic forecasting. The frequent occurrences of structural changes and regime shifts tilt the balance of the argument back in favour of dVARs. One reason is if key parameters like, for example, the means of the cointegrating relationships change after the forecast is made, then forecasts of the EqCM are damaged while the dVAR forecasts are robust (since the affected relationships are omitted from the forecasting mechanism in the first place). Hence, in practice, EqCM forecasts may turn out to be less accurate than forecasts derived from a dVAR. Nevertheless, the EqCM may be the right model to use for policy simulations (e.g. the analysis of the transmission mechanism). Specifically, this is true if the source of forecast failure turns out to be location shifts in, for example, the means of cointegration relationships or in autonomous growth rates, rather than in the model's 'derivative' coefficients, which are the parameters of interest in policy analysis. Theoretical and empirical research indicate that this is a typical situation. Conversely, the 'best model' in terms of economic interpretation and econometrics, may not be the best model for forecasts. In Chapter 11, we investigate the practical relevance of these theoretical developments for forecasts of the Norwegian economy in the 1990s. The model that takes the role of the EqCM is the RIMINI model mentioned earlier.

2

Methodological issues of large-scale macromodels

The chapter focuses on methodology and describes the roles of statistics and of economic theory in macroeconomic modelling. Building on a long tradition, we suggest an approach to macroeconometric modelling which is based on fundamental statistical concepts like the joint distribution function of all observable variables. Users of macroeconomic models often demand a detailed description of the economy, and in order to accommodate that demand, realistic macroeconomic models invariably become too large to be specified simultaneously. The suggested methodology therefore relies on valid conditioning and marginalisation of the joint distribution function in order to arrive at tractable subsystems, which can be analysed with statistical methods.

2.1   Introduction: small vs. large models

Macroeconometric modelling aims at explaining the empirical behaviour of an actual economic system. Such models will be systems of inter-linked equations estimated from time-series data using statistical or econometric techniques.

A conceptual starting point is the idea of a general stochastic process that has generated all data we observe for the economy, and that this process can be summarised in terms of the joint probability distribution of random observable variables in a stochastic equation system: see Section 2.3. For a modern economy, the complexity of such a system, and of the corresponding joint probability distribution, is evident. Nevertheless, it is always possible to take a highly aggregated approach in order to represent the behaviour of a few 'headline' variables (e.g. inflation, GDP growth, unemployment) in a small-scale model. If small enough, the estimation of such econometric models can be

17

based on formally established statistical theory (as with low-dimensional vector autoregressive models [VARs]), where the statistical theory has recently been extended to cointegrated variables.

However, it takes surprisingly little in terms of user-instigated detailing of model features—for example, more than one production sector, separate modelling of consumption and investment—to render simultaneous modelling of all equations impossible in practice. Hence, models that are used for analysing the impact of the governmental budget on the economy are typically very large systems of equations. Even in the cases where the model user from the outset targets only one variable, as with the recently contrived inflation targeting, policy choices are made against the backdrop of a broader analysis of the effects of the interest rate on the economy (the nominal and real exchange rates, output growth, employment and unemployment, housing prices, credit growth, and financial stability). Thus, it has been a long-standing task of model builders to establish good practice and to develop operational procedures for model building which secures that the end product of piecewise modelling is tenable and useful. Important contributions in the literature include Christ (1966), Klein (1983), Fair (1984, 1994), Klein *et al.* (1999), and the surveys in Bodkin *et al.* (1991) and Wallis (1994).

In this book, we supplement the existing literature by suggesting the following operational procedure[1]:

1. By relevant choices of variables we define and analyse subsectors of the economy (by marginalisation).
2. By distinguishing between exogenous and endogenous variables we construct (by conditioning) relevant partial models, which we will call models of type A.
3. Finally, we need to combine these submodels in order to obtain a Model B for the entire economy.

Our thesis is that, given that Model A is a part of Model B, it is possible to learn about Model B from Model A. The alternative to this thesis amounts to a kind of creationism,[2] that is, unless of course macroeconometrics should be restricted to aggregate models.

Examples of properties that can be discovered using our procedure include cointegration in Model B. This follows from a corollary of the theory of cointegrated systems: any nonzero linear combination of cointegrating vectors is also a cointegrating vector. In the simplest case, if there are two cointegrating vectors in Model B, there always exists a linear combination of those cointegrating vectors that 'nets out' one of the variables. Cointegration analysis

---

[1] See Jansen (2002), reply to Søren Johansen (Johansen 2002).

[2] Theory that attributes the origin of matter and species to a special creation (or act of God), as opposed to the evolutionary theory of Darwin.

of the subset of variables (i.e. Model A) excluding that variable will result in a cointegrating vector corresponding to that linear combination. Thus, despite being a property of Model B, cointegration analysis of the subsystem (Model A) identifies one cointegration vector. Whether that identification is economically meaningful or not remains in general an open issue, and any such claim must be substantiated in each separate case. We provide several examples in this book: in Section 2.4 we discuss the identification of a consumption function as a cointegrating relationship, and link that discussion to the concept of partial structure. In Chapter 5, the identification of cointegrating relationships corresponding to price- and wage-setting is discussed in detail.

Other important properties of the full model that can be tested from subsystems include the existence of a natural rate of unemployment (see Chapter 6), and the relevance of forward looking terms in wage- and price-setting (see Chapter 7).

Nevertheless, as pointed out by Johansen (2002), there is a Catch 22 to the above procedure: a general theory for the three steps will contain criteria and conditions which are formulated for the full system. However, sophisticated piecewise modelling can be seen as a sort of gradualism—seeking to establish submodels that represent *partial structure*: that is, partial models that are invariant to extensions of the sample period, to changes elsewhere in the economy (e.g. due to regime shifts) and remain the same for extensions of the information set. However, gradualism also implies a readiness to revise a submodel. Revisions are sometimes triggered by forecast failure, but perhaps less often than believed in academic circles: see Section 2.3.2. More mundane reasons include data revisions and data extensions which allow more precise and improved model specifications. The dialogue between model builders and model users often result in revisions too. For example, experienced model users are usually able to pinpoint unfortunate and unintended implications of a single equation's (or submodel) specification on the properties of the full model.

Obviously, gradualism does not preclude thorough testing of a submodel. On the contrary, the first two steps in the operational procedure above do not require that we know the full model, and testing those conditions has some intuitive appeal since real life provides 'new evidence' through the arrival of new data and by 'natural experiments' through regime shifts like, for example, changes in government or the financial deregulation in many European economies in the recent past. For the last of the three steps, we could in principle think of the full model as the ultimate extension of the information set, and so establishing structure or partial structure represents a way to meet Johansen's observation. In practice, we know that the full model is not attainable. What we do then is to graft the sector model in simplified approximations of Model B, and test the relevant exogeneity assumptions of the partial model within that framework. To the extent that the likelihood function of the simplified Model B is adequately representing or approximating the likelihood function of the full Model B, there is no serious problem left. It is also possible to corroborate the entire procedure, since it is true that Model A can be tested and improved gradually on new information, which is a way of gaining knowledge that parallels modern Darwinism in the natural sciences. We develop these views further in Section 2.5.

A practical reason for focusing on submodels is that the modellers may have good reasons to study some parts of the economy more carefully than other parts. For a central bank that targets inflation, there is a strong case for getting the model of the inflationary process right. This calls for careful modelling of the wage and price formation conditional on institutional arrangements for the wage bargain, the development in financial markets, and the evolving real economy in order to answer a number of important questions: Is there a natural rate (of unemployment) that anchors unemployment as well as inflation? What is the importance of expectations for inflation and how should they be modelled? What is the role of money in the inflationary process?

We find that in order to answer such questions—and to probe the competing hypotheses regarding supply-side economics—a detailed modelling, drawing on information specific to the economy under study—is necessary. Taking account of the simultaneity is to a large extent a matter of estimation efficiency. If there is a tradeoff between such efficiency and the issue of getting the economic mechanisms right, the practitioners of macroeconometric modelling should give priority to the latter.

2.2   The roles of statistics and economic theory in macroeconometrics

Macroeconometrics draws upon and combines two academic disciplines—economics and statistics. There is hardly any doubt that statisticians have had a decisive influence on quantitative economics in general and on modern macroeconometric modelling in particular.

2.2.1   The influx of statistics into economics

The history of macroeconomic modelling starts with the Dutch economist Jan Tinbergen who built and estimated the first macroeconometric models in the mid-1930s (Tinbergen 1937). Tinbergen showed how one could build a system of equations into an econometric model of the business cycle, using economic theory to derive behaviourally motivated dynamic equations and statistical methods (of that time) to test them against data. However, there seems to be universal agreement that statistics entered the discipline of economics and econometrics with the contributions of the Norwegian economist Trygve Haavelmo in his treatise 'The Probability Approach in Econometrics' (Haavelmo 1944; see Royal Swedish Academy of Science 1990, Klein 1988, Morgan 1990, or Hendry and Morgan 1995). Haavelmo was inspired by some of the greatest statisticians of that time. As Morgan (1990, p. 242) points out, he was converted to the usefulness of probability ideas by Jerzy Neyman and he was also influenced by Abraham Wald whom Haavelmo credited as the source of his understanding of statistical theory.

For our purpose, it is central to note that Haavelmo recognised and explained in the context of an economic model, that the joint distribution of all observable variables for the whole sample period provides the most general framework for statistical inference (see Hendry *et al.* 1989). This applies to specification (op. cit., pp. 48–49), as well as identification, estimation, and hypothesis testing:

all come down to one and the same thing, namely to study the properties of the joint probability distribution of random (observable) variables in a stochastic equation system (Haavelmo 1944, p. 85).

Haavelmo's probabilistic revolution changed econometrics. His thoughts were immediately adopted by Jacob Marschak—a Russian-born scientist who had studied statistics with Slutsky—as the research agenda for the Cowles Commision for the period 1943–47, in reconsidering Tinbergen's work on business cycles cited above. Marschak was joined by a group of statisticians, mathematicians, and economists, including Haavelmo himself. Their work was to set the standards for modern econometrics and found its way into the textbooks of econometrics from Tintner (1952) and Klein (1953) onwards.

The work of the Cowles Commision also laid the foundations for the development of macroeconomic models and model building which grew into a large industry in the United States in the next three decades (see Bodkin *et al.* 1991 and Wallis 1994). These models were mainly designed for short (and medium) term forecasting, that is, modelling business cycles. The first model (Klein 1950) was made with the explicit aim of implementing Haavelmo's ideas into Tinbergen's modelling framework for the United States economy. Like Tinbergen's model, it was a small model and Klein put much weight on the modelling of simultaneous equations. Later models became extremely large systems in which more than 1000 equations were used to describe the behaviour of a modern industrial economy. In such models, less care could be taken about each econometric specification, and simultaneity could not be treated in a satisfactory way. The forecasting purpose of these models meant that they were evaluated on their performance. When the models failed to forecast the effects on the industrial economies of the oil price shocks in 1973 and 1979, the macroeconomic modelling industry lost much of its position, particularly in the United States.

In the 1980s, macroeconometric models took advantage of the methodological and conceptual advances in time-series econometrics. Box and Jenkins (1970) had provided and made popular a purely statistical tool for modelling and forecasting univariate time-series. The second influx of statistical methodology into econometrics has its roots in the study of the non-stationary nature of economic data series. Clive Granger—with his background in statistics—has in a series of influential papers shown the importance of an econometric equation being balanced. A stationary variable cannot be explained by a non-stationary variable and vice versa (see, for example, Granger 1990). Moreover, the concept of cointegration (see Granger 1981; Engle and Granger 1987, 1991)—that a linear combination of two or more non-stationary variables can be stationary—has proven useful and important in macroeconometric modelling. Within the framework of a general VAR, the statistician Søren Johansen has provided (see Johansen 1988, 1991, 1995*b*) the most widely used tools for testing for cointegration in a multivariate setting, drawing on the analytical framework of canonical correlation and multivariate reduced rank regression in Anderson (1951).

Also, there has been an increased attention attached to the role of evaluation in modern econometrics (see Granger 1990, 1999). The so-called LSE methodology emphasises the importance of testing and evaluating econometric models (see Hendry 1993*a*, 1995*a*, Mizon 1995, and Ericsson 2005). Interestingly, Hendry *et al.* (1989) claim that many aspects of the Haavelmo research agenda were ignored for a long time. For instance, the joint distribution function for observable variables was recognised by the Cowles Commission as central to solving problems of statistical inference, but the ideas did not influence empirical modelling strategies for decades. By contrast, many developments in econometrics after 1980 are in line with this and other aspects of Haavelmo's research programme. This is also true for the role of economic theory in econometrics:

Theoretical models are necessary tools in our attempts to understand and 'explain' events in real life. (Haavelmo 1944, p. 1)

But whatever 'explanations' we prefer, it is not to be forgotten that they are all our own artificial inventions in a search for an understanding of real life; they are not hidden truths to be 'discovered'. (Haavelmo 1944, p. 3)

With this starting point, one would not expect the facts or the observations to agree with any precise statement derived from a theoretical model. Economic theories must then be formulated as probabilistic statements and Haavelmo viewed probability theory as indispensable in formalising the notion of models being approximations to reality.

2.2.2 Role of economic theory in macroeconometrics

The Cowles Commission research agenda focused on simultaneous equation models (SEMs) and put much weight on the issue of identification. In dealing with these issues, economic theory plays an important part. The prominent representative of this tradition, Lawrence Klein, writes in a very readable survey of the interaction between statistics and economics in the context of macro-econometric modelling (Klein 1988) that the model building approach can be contrasted to pure statistical analysis, which is empirical and not so closely related to received economic theory as is model building.

Still, it is on this score the traditional macroeconomic model building has come under attack (see Favero 2001). Whereas the LSE methodology largely ascribes the failure of those early macroeconomic models to missing dynamics or model mis-specification (omitted energy price effects), other critiques like Robert Lucas and Christopher Sims have claimed that the cause is rather that they had a weak theoretical basis. The Lucas critique (see, for example, Lucas 1976) claims that the failure of conditional models is caused by regime shifts, as a result of policy changes and shifts in expectations. The critique carries over to SEMs if expectations are non-modelled. On the other hand, Sims (1980) argued that SEMs embodied 'incredible' identifying restrictions: the restrictions needed to claim exogeneity for certain variables would not be valid in an environment where agents optimise intertemporally.

Sims instead advocated the use of a low-order vector autoregression to analyse economic time-series. This approach appeared to have the advantage that it did not depend on an 'arbitrary' division between exogenous and endogenous variables and also did not require 'incredible' identifying restrictions. Instead Sims introduced identifying restrictions on the error structure of the model, and this approach has been criticised for being equally arbitrary. Later developments have led to structural VAR models in which cointegration defines long-run relationships between non-stationary variables and where exogenous variables are reintroduced (see Pesaran and Smith 1998 for a survey in which they reanalyse an early model by King *et al.* 1991).[3]

Ever since the Keynes–Tinbergen controversy (see Morgan 1990 and Hendry and Morgan 1995), the role of theory in model specification has represented a major controversy in econometrics (cf. Granger 1990, 1999 for recent surveys). At one end of the theory–empiricism line we have theory-driven models that take the received theory for granted, and do not test it. Prominent examples are the general equilibrium models, dubbed real business cycle models, that have gained a dominating position in academia (see, for example, Kydland and Prescott 1991). There is also a new breed of macroeconometric models which assume intertemporally optimising agents endowed with rational forward-looking expectations, leading to a set of Euler equations (see Poloz *et al.* 1994, Willman *et al.* 2000, Hunt *et al.* 2000, and Nilsson 2002 for models from the central banks of Canada, Finland, New Zealand, and Sweden, respectively). At another extreme we have data based VAR models which initially were statistical devices that made only minimal use of economic theory. As noted above, in the less extreme case of structural VARs, theory restrictions can be imposed as testable cointegrating relationships in levels or they can be imposed on the error structure of the model.

The approach we are advocating has much in common with the LSE methodology referred to above, and it focuses on evaluation as recommended

---

[3] Jacobson *et al.* (2001) use a structural VAR with emphasis on the common trends to analyse the effect of monetary policy under an inflation targeting regime in a small open economy.

in Granger (1999). It represents a compromise between data based (purely stat-istical) models and economic theory: on the one hand learning from the process of trying to take serious account of the data, while on the other hand avoiding making strong theoretical assumptions—needed to make theories 'complete'—which may not make much sense empirically, that is, that are not supported by the data.[4] Moreover, there are common-sense arguments in favour of not adopt-ing a theory-driven model as a basis for policy decisions, which indeed affect reality, until it has gained convincing empirical support (see Granger 1992).

2.3   Identifying partial structure in submodels

Model builders often face demands from model users that are incompatible with a 3–5 equations closed form model. Hence, modellers often find themselves dealing with submodels for the different sectors of the economy. Thus it is often useful to think in terms of a simplification of the joint distribution of all observable variables in the model through sequential factorisation, conditioning, and marginalisations.

2.3.1   The theory of reduction

Consider the joint distribution of $x_t = (x_{1t}, x_{2t}, \ldots, x_{nt})'$, $t = 1, \ldots, T$, and let $x_T^1 = \{x_t\}_{t=1}^T$. Sequential factorisation means that we factorise the joint density function $D_x(x_T^1 \mid x_0, \lambda_x)$ into

$$D_x(x_T^1 \mid x_0; \lambda_x) = D_x(x_1 \mid x_0; \lambda_x) \prod_{t=2}^T D_x(x_t \mid x_{t-1}^1, x_0; \lambda_x), \qquad (2.1)$$

which is what Spanos (1989) named the *Haavelmo distribution*. It explains the present x_t as a function of the past x_{t-1}^1, initial conditions x_0, and a time invariant parameter vector λ_x. This is—by assumption—as close as we can get to representing what Hendry (1995a) calls the data generating process (DGP), which requires the error terms, $\varepsilon_t = x_t - E(x_t \mid x_{t-1}^1, x_0; \lambda_x)$, to be an innovation process. The ensuing approach has been called 'the theory of reduction' as it seeks to explain the origin of empirical models in terms of reduction operations conducted implicitly on the DGP to induce the relevant empirical model (see Hendry and Richard 1982, 1983).

---

[4] As is clear from the discussion above, econometric methodology lacks a consensus, and thus the approach to econometric modelling we are advocating is controversal. Heckman (1992) questions the success, but not the importance, of the probabilistic revolution of Haavelmo. Also, Keuzenkamp and Magnus (1995) offer a critique of the Neyman–Pearson paradigm for hypothesis testing and they claim that econometrics has exerted little influence on the beliefs of economists over the past 50 years; see also Summers (1991). For sceptical accounts of the LSE methodology, see Hansen (1996) and Faust and Whiteman (1995, 1997), to which Hendry (1997b) replies.

The second step in data reduction is further conditioning and simplification. We consider the partitioning $x_t = (y_t', z_t')$ and factorise the joint density function into a conditional density function for $y_t \mid z_t$ and a marginal density function for z_t:

$$D_x(x_t \mid x_{t-1}^1, x_0; \lambda_x) = D_{y|z}(y_t \mid z_t, x_{t-1}^1, x_0; \lambda_{y|z}) \cdot D_z(z_t \mid x_{t-1}^1, x_0; \lambda_z).$$

(2.2)

In practice we then simplify by using approximations by kth order Markov processes and develop models for

$$D_x(x_t \mid x_{t-1}^1, x_0; \lambda_x) \approx D_x(x_t \mid x_{t-1}^{t-k}; \theta_x)$$

(2.3)

$$D_{y|z}(y_t \mid z_t, x_{t-1}^1, x_0, \lambda_{y|z}) \approx D_{y|z}(y_t \mid z_t, x_{t-1}^{t-k}; \theta_{y|z})$$

(2.4)

for $t > k$. The validity of this reduction requires that the residuals remain innovation processes.

A general linear dynamic class of models with a finite number of lags which is commonly used to model the n-dimensional process x_t is the kth order VAR with Gaussian error, that is,

$$x_t = \mu + \sum_{i=1}^{k} \Pi_i x_{t-i} + \varepsilon_t,$$

where ε_t is normal, independent and identically distributed, $N.i.i.d.$ $(0, \Lambda_\varepsilon)$. A VAR model is also the starting point for analysing the cointegrating relationships that may be identified in the x_t-vector (see Johansen 1988, 1991, 1995b). Economic theory helps in determining which information sets to study and in interpreting the outcome of the analysis. In the following, we assume for simplicity that the elements of x_t are non-stationary I(1)-variables that become stationary after being differenced once. Then, if there is cointegration, it is shown in Engle and Granger (1987) that the VAR system always has a vector equilibrium-correcting model (VEqCM) representation, which can be written in differences and levels (disregarding the possible presence of deterministic variables like trends) in the following way:

$$\Delta x_t = \sum_{i=1}^{k-1} A_i \Delta x_{t-i} + \alpha(\beta' x_{t-1}) + \varepsilon_t,$$

(2.5)

where α and β are $n \times r$ matrices of rank r $(r < n)$ and $(\beta' x_{t-1})$ comprises r cointegrating I(0) relationships. Cointegrated processes are seen to define a long-run equilibrium trajectory and departures from this induce 'equilibrium correction' which moves the economy back towards its steady-state path. These models are useful as they often lend themselves to an economic interpretation of model properties and their long-run (steady-state) properties may be given an interpretation as long-run equilibria between economic variables that are derived from economic theory. Theoretical consistency, that is, that the model contains identifiable structures that are interpretable in the light of economic theory, is but one criterion for a satisfactory representation of the economy.

2.3.2 Congruence

If one considers all the reduction operations involved in the process of going from the hypothetical DGP to an empirical model, it is evident that any econometric model is unlikely to coincide with the DGP. An econometric model may however, possess certain desirable properties, which will render it a valid representation of the DGP. According to the LSE methodology (see Mizon 1995 and Hendry 1995a), such a model should satisfy the following six criteria:

1. The model contains identifiable structures that are interpretable in the light of economic theory.
2. The errors should be homoscedastic innovations in order for the model to be a valid simplification of the DGP.
3. The model must be data admissible on accurate observations.
4. The conditioning variables must be (at least) *weakly exogenous* for the parameters of interest in the model.
5. The parameters must be constant over time and remain invariant to certain classes of interventions (depending on the purpose for which the model is to be used).
6. The model should be able to encompass rival models. A model M_i encompasses other models $(M_j, j \neq i)$ if it can explain the results obtained by the other models.

Models that satisfy the first five criteria are said to be *congruent*, whereas an *encompassing congruent* model satisfies all six. Below, we comment on each of the requirements.

Economic theory (item 1) is a main guidance in the formulation of econometric models. Clear interpretation also helps communication of ideas and results among researchers and it structures the debate about economic issues. However, since economic theories are necessarily abstract and build on simplifying assumptions, a direct translation of a theoretical relationship to an econometric model will generally not lead to a satisfactory model. Notwithstanding their structural interpretation, such models will lack structural properties.

There is an important distinction between seeing theory as representing *the* correct specification (leaving parameter estimation to the econometrician), and viewing theory as a guideline in the specification of a model which also accommodates institutional features, attempts to accommodate heterogeneity among agents, addresses the temporal aspects for the data set and so on (see, for example, Granger 1999). Likewise, there is a huge methodological difference between a procedure of sequential simplification while controlling for innovation errors as in Section 2.3.1 and the practice of adopting an axiom of a priori correct specification which by assumption implies white noise errors.

Homoscedastic innovation errors (item 2) mean that residuals cannot be predicted from the model's own information set. Hence they are relative to that set. This is a property that follows logically from the reduction process and it is a necessary requirement for the empirical model to be one that is derived from the DGP. If the errors do not have this property, for example, if they are not white noise, some regularity in the data has not yet been included in the specification.

The requirement that the model must be data admissible (item 3) entails that the model must not produce predictions that are not logically possible. For example, if the data to be explained are proportions, the model should force all outcomes into the zero to one range.

Criterion 4 (weak exogeneity) holds if the parameters of interest are functions of $\theta_{y|z}$ (see (2.4)), which vary independently of θ_x (see equation (2.3) and Engle *et al.* 1983 for a formal definition). This property relates to estimation efficiency: weak exogeneity of the conditioning variables z_t is required for estimation of the conditional model for y_t without loss of information relative to estimation of the joint model for y_t and z_t. In order to make conditional forecasts from the conditional model without loss of information, strong exogeneity is required. This is defined as the joint occurrence of weak exogeneity and *Granger noncausality*, which is absence of feedback from y_t to z_t, that is x_{t-1}^1 in the marginal density function for z_t, $D_z(z_t \mid x_{t-1}^1, x_0; \lambda_z)$ in equation (2.2), does not include lagged values of y_t.

Item 5 in the list is spelt out in greater detail in Hendry (1995a: pp. 33–4), where he gives a formal and concise definition. He defines structure as the set of basic permanent features of the economic mechanism. A vector of parameters defines a structure if it is invariant and directly characterises the relations under analysis, that is, it is not derived from more basic parameters. A parameter can be structural only if it is

- constant and so is invariant to an extension of the sample period;
- unaltered by changes elsewhere in the economy and so is invariant to regime shifts, etc.;
- remains the same for extensions of the information set and so is invariant to adding more variables to the analysis.

This invariance property is of particular importance for a progressive research programme: ideally, empirical modelling is a cumulative process where models continuously become overtaken by new and more useful ones. By useful, we understand models that possess structural properties (items 1–5), in particular models that are relatively invariant to changes elsewhere in the economy, that is, they contain autonomous parameters (see Frisch 1938, Haavelmo 1944,

Johansen 1977, and Aldrich 1989). Models with a high degree of autonomy represent structure: they remain invariant to changes in economic policies and other shocks to the economic system, as implied by the definition above.

However, structure is partial in two respects: first, autonomy is a relative concept, since an econometric model cannot be invariant to every imaginable shock; second, all parameters of an econometric model are unlikely to be equally invariant. Parameters with the highest degree of autonomy represent *partial structure* (see Hendry 1993*b*, 1995*b*). Examples are elements of the β-vector in a cointegrating equation, which are often found to represent partial structure, as documented by Ericsson and Irons (1994). Finally, even though submodels are unlikely to contain partial structure to the same degree, it seems plausible that very aggregated models are less autonomous than the submodels, simply because the submodels can build on a richer information set.

Data congruence, that is, ability to characterise the data, remains an essential quality of useful econometric models (see Granger 1999 and Hendry 2002). In line with this, our research strategy is to check any hypothesised general model which is chosen as the starting point of a specification search for data congruence, and to decide on a final model after a general-to-specific (Gets) specification search. Due to recent advances in the theory and practice of data based model building, we know that by using Gets algorithms, a researcher stands a good chance of finding a close approximation to the data generating process (see Hoover and Perez 1999 and Hendry and Krolzig 1999), and that the danger of over-fitting is in fact surprisingly low.[5]

A congruent model is not necessarily a true model. Hendry (1995*a*: ch. 4) shows that an innovation is relative to its information set but may be predictable from other information. Hence, a sequence of congruent models could be developed and each of them encompassing all previous models. So satisfying all six criteria provides a recipe for a progressive research strategy. Congruency and its absence can be tested against available information, and hence, unlike *truth*, it is an operational concept in an empirical science (see Bontemps and Mizon 2003).

Finally, it should be noted that a strategy that puts a lot of emphasis on forecast behaviour, without a careful evaluation of the causes of forecast failure *ex post*, runs a risk of discarding models that actually contain important elements of structure. Hence, for example, Doornik and Hendry (1997*a*)

---

[5] Naturally, with a very liberal specification strategy, overfitting will result from Gets modelling, but with 'normal' requirements of levels of significance, robustness to sample splits, etc., the chance of overfitting is small. Thus the documented performance of Gets modelling now refutes the view that the *axiom of correct specification* must be invoked in applied econometrics (Leamer 1983). The real problem of empirical modelling may instead be to keep or discover an economically important variable that has yet to manifest itself strongly in the data (see Hendry and Krolzig 2001). Almost by implication, there is little evidence that Gets leads to models that are prone to forecast failure: see Clements and Hendry (2002).

and Clements and Hendry (1999a: ch.3) show that the main source of forecast failure is deterministic shifts in means (e.g. the equilibrium savings rate), and not shifts in such coefficients (e.g. the propensity to consume) that are of primary concern in policy analysis. Structural breaks are a main concern in econometric modelling, but like any hypothesis of theory, the only way to judge the quality of a hypothesised break is by confrontation with the evidence in the data. Moreover, given that an encompassing approach is followed, a forecast failure is not merely destructive but represents a potential for improvement, since respecification follows in its wake: see Section 2.4.2.

2.4 An example: modelling the household sector

The complete Haavelmo distribution function—for example, the joint distribution (2.1) of all variables of the macro model—is not tractable and hence not an operational starting point for empirical econometric analysis. In practice, we have to split the system into subsystems of variables and to analyse each of them separately. Joint modelling is considered only within subsystems. But by so doing, one risks ignoring possible influences across the subsystems. This would translate into invalid conditioning (the weak exogeneity assumption is not satisfied) and invalid marginalisation (by omitting relevant explanatory variables from the analysis), which are known to imply inefficient statistical estimation and inference. The practical implementation of these principles is shown in an example drawn from the modelling of the household sector of the RIMINI model (see Chapter 1).

The process of sequential decomposition into conditional and marginal models is done repeatedly within the subsystems of RIMINI. In the household sector subsystem, total consumer expenditure, ch_t, is modelled as a function of real household disposable income, yh_t, and real household wealth, wh_t. (Here and in the rest of the book, small letters denote logs of variables.) Total wealth consists of the real value of the stock of housing capital plus net financial wealth. The volume of the residential housing stock is denoted H_t and the real housing price is $(PH)_t/P_t$, where P_t is the national accounts price deflator for total consumption expenditure. The sum of net real financial assets is equal to the difference between real gross financial assets and real loans ($M_t - L_t$), yielding

$$wh_t = \ln WH_t = \ln \left[\left(\frac{PH_t}{P_t} \right) H_{t-1} + M_t - L_t \right].$$

The joint distribution function for this subsystem can be written as (2.1) with $x_t = (ch_t, yh_t, wh_t)$. The conditional submodel for total real consumer expenditure ch_t (Brodin and Nymoen 1992—B&N hereafter), is

$$D_{c|y,w}(ch_t \mid yh_t, wh_t; \lambda_c),$$

relying on the corresponding conditional density function, (2.4), to be a valid representation of the DGP. RIMINI contains submodels for yh_t and for all individual components in wh_t. For example, the conditional submodel for simultaneous determination of housing prices, ph_t, and real household loans, l_t, is

$$D_{w|y}(ph_t, l_t \mid RL_t, yh_t, h_{t-1}; \lambda_w),$$

where RL_t denotes the interest rate on loans, and conditional submodels for the net addition to housing capital stock Δh_t, and the price of new housing capital, phn_t

$$D_{\Delta h|\cdot}(\Delta h_t \mid ph_t, phn_t, RL_t, yh_t, h_{t-1}; \lambda_{\Delta h})$$

$$D_{phn|\cdot}(phn_t \mid ph_t, pj_t, h_{t-1}; \lambda_{phn}),$$

where pj_t is the deflator of gross investments in dwellings.

2.4.1 The aggregate consumption function

The model for aggregate consumption in B&N satisfies the criteria we listed in Section 2.3. They provide a model in which cointegration analysis establishes that the linear relationship

$$ch_t = \text{constant} + 0.56yh_t + 0.27wh_t, \qquad (2.6)$$

is a cointegrating relationship and that the cointegration rank is one. Hence, while the individual variables in (2.6) are assumed to be non-stationary and integrated, the linear combination of the three variables is stationary with a constant mean showing the discrepancy between consumption and its long-run equilibrium level $0.56yh_t + 0.27wh_t$. Moreover, income and wealth are *weakly exogenous* for the cointegration parameters. Hence, the equilibrium correction model for Δch_t satisfies the requirements of valid conditioning. Finally, the cointegration parameters appears to be *invariant*. The estimated marginal models for income and wealth show evidence of structural breaks. The joint occurrence of a stable conditional model (the consumption function) and unstable marginal models for the conditional variables is evidence of within sample invariance of the coefficients of the conditional model and hence super exogenous conditional variables (income and wealth). The result of invariance is corroborated by Jansen and Teräsvirta (1996), using an alternative method based on smooth transition models.

The empirical consumption function in B&N has proven to be relatively stable for more than a decade, in particular this applies to the cointegration part of the equation. Thus, it is of particular interest to compare it with rival models in the literature.

2.4.2 Rival models

Financial deregulation in the mid-1980s led to a strong rise in aggregate consumption relative to income in several European countries. The pre-existing empirical macroeconometric consumption functions in Norway, which typically explained aggregate consumption by income, all broke down—that is, they failed in forecasting, and failed to explain the data *ex post*.

As stated in Eitrheim *et al.* (2002*b*), one view of this forecast failure is that it provided direct evidence in favour of the rival rational expectations, permanent income hypothesis: in response to financial deregulation, consumers revised their expected permanent income upward, thus creating a break in the conditional relationship between consumption and income. The breakdown has also been interpreted as a confirmation of the relevance of the Lucas critique, in that it was a shock to a non-modelled expectation process that caused the structural break in the existing consumption functions.

Eitrheim *et al.* (2002*b*) compare the merits of the two competing models: the empirical consumption function (CF), conditioning on income in the long run, and an Euler equation derived from a model for expectation formation. We find that while the conditional consumption function encompasses the Euler equation (EE) on a sample from 1968(2) to 1984(4), both models fail to forecast the annual consumption growth in the next years. In the paper, we derive the theoretical properties of forecasts based on the two models. Assuming that the EE is the true model and that the consumption function is a mis-specified model, we show that both sets of forecasts are immune to a break (i.e. shift in the equilibrium savings rate) that occurs after the forecast have been made. Moreover, failure in 'before break' CF-forecasts is only (logically) possible if the consumption function is the true model *within* the sample. Hence, the observed forecast failure of the CF is corroborating evidence in favour of the conditional consumption function for the period before the break occurred.

However, a respecified consumption function—B&N of the previous section—that introduced wealth as a new variable was successful in accounting for the breakdown *ex post*, while retaining parameter constancy in the years of financial consolidation that followed after the initial plunge in the savings rate. The respecified model was able to adequately account for the observed high variability in the savings rate, whereas the earlier models failed to do so.

B&N noted the implication that the respecification explained why the Lucas critique lacked power in this case: first, while the observed breakdown of conditional consumption functions in 1984–85 is consistent with the Lucas critique, that interpretation is refuted by the finding of a conditional model with constant parameters. Second, the invariance result shows that an Euler equation type model (derived from, for example, the stochastic permanent income model) cannot be an encompassing model. Even if the Euler approach is supported by empirically constant parameters, such a finding cannot explain why a conditional model is also stable. Third, finding that invariance holds, at least as an empirical approximation, yields an important basis for the use of the dynamic consumption function in forecasting and policy analysis, the main practical usages of empirical consumption functions.

Eitrheim *et al.* (2002*b*) extend the data set by nine years of quarterly observations, that is, the sample is from 1968(3) to 1998(4). The national accounts were heavily revised for that period. We also extended the wealth measure to include non-liquid financial assets. Still we find that the main results of B&N are confirmed. Empirical support for one cointegrating vector between ch_t, yh_t, and wh_t, and valid conditioning in the consumption function is reconfirmed on the new data. In fact, full information maximum likelihood estimation of a four equation system explaining (the change in) ch_t, yh_t, wh_t, and $(ph_t - p_t)$ yields the same empirical results as estimation based on the conditional model. These findings thus corroborate the validity of the conditional model of B&N.

2.5   Is modelling subsystems and combining them to a global model a viable procedure?

The traditional approach to building large-scale macroeconometric models has been to estimate one equation (or submodel) at a time and collect the results in the simultaneous setting. Most often this has been done without testing for the adequacy of that procedure. The approach could, however, be defended from the estimation point of view. By adopting limited information maximum likelihood (LIML) methods, one could estimate the parameters of one equation, while leaving the parameters of other equations unrestricted: see Anderson and Rubin (1949)[6] and Koopmans and Hood (1953).[7] It has, however, also been argued that the limited information methods were more robust against mis-specified equations elsewhere in the system in cases where one had better theories or more reliable information about a subset of variables than about the rest (cf. Christ 1966, p. 539). Historically, there is little doubt that limited information methods—like LIML—were adopted out of practical considerations, to avoid the computational burden of full information methods—like full information maximum likelihood (FIML). The problem of sorting out the properties of the system that obtained when the bits and pieces were put together, remained unsolved.

That said, it is no doubt true that we run into uncharted territory when we—after constructing relevant submodels by marginalisation and conditioning—combine the small models of subsectors to a large macroeconometric

---

[6] Interestingly, the papers that introduced the limited information methods also introduced the first tests of overidentifying restrictions in econometrics.

[7] Johansen (2002) has pointed out that LIML does not work with cointegrated systems, where relaxing cross equation restrictions (implied by cointegration) changes the properties of the system.

model. As we alluded to in the Section 2.1, Johansen (2002) points out that a
general theory for the validity of the three steps will invariably contain criteria
and conditions which are formulated for the full system. The question thus is:
given that the full model is too large to be modelled simultaneously, is there a
way out?

One solution might be to stay with very aggregated models that are small
enough to be analysed as a complete system. Such an approach will necessarily leave out a number of economic mechanisms which we have found to be
important and relevant in order to describe the economy adequately.

Our general approach can be seen as one of gradualism—seeking to establish
structure (or *partial structure*) in the submodels. In Section 2.3.2 we gave a
formal definition of partial structure as partial models that are (1) invariant
to extensions of the sample period; (2) invariant to changes elsewhere in the
economy (e.g. due to regime shifts); and (3) remains the same for extensions of
the information set.

The first two of these necessary conditions do not require that we know
the full model. The most common cause for them to be broken is that there
are important omitted explanatory variables. This is detectable within the
frame of the submodel once the correlation structure between included and
excluded variables changes.

For the last of these conditions we can, at least in principle, think of the
full model as the ultimate extension of the information set, and so establishing structure or partial structure represents a way to break free of Søren
Johansen's Catch 22. In practice, however, we know that the full model is
not attainable. Nevertheless, we note that the conditional consumption function of Section 2.4.2 is constant when the sample is extended with nine years
of additional quarterly observations; it remains unaltered through the period
of financial deregulation and it also sustains the experiment of simultaneous
modelling of private consumption, household disposable income, household
wealth and real housing prices. We have thus found corroborating *inductive*
evidence for the conditional consumption function to represent partial structure. The simultaneous model is in this case hardly an ideal substitute for
a better model of the supply-side effects that operate through the labour
market, nonetheless it offers a safeguard against really big mistakes of the type
that causation 'goes the other way', for example, income is in fact equilibrium
correcting, not consumption.

There may be an interesting difference in focus between statisticians and
macroeconomic modellers. A statistician may be concerned about the estimation perspective, that is, the lack of efficiency by analysing a sequence of
submodels instead of a full model, whereas a macroeconomic modeller primarily wants to avoid mis-specified relationships. The latter is due to pragmatic
real-world considerations as macroeconomic models are used as a basis for
policy-making. From that point of view it is important to model the net coefficients of all relevant explanatory variables by also conditioning on all relevant

and applicable knowledge about institutional conditions in the society under study. Relying on more aggregated specifications where gross coefficients pick up the combined effects of the included explanatory variables and correlated omitted variables may lead to misleading policy recommendations. Our conjecture is that such biases are more harmful for policy makers than the simultaneity bias one may incur by combining submodels. Whether this holds true or not is an interesting issue which is tempting to explore by means of Monte Carlo simulations on particular model specifications.

That said, it is of particular importance to get the long-run properties of the submodel right. We know that once a cointegrating equation is found, it is invariant to extensions of the information set. On the other hand, this is a property that needs to be established in each case. We do not know what we do not know. One line of investigation that may shed light on this is associated with the notion of separation in cointegrated systems as described in Granger and Haldrup (1997). Their idea is to decompose each variable into a persistent (long-memory) component and a transitory (short-memory) component. Within the framework of a vector equilibrium correcting model like (2.5), the authors consider two subsystems, where the variables of one subsystem do not enter the cointegrating equations of the other subsystem (cointegration separation). Still, there may be short-term effects of the variables in one subsystem on the variables in the other and the cointegrating equations of one system may also affect the short-term development of the variables in the other. Absence of both types of interaction is called complete separation while if only one of these is present it is referred to as partial separation. These concepts are of course closely related to strong and weak exogeneity of the variables in one subsystem with respect to the parameters of the other. Both partially and completely separated submodels are testable hypotheses, which ought to be tested as part of the cointegration analysis. Hecq *et al.* (2002) extend the results of Granger and Haldrup (1997). The conclusion of Hecq *et al.* (2002) is, however, that testing of separation requires that the full system is known, which is in line with Søren Johansen's observation earlier.

In Chapter 9 we introduce a stepwise procedure for assessing the validity of a submodel for wages and prices for the economy at large. A detailed and carefully modelled core model for wage and price determination (a Model A of Section 2.1) is supplemented with marginal models for the conditioning variables in the core model. The extended model is cruder and more aggregated than the full Model B of Section 2.1. Notwithstanding this, it enables us to test valid conditioning (weak exogeneity) as well as invariance (which together with weak exogeneity defines super exogeneity) of the core model on criteria and conditions formulated within the extended model. The approach features a number of ingredients that are important for establishing an econometrically relevant submodel, and—as in the case of the consumption function—this points to a way to avoid the Catch 22 by establishing partial structure.

3

Inflation in open economies: the main-course model

The chapter introduces Aukrust's main-course model of wage- and price-setting. Our reconstruction of Aukrust's model will use elements both from rational reconstructions, which present past ideas with the aid of present-day concepts and methods, and historical reconstructions, which understand older theories in the context of their own times. Our excursion into the history of macroeconomic thought is both traditional and pluralistic. The aim with the appraisal is to communicate the modernity of the set of testable hypotheses emerging from Aukrust's model to interested practitioners.

3.1   Introduction

As noted in the introductory chapter, an important development of macro-econometric models has been the representation of the supply side of the economy, and wage–price dynamics in particular. This chapter and the next three (Chapters 4–6) present four frameworks for wage–price modelling, which all have played significant roles in shaping macroeconometric models in Norway, as well as in several other countries. We start in this chapter with a reconstruction of Aukrust's main-course model of inflation, using the modern econometric concepts of cointegration and causality. This rational reconstruction shows that, despite originating back in the mid-1960s, the main-course model resembles present day theories of wage formation with unions and price-setting firms, and markup pricing by firms.

In its time, the main-course model of inflation was viewed as a contender to the Phillips curve, and in retrospect it is easy to see that the Phillips curve won. However, the Phillips curve and the main-course model are in fact not mutually exclusive. A conventional open economy version of the Phillips curve can be incorporated into the main-course model, and in Chapter 4, we approach the Phillips curve from that perspective.

The main-course model of inflation was formulated in the 1960s.[1] It became the framework for both medium-term forecasting and normative judgements about 'sustainable' centrally negotiated wage growth in Norway.[2] In this section we show that Aukrust's (1977) version of the model can be reconstructed as a set of propositions about cointegration properties and causal mechanisms. The reconstructed main-course model serves as a reference point for, and in some respects also as a corrective to, the modern models of wage formation and inflation in open economies, for example, the open economy Phillips curve and the imperfect competition model of, for example, Layard and Nickell (1986: Sections 4.2 and 5). It also motivates our generalisation of these models in Section 6.9.2.

Central to the model is the distinction between a sector where strong competition makes it reasonable to model firms as price takers, and another sector (producing non-traded goods) where firms set prices as markups on wage costs. Following convention, we refer to the price taking sector as the *exposed* sector, and the other as the *sheltered* sector. In equations (3.1)–(3.7), $w_{e,t}$ denotes the nominal wage in the exposed (e) industries in period t. Foreign prices in domestic currency are denoted by pf_t, while $q_{e,t}$ and $a_{e,t}$ are the product price and average labour productivity of the exposed sector. $w_{s,t}$, $q_{s,t}$, and $a_{s,t}$ are the corresponding variables of the sheltered (s) sector.[3] All variables are measured

---

[1] In fact there were two models, a short-term multisector model and the long-term two sector model that we reconstruct using modern terminology in this chapter. The models were formulated in 1966 in two reports by a group of economists who were called upon by the Norwegian government to provide background material for that year's round of negotiations on wages and agricultural prices. The group (Aukrust, Holte, and Stoltz) produced two reports. The second (dated 20 October 1966, see Aukrust 1977) contained the long-term model that we refer to as the main-course model. Later, there were similar developments in, for example, Sweden (see Edgren *et al.* 1969) and the Netherlands (see Driehuis and de Wolf 1976).

In later usage the distinction between the short- and long-term models seems to have become blurred, in what is often referred to as the Scandinavian model of inflation. Rødseth (2000: ch. 7.6) contains an exposition and appraisal of the Scandinavian model in terms of current macroeconomic theory. We acknowledge Aukrust's clear exposition and distinction in his 1977 paper, and use the name main-course model for the long-run version of his theoretical framework.

[2] On the role of the main-course model in Norwegian economic planning, see Bjerkholt (1998).

[3] In France, the distinction between sheltered and exposed industries became a feature of models of economic planning in the 1960s, and quite independently of the development in Norway. In Courbis (1974), the main-course theory is formulated in detail and illustrated with data from French post-war experience (we are grateful to Odd Aukrust for pointing this out to us).

in natural logarithms, so, for example, $w_{i,t} = \log(W_{i,t})$ for the wage rates $(i = e, s)$.

$$q_{e,t} = pf_t + v_{1,t}, \tag{3.1}$$

$$pf_t = g_f + pf_{t-1} + v_{2,t}, \tag{3.2}$$

$$a_{e,t} = g_{a_e} + a_{e,t-1} + v_{3,t}, \tag{3.3}$$

$$w_{e,t} - q_{e,t} - a_{e,t} = m_e + v_{4,t}, \tag{3.4}$$

$$w_{s,t} = w_{e,t} + v_{5,t}, \tag{3.5}$$

$$a_{s,t} = g_{a_s} + a_{s,t-1} + v_{6,t}, \tag{3.6}$$

$$w_{s,t} - q_{s,t} - a_{s,t} = m_s + v_{7,t}. \tag{3.7}$$

The parameters g_i $(i = f, a_e, a_s)$ are constant growth rates, while m_i $(i = e, s)$ are means of the logarithms of the wage shares in the two industries.

The seven stochastic processes $v_{1,t}$ $(i = 1, \ldots, 7)$ play a key role in our reconstruction of Aukrust's theory. They represent separate ARMA processes. The roots of the associated characteristic polynomials are assumed to lie on or outside the unit circle. Hence, $v_{1,t}$ $(i = 1, \ldots, 7)$ are causal ARMA processes, cf. Brockwell and Davies (1991).

Before we turn to the interpretation of the model, we follow convention and define p_t, the log of the consumer price index (CPI), as a weighted average of $q_{s,t}$ and $q_{e,t}$:

$$p_t = \phi q_{s,t} + (1 - \phi) q_{e,t}, \qquad 0 < \phi < 1, \tag{3.8}$$

where ϕ is a coefficient that reflects the weight of non-traded goods in private consumption.[4]

3.2 Cointegration

Equation (3.1) captures the price taking behaviour characterising the exposed industries, and (3.2)–(3.3) define foreign prices of traded goods (pf_t) and labour productivity as random walks with drifts. Equation (3.4) serves a double function: first, it defines the exposed sector wage share $w_{e,t} - q_{e,t} - a_{e,t}$ as a stationary variable since $v_{4,t}$ on the right-hand side is I(0) by assumption; second, since both $q_{e,t}$ and $a_{e,t}$ are I(1) variables, the nominal wage $w_{e,t}$ is also non-stationary I(1).

The sum of the technology trend and the foreign prices plays an important role in the theory since it traces out a central tendency or long-run sustainable scope for wage growth. Aukrust (1977) refers to this as the main course for wages in the exposed industries. Thus, for later use, we define the main-course variable: $mc_t = a_{e,t} + q_{e,t}$. The essence of the statistical interpretation of the theory is captured by the assumption that $v_{1,t}$ is ARMA, and thus I(0).

---

[4] Note that, due to the log-form, $\phi = x_s/(1 - x_s)$ where x_s is the share of non-traded goods in consumption.

It follows that $w_{e,t}$ and mc_t are cointegrated, that the difference between $w_{e,t}$ and mc_t has a finite variance, and that deviations from the main course will lead to equilibrium correction in $w_{e,t}$ (see Nymoen 1989a and Rødseth and Holden 1990).

Hypothetically, if shocks were switched off from period 0 onwards, the wage level would follow the deterministic function

$$\mathsf{E}[w_{e,t} \mid mc_0] = m_e + (g_f + g_{a_e})t + mc_0, \qquad mc_0 = pf_0 + a_{e,0}, \quad (t = 1, 2, \ldots). \tag{3.9}$$

The variance of $w_{e,t}$ is unbounded, reflecting the stochastic trends in productivity and foreign prices, thus $w_{e,t} \sim \mathsf{I}(1)$.

In his 1977 paper, Aukrust identifies the 'controlling mechanism' in equation (3.4) as fundamental to his theory:

The profitability of the E industries is a key factor in determining the wage level of the E industries: mechanism are assumed to exist which ensure that the higher the profitability of the E industries, the higher their wage level; there will be a tendency of wages in the E industries to adjust so as to leave actual profits within the E industries close to a 'normal' level (for which, however, there is no formal definition). (Aukrust 1977, p. 113)

In our reconstruction of the theory, the normal rate of profit is simply $1 - m_e$. Aukrust also carefully states the long-term nature of his hypothesised relationship:

The relationship between the 'profitability of E industries' and the 'wage level of E industries' that the model postulates, therefore, is certainly not a relation that holds on a year-to-year basis. At best it is valid as a long-term tendency and even so only with considerable slack. It is equally obvious, however, that the wage level in the E industries is not completely free to assume any value irrespective of what happens to profits in these industries. Indeed, if the actual profits in the E industries deviate much from normal profits, it must be expected that sooner or later forces will be set in motion that will close the gap. (Aukrust 1977, pp. 114–15)

Aukrust goes on to specify 'three corrective mechanisms', namely wage negotiations, market forces (wage drift, demand pressure) and economic policy. If we in these quotations substitute 'considerable slack' with '$v_{1,t}$ being autocorrelated but $\mathsf{I}(0)$', and 'adjustment' and 'corrective mechanism' with 'equilibrium correction', it is seen how well the concepts of cointegration and equilibrium correction match the gist of Aukrust's original formulation. Conversely, the use of growth rates rather than levels, which became common in both text book expositions of the theory and in econometric work claiming to test it (see Section 3.2.3) misses the crucial point about a low frequency, long-term relationship between foreign prices, productivity, and exposed sector wage-setting.

Aukrust coined the term 'wage corridor' to represent the development of wages through time and used a graph similar to Figure 3.1 to illustrate his ideas. The main course defined by equation (3.9) is drawn as a straight line since the wage is measured in logarithmic scale. The two dotted lines represent

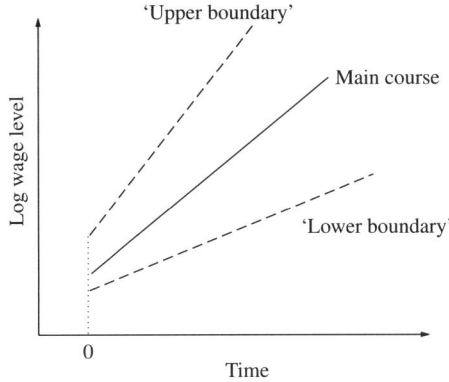

Figure 3.1. The 'wage corridor' in the Norwegian model of inflation

what Aukrust called the 'elastic borders of the wage corridor'. In econometric terminology, the vertical distance between the lines represents a confidence interval, for example, $\mathsf{E}[w_t|mc_0] \pm 2$ standard errors, where the standard errors are conditional on an initial value mc_0. The unconditional variance does not exist, so the wage corridor widens up as we move away from mc_0.

Equation (3.5) incorporates two other substantive hypotheses in the Norwegian model of inflation: stationarity of the relative wage between the two sectors (normalised to unity), and wage leadership of the exposed sector. Thus, the sheltered sector is a wage follower, with exposed sector wage determinants also in effect shaping sheltered sector wage development.

Equation (3.6) allows a separate trend in labour productivity in the sheltered sector and equation (3.7) contains the stationarity hypothesis of the sheltered sector wages share. Given the nature of wage-setting and the exogenous technology trend, equation (3.7) implies that sheltered sector price-setters mark up their prices on average variable costs. Thus sheltered sector price formation adheres to so-called *normal cost pricing*.

To summarise, the three cointegration propositions of the reconstructed main-course model are:

$$\text{H1}_{mc} \quad w_{e,t} - q_{e,t} - a_{e,t} = m_e + v_{4,t}, v_{4,t} \sim \mathsf{I}(0),$$
$$\text{H2}_{mc} \quad w_{e,t} = w_{s,t} + v_{5,t}, v_{5,t} \sim \mathsf{I}(0),$$
$$\text{H3}_{mc} \quad w_{s,t} - q_{s,t} - a_{s,t} = m_s + v_{7,t}, v_{7,t} \sim \mathsf{I}(0).$$

H1_{mc} states that the exposed sector wage level cointegrates with the sectorial price and productivity levels, with unit coefficients and for a constant mean of the wage share, m_e. However, the institutional arrangements surrounding wage-setting change over time, so heuristically m_e may be time dependent. For example, bargaining power and unemployment insurance systems are not constant factors but evolve over time, sometimes abruptly too. In his 1977 paper, Aukrust himself noted that the assumption of a completely constant mean wage share over long time spans was probably not tenable. However, no internal inconsistency is caused by replacing the assumption of unconditionally stationary wage shares with the weaker assumption of conditional stationarity. Thus, we consider in the following an extended main-course model where the mean of the wage share is a linear function of exogenous I(0) variables and of deterministic terms.

For example, a plausible generalisation of $H1_{mc}$ is represented by

$$H1_{gmc} \quad w_{e,t} - q_{e,t} - a_{e,t} = m_{e,0} + \beta_{e,1} u_t + \beta_{e,2} D_t + v_{4,t},$$

where u_t is the log of the rate of unemployment and D_t is a dummy (vector) that along with u_t help explain shifts in the mean of the wage share, thus in $H1_{gmc}$, $m_{e,0}$ denotes the mean of the cointegration relationship, rather than of the wage share itself. Consistency with the main-course theory requires that the rate of unemployment is interpreted as I(0), but not necessarily stationary, since u_t may in turn be subject to changes in its mean, that is, structural breaks. Graphically, the main course in Figure 3.1 is no longer necessarily a straight unbroken line (unless the rate of unemployment and D_t stay constant for the whole time period considered).

Other candidate variables for inclusion in an extended main-course hypothesis are the ratio between unemployment insurance payments and earnings (the so-called *replacement ratio*) and variables that represent unemployment composition effects (unemployment duration, the share of labour market programmes in total unemployment); see Nickell (1987), Calmfors and Forslund (1991). In Section 3.5 we shall see that in this extended form, the cointegration relationship implied by the main-course model is fully consistent with modern wage bargaining theory.

Following the influence of trade union and bargaining theory, it has also become popular to estimate real-wage equations that include a so-called *wedge* between real wages and the consumer real wage, that is, $p_t - q_{e,t}$ in the present framework. However, inclusion of a wedge variable in the cointegrating wage equation of an exposed sector is inconsistent with the main-course hypothesis, and finding such an effect empirically may be regarded as evidence against the framework. On the other hand, there is nothing in the main-course theory that rules out substantive short-run influences of the CPI, that is, of Δp_t in a dynamic wage equation. In Chapter 6 we analyse a model that contains this form of realistic short-run dynamics.

The other two cointegration propositions ($H2_{mc}$ and $H3_{mc}$) in Aukrust's model have not received nearly as much attention as $H1_{mc}$ in empirical research, but exceptions include Rødseth and Holden (1990) and Nymoen (1991). In part, this is due to lack of high quality wage and productivity data for the private service and retail trade sectors. Another reason is that both economists and policy makers in the industrialised countries place most emphasis on understanding and evaluating wage-setting in manufacturing, because of its continuing importance for the overall economic performance.

3.2.1 Causality

The main-course model specifies the following three hypothesis about causation:

$$H4_{mc} \quad mc_t \rightarrow w_{e,t},$$
$$H5_{mc} \quad w_{e,t} \rightarrow w_{s,t},$$
$$H6_{mc} \quad w_{s,t} \rightarrow p_t,$$

where $\rightarrow$ denotes one-way causation. Causation may be contemporaneous or of the Granger-causation type. In any case the defining characteristic of the main-course model is that there is no feedback between, for example, domestic cost of living (p_t) and the wage level in the exposed sector. In his 1977 paper, Aukrust sees the causation part of the theory ($H4_{mc}$–$H6_{mc}$) as just as important as the long-term 'controlling mechanism' ($H1_{mc}$–$H3_{mc}$). If anything, Aukrust seems to put extra emphasis on the causation part. For example, he argues that exchange rates must be controlled and not floating, otherwise pf_t (foreign prices denoted in domestic currency) is not a pure causal factor of the domestic wage level in equation (3.2), but may itself reflect deviations from the main course, thus

In a way, ..., the basic idea of the Norwegian model is the 'purchasing power doctrine' in reverse: whereas the purchasing power doctrine assumes floating exchange rates and explains exchange rates in terms of relative price trends at home and abroad, this model assumes controlled exchange rates and international prices to explain trends in the national price level. If exchange rates are floating, the Norwegian model does not apply. (Aukrust 1977, p. 114)

From a modern viewpoint this seems to be unduly restrictive since the cointegration part of the model can be valid even if Aukrust's one-way causality is untenable. Consider for example, $H1_{mc}$, the main-course proposition for the exposed sector, which in modern econometric methodology implies rank reduction in the system made up of $w_{e,t}$, $q_{e,t}$, and $a_{e,t}$, but not necessarily one-way causation. Today, we would regard it as both meaningful and significant if an econometric study showed that $H1_{mc}$ (or more realistically $H1_{gmc}$) constituted a single cointegrating vector between the three I(1) variables $\{w_{e,t}, q_{e,t}, a_{e,t}\}$, even if $q_{e,t}$ and $a_{e,t}$, not only $w_{e,t}$, contribute to the correction of deviations from the main course. Clearly, we would no longer have a 'wage model' if w_t was found to be weakly exogenous with respect to the parameters of the cointegrating vector, but that is a very special case, just as $H4_{mc}$ is a very strict hypothesis. Between these polar points there are many constellations with two-way causation that make sense in a dynamic wage–price model.

In sum, although care must be taken when we attempt to estimate a long-run wage equation with data from different exchange rate regimes, it seems unduly restrictive a priori to restrict the relevance of Aukrust's model to a fixed exchange rate regime.

3.2.2 Steady-state growth

In a hypothetical steady-state situation, with all shocks represented by $(i = 1, 2, \ldots, 7)$ switched off, the model can be written as a set of (deterministic) equations between growth rates

$$\Delta w_{e,t} = g_f + g_{a_e}, \tag{3.10}$$
$$\Delta w_{s,t} = \Delta w_{e,t}, \tag{3.11}$$
$$\Delta q_{s,t} = \Delta w_{s,t} - g_{a_s}, \tag{3.12}$$
$$\Delta p_t = \phi g_f + (1 - \phi)\Delta q_{s,t}. \tag{3.13}$$

Most economist are familiar with this 'growth rate' version of the model, often referred to as the 'Scandinavian model of inflation'. The model can be solved for the domestic rate of inflation:

$$\Delta p_t = g_f + (1 - \phi)(g_{a_e} - g_{a_s}),$$

implying a famous result of the Scandinavian model, namely that a higher productivity growth in the exposed sector *ceteris paribus* implies increased domestic inflation.

3.2.3 Early empiricism

In the reconstruction of the model that we have undertaken above, no inconsistencies exist between Aukrust's long-term model and the steady-state model in growth-rate form. However, economists and econometricians have not always been precise about the steady-state interpretation of the system (3.10)–(3.13). For example, it seems to have inspired the use of differenced-data models in empirical tests of the Scandinavian model—Nordhaus (1972) is an early example.[5]

With the benefit of hindsight, it is clear that growth rate regressions only superficially capture Aukrust's ideas about long-run relationships between price and technology trends: by differencing, the long-run frequency is removed from the data used in the estimation; see, for example, Nymoen (1990: ch. 1). Consequently, the regression coefficient of, for example, $\Delta a_{e,t}$ in a model of $\Delta w_{e,t}$ does not represent the long-run elasticity of the wage with respect to productivity. The longer the adjustment lags, the larger the bias caused by wrongly identifying coefficients on growth-rate variables with true long-run elasticities. Since there are typically long adjustment lags in wage-setting, even studies that use annual data typically find very low coefficients on the productivity growth terms.

The use of differenced data clearly reduced the chances of finding formal evidence of the long-term propositions of Aukrust's theory. However, at the same time, the practice of differencing the data also meant that one

---

[5] See Hendry (1995a: ch. 7.4) on the role of differenced data models in econometrics.

avoided the pitfall of *spurious* regressions (see Granger and Newbold 1974). For example, using conventional tables to evaluate '*t*-values' from levels regression, it would have been all too easy to find support for a relationship between the main course and the level of wages, even if no such relationship existed. Statistically valid testing of the Norwegian model had to await the arrival of cointegration methods and inference procedures for integrated data (see Nymoen 1989*a* and Rødseth and Holden 1990). Our evaluation of the validity of the extended main-course model for Norwegian manufacturing is found in Section 5.5, where we estimate a cointegrating relationship for Norwegian manufacturing wages, and in Section 6.9.2, where a dynamic model is formulated.

3.2.4 Summary

Unlike the other approaches to modelling wages and prices that we discuss in the next chapters, Aukrust's model (or the Scandinavian model for that matter) is seldom cited in the current literature. There are two reasons why this is unfortunate. First, Aukrust's theory is a rare example of a genuinely macroeconomic theory that deals with aggregates which have precise and operational definitions. Moreover, Aukrust's explanation of the hypothesised behavioural relationships is 'thick', that is, he relies on a broad set of formative forces which are not necessarily reducible to specific ('thin') models of individual behaviour. Second, the Norwegian model of inflation sees inflation as a many-faceted system property, thus avoiding the one-sidedness of many more recent theories that seek to pinpoint one (or a few) factors behind inflation (e.g. excess money supply, excess product demand, too low unemployment, etc.).

In the typology of Rorty (1984), our reconstruction of Aukrust's model has used elements both from rational reconstructions, which present past ideas with the aid of present-day concepts and methods, and historical reconstructions, which understand older theories in the context of their own times. Thus, our brief excursion into the history of macroeconomic thought is traditional and pluralistic as advocated by Backhouse (1995: ch. 1). Appraisal in terms of modern concepts hopefully communicates the set of testable hypotheses emerging from Aukrust's model to interested practitioners. On the other hand, Aukrust's taciturnity on the relationship between wage-setting and the determination of long-run unemployment is clearly conditioned by the stable situation of near full employment in the 1960s. In Chapter 4, we show how a Phillips curve can be combined with Aukrust's model so that unemployment is endogenised. We will also show that later models of the bargaining type, can be viewed as extensions (and new derivations) rather than contradictory to Aukrust's contribution.

4

The Phillips curve

The Phillips curve ranges as the dominant approach to wage and price modelling in macroeconomics. In the United States, in particular, it retains its role as the operational framework for both inflation forecasting and for estimating the NAIRU. In this chapter, we will show that the Phillips curve is consistent with cointegration between prices, wages, and productivity, and a stationary rate of unemployment, and hence there is common ground between the Phillips curve and the Norwegian model of inflation of the previous chapter.

4.1 Introduction

The Norwegian model of inflation and the Phillips curve are rooted in the same epoch of macroeconomics. But while Aukrust's model dwindled away from the academic scene, the Phillips curve literature 'took off' in the 1960s and achieved immense impact over the next four decades. Section 4.1.1 records some of the most noteworthy steps in the developments of the Phillips curve. In the 1970s, the Phillips curve and Aukrust's model were seen as alternative, representing 'demand' and 'supply' model of inflation respectively (see Frisch 1977). However, as pointed out by Aukrust (1977), the difference between viewing the labour market as the important source of inflation and the Phillips curve's focus on product market, is more a matter of emphasis than of principle, since both mechanisms may be operating together.[1] In Section 4.2, we show formally how the two approaches can be combined by letting the Phillips curve take the role of a short-run relationship of nominal wage growth, while the main-course thesis holds in the long run.

This chapter also addresses issues which are central to modern applications of the Phillips curve: its representation in a system of cointegrated variables;

---

[1] See Aukrust (1977, p. 130).

consistency or otherwise with hysteresis and mean shifts in the rate of unemployment (Section 4.3); the uncertainty of the estimated Phillips curve NAIRU (Section 4.4); and the status of the inverted Phillips curve, that is, Lucas's supply curve (Section 4.5.2). Sections 4.1.1–4.5 cover these theoretical and methodological issues while Section 4.6 shows their practical relevance in a substantive application to the Norwegian Phillips curve.

4.1.1   Lineages of the Phillips curve

Following Phillips' (1958) discovery of an empirical regularity between the rate of unemployment and money wage inflation in the United Kingdom, the Phillips curve was integrated in macroeconomics through a series of papers in the 1960s. Samuelson and Solow (1960) interpreted it as a tradeoff facing policy makers, and Lipsey (1960) was the first to estimate Phillips curves with multivariate regression techniques. Lipsey interpreted the relationship from the perspective of classical price dynamics, with the rate of unemployment acting as a proxy for excess demand and friction in the labour market. Importantly, Lipsey included consumer price growth as an explanatory variable in his regressions, and thus formulated what has become known as the expectations augmented Phillips curve. Subsequent developments include the distinction between the short-run Phillips curve, where inflation deviates from expected inflation, and the long-run Phillips curve, where inflation expectations are fulfilled. Finally, the concept of a natural rate of unemployment was defined as the steady-state rate of unemployment corresponding to a vertical long-run curve (see Phelps 1968 and Friedman 1968).

The relationship between money wage growth and economic activity also figures prominently in new classical macroeconomics; see, for example, Lucas and Rapping (1969, 1970), Lucas (1972). However, in new classical economics the causality in Phillips' original model was reversed: if a correlation between inflation and unemployment exists at all, the causality runs from inflation to the level of activity and unemployment. Lucas's and Rapping's inversion is based on the thesis that the level of prices is anchored in a quantity theory relationship and an autonomous money stock. Price and wage growth is then determined from outside the Phillips curve, so the correct formulation would be to have the rate of unemployment on the left-hand side and the rate of wage growth (and/or inflation) on the right-hand side.

Lucas's 1972 paper provides another famous derivation based on rational expectations about uncertain relative product prices. If expectations are fulfilled (on average), aggregate supply is unchanged from last period. However, if there are price surprises, there is a departure from the long-term mean level of output. Thus, we have the 'surprise only' supply relationship.

The Lucas supply function is the counterpart to the vertical long-run curve in Lipsey's expectations augmented Phillips curve, but derived with the aid of microeconomic theory and the rational expectations hypothesis. Moreover, for conventional specifications of aggregate demand (see, for example, Romer 1996: ch. 6.4), the model implies a positive association between output and inflation, or a negative relationship between the rate of unemployment and inflation. Thus, there is also a new classical correspondence to the short-run Phillips curve. However, the *Lucas supply curve* when applied to data and estimated by ordinary least squares (OLS), does not represent a causal relationship that can be exploited by economic policy makers. On the contrary, it will change when, for example, the money supply is increased in order to stimulate output, in a way that leaves the policy without an effect on real output or unemployment. This is the *Lucas critique* (Lucas 1976), which was formulated as a critique of the Phillips curve inflation–unemployment tradeoff, which figured in the academic literature, as well as in the macroeconometric models of the 1970s (see Wallis 1995). The force of the critique, however, stems from its generality: it is potentially damaging for all conditional econometric models; see Section 4.5.

The causality issue also crops up in connection with the latest versions of the Phillips curve—the forward looking New Keynesian variety—which we return to in Chapter 7.[2] In the United States, an empirical Phillips curve version, dubbed 'the triangle model of inflation', has thrived in spite of the Lucas critique—see Gordon (1983, 1997) and Staiger *et al.* (2001) for recent contributions. As we will argue below, one explanation of the viability of the US Phillips curve is that the shocks to the rate of unemployment have been of an altogether smaller order of magnitude than in European countries.

4.2 Cointegration, causality, and the Phillips curve natural rate

As indicated earlier, there are many ways that a Phillips curve for an open economy can be derived from economic theory. Our appraisal of the Phillips curve in this section builds on Calmfors (1977), who reconciled the Phillips curve with the Scandinavian model of inflation. We want to go one step further, however, and incorporate the Phillips curve in a framework that allows for integrated wage and price series. Reconstructing the model in terms of cointegration and causality reveals that the Phillips curve version of the main-course model forces a particular equilibrium correction mechanism on the system. Thus, while it is consistent with Aukrust's main-course theory, the Phillips curve is also a special model thereof, since it includes only one of the many wage stabilising mechanisms discussed by Aukrust.

---

[2] The main current of theoretical work is definitively guided by the search for 'microfoundations for macro relationships' and imposes an isomorphism between micro and macro. An interesting alternative approach is represented by Ferri (2000) who derives the Phillips curve as a system property.

Without loss of generality we concentrate on the wage Phillips curve and recall that, according to Aukrust's theory, it is assumed that (using the same symbols as in Chapter 3):

1. $(w_{e,t} - q_{e,t} - a_{e,t}) \sim \mathsf{I}(0)$ and $u_t \sim \mathsf{I}(0)$, possibly after removal of deterministic shifts in their means; and
2. the causal structure is 'one way' as represented by H4$_{mc}$ and H5$_{mc}$ in Chapter 3.

Consistency with the assumed cointegration and causality requires that there exists an equilibrium-correction model (EqCM hereafter) for the nominal wage rate in the exposed sector. Assuming first-order dynamics for simplicity, a Phillips curve EqCM system is defined by the following two equations:

$$\Delta w_t = \beta_{w0} - \beta_{w1} u_t + \beta_{w2} \Delta a_t + \beta_{w3} \Delta q_t + \varepsilon_{w,t},$$
$$0 \le \beta_{w1}, \quad 0 < \beta_{w2} < 1, \quad 0 < \beta_{w3} < 1, \tag{4.1}$$
$$\Delta u_t = \beta_{u0} - \beta_{u1} u_{t-1} + \beta_{u2}(w - q - a)_{t-1} - \beta_{u3} z_{u,t} + \varepsilon_{u,t},$$
$$0 < \beta_{u1} < 1, \quad \beta_{u2} > 0, \quad \beta_{u3} \ge 0, \tag{4.2}$$

where we have simplified the notation somewhat by dropping the 'e' subscript.[3] Δ is the difference operator. $\varepsilon_{w,t}$ and $\varepsilon_{u,t}$ are innovations with respect to an information set available in period $t-1$, denoted $\mathcal{I}_{t-1}$.[4] Equation (4.1) is the short-run Phillips curve, while (4.2) represents the basic idea that profitability (in the e-sector) is a factor that explains changes in the economy-wide rate of unemployment. $z_{u,t}$ represents (a vector of) other $\mathsf{I}(0)$ variables (and deterministic terms) which *ceteris paribus* lower the rate of unemployment. $z_{u,t}$ will typically include a measure of the growth rate of the domestic economy, and possibly factors connected with the supply of labour. Insertion of (4.2) into (4.1) is seen to give an explicit EqCM for wages.

To establish the main-course rate of equilibrium unemployment, we rewrite (4.1) as

$$\Delta w_t = -\beta_{w1}(u_t - \breve{u}) + \beta_{w2} \Delta a_t + \beta_{w3} \Delta q_t + \varepsilon_{w,t}, \tag{4.3}$$

where

$$\breve{u} = \frac{\beta_{w0}}{\beta_{w1}} \tag{4.4}$$

is the rate of unemployment which does not put upward or downward pressure on wage growth. Taking unconditional means, denoted by E, on both sides

---

[3] Alternatively, given H2$_{mc}$, Δw_t represents the average wage growth of the two sectors.

[4] The rate of unemployment enters linearly in many US studies; see, for example, Fuhrer (1995). For most other datasets, however, a concave transform improves the fit and the stability of the relationship; see, for example, Nickell (1987) and Johansen (1995a).

of (4.3) gives

$$\mathsf{E}[\Delta w_t] - g_f - g_a = -\beta_{w1}\mathsf{E}[u_t - \breve{u}] + (\beta_{w2} - 1)g_a + (\beta_{w3} - 1)g_f.$$

Using the assumption of a stationary wage share, the left-hand side is zero. Thus, using g_a and g_f to denote the constant steady-state growth rates of productivity and foreign prices, we obtain

$$\mathsf{E}[u_t] \equiv u^{\mathrm{phil}} = \left(\breve{u} + \frac{\beta_{w2} - 1}{\beta_{w1}} g_a + \frac{\beta_{w3} - 1}{\beta_{w1}} g_f \right), \qquad (4.5)$$

as the solution for the main-course equilibrium rate of unemployment which we denote u^{phil}. The long-run mean of the wage share is consequently

$$\mathsf{E}[w_t - q_t - a_t] \equiv wsh^{\mathrm{phil}} = -\frac{\beta_{u0}}{\beta_{u2}} + \frac{\beta_{u1}}{\beta_{u2}} u^{\mathrm{phil}} + \frac{\beta_{u3}}{\beta_{u2}} \mathsf{E}[z_{u,t}]. \qquad (4.6)$$

Moreover, u^{phil} and wsh^{phil} represent the unique and stable steady state of the corresponding pair of deterministic difference equations.

The well-known dynamics of the Phillips curve is illustrated in Figure 4.1. Assume that the economy is initially running at a low level of unemployment, that is, u_0 in the figure. The short-run Phillips curve (4.1) determines the rate of wage inflation Δw_0. The wage share consistent with equation (4.2) is above its long-run equilibrium, implying that unemployment starts to rise and wage growth is reduced along the Phillips curve. The steep Phillips curve is defined for the case of $\Delta w_t = \Delta q_t + \Delta a_t$. The slope of this curve is given by $-\beta_{w1}/(1 - \beta_{w3})$, and it has been dubbed the long-run Phillips curve in the literature. The stable equilibrium is attained when wage growth is equal to the steady-state growth of the main course, that is, $g_f + g_a$ and the corresponding

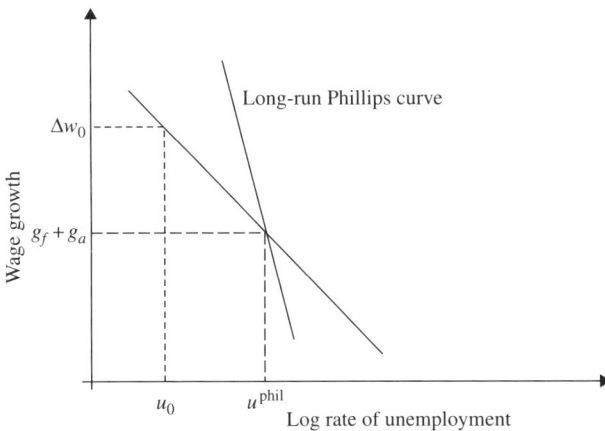

Figure 4.1. Open economy Phillips curve dynamics and equilibrium

level of unemployment is given by u^{phil}. The issue about the slope of the long-run Phillips curve is seen to hinge on the coefficient β_{w3}, the elasticity of wage growth with respect to the product price. In the figure, the long-run curve is downward sloping, corresponding to $\beta_{w3} < 1$ which is conventionally referred to as dynamic inhomogeneity in wage-setting. The converse, dynamic homogeneity, implies $\beta_{w3} = 1$ and a vertical Phillips curve. Subject to dynamic homogeneity, the equilibrium rate u^{phil} is independent of world inflation g_f.

The slope of the long-run Phillips curve represented one of the most debated issues in macroeconomics in the 1970s and 1980s. One argument in favour of a vertical long-run Phillips curve is that it is commonly observed that workers are able to obtain full compensation for CPI-inflation. Hence $\beta_{w3} = 1$ is a reasonable restriction on the Phillips curve, at least if Δq_t is interpreted as an expectations variable. The downward sloping long-run Phillips curve has also been denounced on the grounds that it gives a too optimistic picture of the powers of economic policy: namely that the government can permanently reduce the level of unemployment below the natural rate by 'fixing' a suitably high level of inflation (see, for example, Romer 1996, ch. 5.5). In the context of an open economy this discussion appears to be somewhat exaggerated, since a long-run tradeoff between inflation and unemployment in any case does not follow from the premise of a downward-sloping long-run curve. Instead, as shown in Figure 4.1, the steady-state level of unemployment is determined by the rate of imported inflation g_f and exogenous productivity growth, g_a. Neither of these are normally considered as instruments (or intermediate targets) of economic policy.[5]

In the real economy, cost-of-living considerations plays a significant role in wage-setting; see, for example, Carruth and Oswald (1989, ch. 3) for a review. Thus, in applied econometric work, one usually includes current and lagged CPI-inflation, reflecting the weight put on cost-of-living considerations in actual wage bargaining situations. To represent that possibility, consider the following system (4.7)–(4.9):

$$\Delta w_t = \acute{\beta}_{w0} - \acute{\beta}_{w1} u_t + \acute{\beta}_{w2} \Delta a_t + \acute{\beta}_{w3} \Delta q_t + \acute{\beta}_{w4} \Delta p_t + \acute{\varepsilon}_{wt}, \qquad (4.7)$$

$$\Delta u_t = \beta_{u0} - \beta_{u1} u_{t-1} + \beta_{u2} (w - q - a)_{t-1} - \beta_{u3} z_t + \varepsilon_{ut}, \qquad (4.8)$$

$$\Delta p_t = \beta_{p1} (\Delta w_t - \Delta a_t) + \beta_{p2} \Delta q_t + \varepsilon_{p,t}. \qquad (4.9)$$

The first equation augments (4.1) with the change in consumer prices Δp_t, with coefficient $0 \le \acute{\beta}_{w4} \le 1$. To distinguish formally between this equation and (4.1), we use an accent above the other coefficients as well (and above the disturbance term). The second equation is identical to the unemployment equation (4.2). The last stochastic price equation combines the stylised definition of consumer prices in (3.8) with the twin assumption of stationarity

---

[5] To affect u^{phil}, policy needs to incur a higher or lower permanent *rate* of currency depreciation.

of the sheltered sector wages share and wage leadership of the exposed sector.[6]

Using (4.9) to eliminate Δp_t in (4.7) brings us back to (4.1), with coefficients and ε_{wt} suitably redefined. Thus, the expression for the equilibrium rate u^{phil} in (4.5) applies as before. However, it is useful to express u^{phil} in terms of the coefficients of the extended system (4.7)–(4.9):

$$u^{\text{phil}} = \breve{u} + \frac{\acute{\beta}_{w2} - 1}{\acute{\beta}_{w1}} g_a + \frac{\acute{\beta}_{w3} + \acute{\beta}_{w4}(\beta_{p1} + \beta_{p2}) - 1}{\acute{\beta}_{w1}} g_f, \qquad (4.10)$$

since there are now two homogeneity restrictions needed to ensure a vertical long-run Phillips curve: namely $\acute{\beta}_{w3} + \acute{\beta}_{w4} = 1$ and $\beta_{p1} + \beta_{p2} = 1$.

Compared to the implicit dynamics of Chapter 3, the open economy wage Phillips curve system represents a full specification of the dynamics of the Norwegian model of inflation. Clearly, the dynamic properties of the model apply to other versions of the Phillips curve as well. In particular, all Phillips curve systems imply that the natural rate (or NAIRU) of unemployment is a stable stationary solution. As a single equation, the Phillips curve *equation* itself is dynamically unstable for a given rate of unemployment. Dynamic stability of the wage share and the rate of unemployment hinges on the equilibrating mechanism embedded in the equation for the rate of unemployment. In that sense, a Phillips curve specification of wage formation cannot logically accommodate an economic policy that targets the level of (the rate of) unemployment, since only the natural rate of unemployment is consistent with a stable wage share. Any other (targeted) level leads to an ever increasing or ever declining wage share.

The question about the dynamic stability of the natural rate (or NAIRU) is of course of great interest, and cannot be addressed in the incomplete Phillips curve system, that is, by estimating a single-equation Phillips curve model. Nevertheless, as pointed out by Desai (1995), there is a long-standing practice of basing the estimation of the NAIRU on the incomplete system. For United States, the question of correspondence with a steady state may not be an issue: Staiger *et al.* (1997) is an example of an important study that follows the tradition of estimating only the Phillips curve (leaving the equilibrating mechanism, for example, (4.2) implicit). For other countries, European in particular, where the stationarity of the rate of unemployment is less obvious, the issue about the correspondence between the estimated NAIRUs and the steady state is a more pressing issue.

In the following sections, we turn to two separate aspects of the Phillips curve NAIRU. First, Section 4.3 discusses how much flexibility and time

---

[6] Hence, the first term in (4.9) reflects normal cost pricing in the sheltered sector. Also, as a simplification, we have imposed identical productivity growth in the two sectors, $\Delta a_{e,t} = \Delta a_{s,t} \equiv \Delta a_t$.

dependency one can allow to enter into NAIRU estimates, while still claiming consistency with the Phillips curve framework. Second, in Section 4.4 we
discuss the statistical problems of measuring the uncertainty of an estimated
time independent NAIRU.

4.3   Is the Phillips curve consistent with persistent changes in unemployment?

In the expressions for the main-course NAIRU (4.5) and (4.10), u^{phil} depends
on parameters of the wage Phillips curve (4.1) and exogenous growth rates. The
coefficients of the unemployment equation do not enter into the natural rate
NAIRU expression. In the other version of the Phillips curve, the expression
for the NAIRU depends on parameters of price-setting as well as wage-setting,
that is, the model is specified as a *price* Phillips curve rather than a wage
Phillips curve. But the NAIRU expression from a price Phillips curve remains
independent of parameters from equation (4.2) (or its counterpart in other
specifications).

The fact that an important system property (the equilibrium of unemployment) can be estimated from a single equation goes some way towards
explaining the popularity of the Phillips curve model. Nevertheless, results
based on analysis of the incomplete system gives limited information. In particular, a single-equation analysis gives insufficient information of the dynamic
properties of the system. First, unless the Phillips curve is estimated jointly
with equation (4.2), dynamic stability cannot be tested, and the correspondence between u^{phil} and the steady state of the system cannot be asserted. Thus,
single equation estimates of the NAIRU are subject to the critique that the
correspondence principle may be violated (see Samuelson 1941). Second, even
if one is convinced a priori that u^{phil} corresponds to the steady state of the
system, the speed of adjustment towards the steady state is clearly of interest
and requires estimation of equation (4.2) as well as of the Phillips curve (4.1).

During the last 20–25 years of the previous century, European rates of
unemployment rose sharply and showed no sign of reverting to the levels of the
1960s and 1970s. Understanding the stubbornly high unemployment called for
models that (1) allow for long adjustments lags around a constant natural rate,
or (2) allow the equilibrium to change. A combination of the two is of course
also possible.

Simply by virtue of being a dynamic system, the Phillips curve model
accommodates *slow* dynamics. In principle, the adjustment coefficient β_{u1} in
the unemployment equation (4.2) can be arbitrarily small—as long as it is
not zero the u^{phil} formally corresponds to the steady state of the system.
However, there is a question of *how* slow the speed of adjustment can be before
the concept of equilibrium becomes undermined 'from within'. According to
the arguments of Phelps and Friedman, the natural rate ought to be quite

stable, and it should be a strong attractor of the actual rate of unemployment (see Phelps 1995). However, the experience of the 1980s and 1990s has taught us that the natural rate is at best a weak attractor. There are important practical aspects of this issue too: policy makers, pondering the prospects after a negative shock to the economy, will find small comfort in learning that eventually the rate of unemployment will return to its natural rate, but only after 40 years or more! In Section 4.6 we show how this kind of internal inconsistency arises in an otherwise quite respectable empirical version of the Phillips curve system (4.1) and (4.2).

Moreover, the Phillips curve framework offers only limited scope for an economic explanation of the regime shifts that sometimes occur in the mean of the rate of unemployment. True, expression (4.10) contains a long-run Okun's law type relationship between the rate of unemployment and the rate of productivity growth. However, it seems somewhat incredible that changes in the real growth rate g_a alone should account for the sharp and persistent rises in the rate of unemployment experienced in Europe. A nominal growth rate like g_f can of course undergo sharp and large rises, but for those changes to have an impact on the equilibrium rate requires a downward sloping long-run Phillips curve—which many macroeconomists will not accept.

Thus, the Phillips curve is better adapted to a stable regime characterised by a modest adjustment lag around a fairly stable mean rate of unemployment, than to the regime shift in European unemployment of the 1980s and 1990s. This is the background against which the appearance of new models in the 1980s must be seen, that is, models that promised to be able to explain the shifts in the equilibrium rate of unemployment (see Backhouse 2000), and there is now a range of specifications of how the structural characteristics of labour and commodity markets affect the equilibrium paths of unemployment (see Nickell 1993 for a survey and Chapter 5 of this book). Arguably however, none of the new models have reached the status of being an undisputed consensus model that once was the role of the Phillips curve.

So far we have discussed permanent changes in unemployment as being due to large deterministic shifts that occur intermittently, in line with our maintained view of the rate of unemployment as I(0) but subject to (infrequent) structural breaks. An alternative view, which has become influential in the United States, is the so-called time varying NAIRU: cf. Gordon (1997), Gruen *et al.* (1999), and Staiger *et al.* (1997). The basic idea is that the NAIRU reacts to small supply-side shocks that occur frequently. The following modifications of equation (4.3) defines the time varying NAIRU

$$\Delta w_t = -\beta_{w1}(u_t - \breve{u}_t) + \beta_{w2}\Delta a_t + \beta_{w3}\Delta q_t + \varepsilon_{wt}, \qquad (4.11)$$

$$\breve{u}_t = \breve{u}_{t-1} + \varepsilon_{u,t}. \qquad (4.12)$$

The telling difference is that the natural rate $\breve{u}$ is no longer a time-independent parameter, but a stochastic parameter that follows the random walk (4.12), and a disturbance $\varepsilon_{u,t}$ which in this model represents small supply-side shocks. When estimating this pair of equations (by the Kalman filter) the standard error of $\varepsilon_{u,t}$ typically is limited at the outset, otherwise $\breve{u}_t$ will jump up and down and soak up all the variation in Δw_t left unexplained by the conventional explanatory variables. Hence, time varying NAIRU estimates tend to reflect how much variability a researcher accepts and finds possible to communicate. Logically, the methodology implies a unit root, both in the observed rate of unemployment and in the NAIRU itself. Finally, the practical relevance of this framework seems to be limited to the United States, where there are few big and lasting shifts in the rate of unemployment.

Related to the time varying NAIRU is the concept of hysteresis. Following Blanchard and Summers (1986), economists have invoked the term unemployment 'hysteresis' for the case of a unit root in the rate of unemployment, in which case the equilibrium rate might be said to become identical to the lagged rate of unemployment. However, Røed (1994) instructively draws the distinction between genuine hysteresis as a non-linear and multiple equilibrium phenomenon, and the linear property of a unit root. Moreover, Cross (1995) have convincingly shown that 'hysteresis' is not actually hysteresis (in its true meaning, as a non-linear phenomenon), and that proper hysteresis creates a time path for unemployment which is inconsistent with the natural rate hypothesis.

4.4   Estimating the uncertainty of the Phillips curve NAIRU

This section describes three approaches for estimation of a 'confidence region' of a (time independent) Phillips curve NAIRU. As noted by Staiger *et al.* (1997) the reason for the absence of confidence intervals in most NAIRU calculations has to do with the fact that the NAIRU (e.g. in (4.4)) is a non-linear function of the regression coefficients. Nevertheless, three approaches can be used to construct confidence intervals for the NAIRU: the Wald, Fieller, and likelihood ratio statistics.[7] The Fieller and likelihood ratio forms appear preferable because of their finite sample properties.

The first and most intuitive approach is based on the associated standard error and t ratio for the estimated coefficients, and thus corresponds to a Wald statistic; see Wald (1943) and Silvey (1975, pp. 115–18). This method may be characterised as follows. A wage Phillips curve is estimated in the form of (4.1) in Section 4.2. In the case of full pass-through of productivity gains on wages, and no 'money illusion', the Phillips curve NAIRU u^{phil} is β_{w0}/β_{w1}, and its estimated value $\hat{\mu}_u$ is $\hat{\beta}_{w0}/\hat{\beta}_{w1}$, where a circumflex denotes estimated values.

---

[7] This section draws on Ericsson *et al.* (1997).

As already noted, (4.1) is conveniently rewritten as:

$$\Delta w_t - \Delta a_t - \Delta q_t = -\beta_{w1}(u_t - u^{\text{phil}}) + \varepsilon_{wt}, \tag{4.13}$$

where u^{phil} may be estimated directly by (say) non-linear least squares. The result is numerically equivalent to the ratio $\hat{\beta}_{w0}/\hat{\beta}_{w1}$ derived from the *linear* estimates $(\hat{\beta}_{w0}, \hat{\beta}_{w1})$ in (4.1). In either case, a standard error for $\hat{\mu}^{\text{phil}}$ can be computed, from which confidence intervals are directly obtained.

More generally, a confidence interval includes the unconstrained/most likely estimate of u^{phil}, which is $\hat{\beta}_{w0}/\hat{\beta}_{w1}$, and some region around that value. Heuristically, the confidence interval contains each value of the ratio that does not violate the hypothesis

$$H_{\text{W}}: \quad \frac{\beta_{w0}}{\beta_{w1}} = u_0^{\text{phil}} \tag{4.14}$$

too strongly in the data. More formally, let $F_{\text{W}}(u_0^{\text{phil}})$ be the Wald-based F-statistic for testing H_{W}, and let $\text{Pr}(\cdot)$ be the probability of its argument. Then, a confidence interval of $(1 - \alpha)\%$ is $[u_{\text{low}}^{\text{phil}}, u_{\text{high}}^{\text{phil}}]$ defined by $\text{Pr}(F_{\text{W}}(u_0^{\text{phil}})) \leq 1 - \alpha$ for $u_0^{\text{phil}} \in [u_{\text{low}}^{\text{phil}}, u_{\text{high}}^{\text{phil}}]$.

If β_{w1}, the elasticity of the rate of unemployment in the Phillips curve, is precisely estimated, the Wald approach is usually quite satisfactory. Small sample sizes clearly endanger estimation precision, but 'how small is small' depends on the amount of information 'per observation' and the effective sample size. However, if β_{w1} is imprecisely estimated (i.e. not very significant statistically), this approach can be highly misleading. Specifically, the Wald approach ignores how $\hat{\beta}_{w0}/\hat{\beta}_{w1}$ behaves for values of $\hat{\beta}_{w1}$ relatively close to zero, where 'relatively' reflects the uncertainty in the estimate of β_{w1}. For European Phillips curves, the β_{w1} estimates are typically insignificant statistically, so this concern is germane to calculating Phillips curve natural rates for Europe. In essence, the problem arises because $\hat{\mu}_u$ is a non-linear function of estimators $(\hat{\beta}_{w0}, \hat{\beta}_{w1})$ that are (approximately) jointly normally distributed; see Gregory and Veall (1985) for details.

The second approach avoids this problem by transforming the non-linear hypothesis (4.14) into a linear one, namely:

$$H_F: \quad \beta_{w0} - \beta_{w1} u_0^{\text{phil}} = 0. \tag{4.15}$$

This approach is due to Fieller (1954), so the hypothesis in (4.15) and corresponding F-statistic are denoted H_F and $F_F(u_0^{\text{phil}})$. Because the hypothesis (4.15) is linear in the parameters β_{w0} and β_{w1}, tests of this hypothesis are typically well-behaved, even if $\hat{\beta}_{w1}$ is close to zero. Determination of confidence intervals is exactly as for the Wald approach, except that the F-statistic is constructed for $\beta_{w0} - \beta_{w1} u_0^{\text{phil}}$. See Kendall and Stuart (1973, pp. 130–2) for a summary.

The third approach uses the likelihood ratio (LR) statistic (see Silvey 1975, pp. 108–12), to calculate the confidence interval for the hypothesis H_{W}.

That is, (4.13) is estimated both unrestrictedly and under the restriction H_W, corresponding likelihoods (or residual sums of squares for single equations) are obtained, and the confidence interval is constructed from values of u_0^{phil} for which the LR statistic is less than a given critical value.

Three final comments are in order. First, if the original model is linear in its parameters, as in (4.1), then Fieller's solution is numerically equivalent to the LR one, giving the former a generic justification. Second, if the estimated Phillips curve does not display dynamic homogeneity, $\hat{\beta}_{w0}/\hat{\beta}_{w1}$ is only a component of the NAIRU estimate that would be consistent with the underlying theory, cf. the general expression (4.10). This complicates the computation of the NAIRU further, in that one should take into account the covariance of terms like $\hat{\beta}_{w0}/\hat{\beta}_{w1}$, and $(\hat{\beta}'_{w3} + \hat{\beta}'_{w4} - 1)/\hat{\beta}'_{w1}$. However, unless the departure from homogeneity is numerically large, $[u_{\text{low}}^{\text{phil}}, u_{\text{high}}^{\text{phil}}]$ may be representative of the degree of uncertainty that is associated with the estimated Phillips curve natural rate. Third, identical statistical problems crop up in other areas of applied macroeconomics too, for example, in the form of a 'Monetary conditions index'; see Eika *et al.* (1996).

Section 4.6 contains an application of the Wald and Fieller/LR methods to the Phillips curve NAIRU of the Norwegian economy.

4.5   Inversion and the Lucas critique

As pointed out by Desai (1984), the reversal of dependent and independent variables represents a continuing controversy in the literature on inflation modelling. Section 4.1.1 recounts how Lucas's supply curve turns the causality of the conventional Phillips curve on its head. Moreover, the Lucas critique states that conditional Phillips curve models will experience structural breaks whenever agents change their expectations, for example, following a change in economic policy. In this section, we discuss both inversion and the Lucas critique, with the aim of showing how the direction of the regression and the relevance of the Lucas critique can be tested in practice.

4.5.1   Inversion

Under the assumption of super exogeneity,[8] the results for a conditional econometric model, for example, a conventional augmented Phillips curve, are not invariant to a re-normalisation. One way to see this is to invoke the well-known formula

$$\hat{\beta} \cdot \hat{\beta}^* = r_{yx}^2, \tag{4.16}$$

[8] Super exogeneity is defined as the joint occurrence of weak exogeneity of the explanatory variables with respect to the parameters of interest and invariance of the parameters in the conditional model with respect to changes in the marginal models for the explanatory variables, see Engle *et al.* (1983).

where r_{yx} denotes the correlation coefficient and $\hat{\beta}$ is the estimated regression coefficient when y is the dependent variable and x is the regressor. $\hat{\beta}^*$ is the estimated coefficient in the reverse regression. By definition, 'regime shifts' entail that correlation structures alter, hence r_{yx} shifts. If, due to super exogeneity, $\hat{\beta}$ nevertheless is constant, then $\hat{\beta}^*$ cannot be constant.

Equation (4.16) applies more generally, with r_{yx} interpreted as the partial correlation coefficient. Hence, if (for example) the Phillips curve (4.1) is estimated by OLS, then finding that $\hat{\beta}_{w1}$ is recursively stable entails that $\hat{\beta}^*_{w1}$ for the re-normalised equation (on the rate of unemployment) is recursively unstable. Thus, finding a stable Phillips curve over a sample period that contains changes in the (partial) correlations, refutes any claim that the model has a Lucas supply curve interpretation. This simple procedure also applies to estimation by instrumental variables (due to endogeneity of, for example, Δq_t and/or Δp_t) provided that the number of instrumental variables is lower than the number of endogenous variables in the Phillips curve.

4.5.2 Lucas critique

Lucas's 1976 thesis states that conditional econometric models will be prone to instability and break down whenever non-modelled expectations change. This section establishes the critique for a simple algebraic case. In the following section we discuss how the Lucas critique can be confirmed or refuted empirically.

Without loss of generality, consider a single time-series variable y_t, which can be split into an explained part y_t^p, and an independent unexplained part, $\epsilon_{y,t}$:

$$y_t = y_t^p + \epsilon_{y,t}. \tag{4.17}$$

Following Hendry (1995a: ch. 5.2) we think of y_t^p as a *plan* attributable to agents, and $\epsilon_{y,t}$ as the difference between the planned and actual outcome of y_t. Thus,

$$\mathsf{E}[y_t \mid y_t^p] = y_t^p, \tag{4.18}$$

and $\epsilon_{y,t}$ is an innovation relative to the plan, hence

$$\mathsf{E}[\epsilon_{y,t} \mid y_t^p] = 0. \tag{4.19}$$

Assume next that agents use an information set $\mathcal{I}_{t-1}$ to form rational expectations for a variable x_t, that is,

$$x_t^e = \mathsf{E}[x_t \mid \mathcal{I}_{t-1}] \tag{4.20}$$

and that expectations are connected to the plan

$$y_t^p = \beta x_t^e, \tag{4.21}$$

which is usually motivated by, or derived from, economic theory.

By construction, $\mathsf{E}[y_t^p \mid \mathcal{I}_{t-1}] = y_t^p$, while we assume that $\epsilon_{y,t}$ in (4.17) is an innovation

$$\mathsf{E}[\epsilon_{y,t} \mid \mathcal{I}_{t-1}] = 0 \tag{4.22}$$

and, therefore

$$\mathsf{E}[y_t \mid \mathcal{I}_{t-1}] = y_t^p. \tag{4.23}$$

Initially, x_t^e is assumed to follow a first-order AR process (non-stationarity is considered below):

$$x_t^e = \mathsf{E}[x_t \mid \mathcal{I}_{t-1}] = \alpha_1 x_{t-1}, \qquad |\alpha_1| < 1. \tag{4.24}$$

Thus $x_t = x_t^e + \epsilon_{x,t}$, or:

$$x_t = \alpha_1 x_{t-1} + \epsilon_{x,t}, \quad \mathsf{E}[\epsilon_{x,t} \mid x_{t-1}] = 0. \tag{4.25}$$

For simplicity, we assume that $\epsilon_{y,t}$ and $\epsilon_{x,t}$ are independent.

Assume next that the single parameter of interest is β in equation (4.21). The reduced form of y_t follows from (4.17), (4.21), and (4.24):

$$y_t = \alpha_1 \beta x_{t-1} + \epsilon_{x,t}, \tag{4.26}$$

where x_t is weakly exogenous for $\xi = \alpha_1 \beta$, but the parameter of interest β is not identifiable from (4.26) alone. Moreover the reduced form equation (4.26), while allowing us to estimate ξ consistently in a state of nature characterised by stationarity, is susceptible to the Lucas critique, since ξ is not invariant to changes in the autoregressive parameter of the marginal model (4.24).

In practice, the Lucas critique is usually aimed at 'behavioural equations' in simultaneous equations systems, for example,

$$y_t = \beta x_t + \eta_t \tag{4.27}$$

with disturbance term:

$$\eta_t = \epsilon_{y,t} - \epsilon_{x,t} \beta. \tag{4.28}$$

It is straightforward (see Appendix A.1) to show that estimation of (4.27) by OLS on a sample $t = 1, 2, \ldots, T$, gives

$$\operatorname*{plim}_{T \to \infty} \hat{\beta}_{\mathrm{OLS}} = \alpha_1^2 \beta, \tag{4.29}$$

establishing that, 'regressing y_t on x_t' does not represent the counterpart to $y_t^p = x_t^e \beta$ in (4.21). Specifically, instead of β, we estimate $\alpha_1^2 \beta$, and changes in the expectation parameter α_1 damage the stability of the estimates, thus confirming the Lucas critique.

However, the applicability of the critique rests on the assumptions made. For example, if we change the assumption of $|\alpha_1| < 1$ to $\alpha_1 = 1$, so that x_t has a unit root but is cointegrated with y_t, the Lucas critique does *not* apply: under cointegration, $\operatorname{plim}_{T \to \infty} \hat{\beta}_{\mathrm{OLS}} = \beta$, since the cointegration parameter is unique and can be estimated consistently by OLS.

As another example of the importance of the *exact* set of assumptions made, consider replacing (4.21) with another economic theory, namely the contingent plan

$$y_t^p = \beta x_t. \tag{4.30}$$

Equations (4.30) and (4.17) give

$$y_t = \beta x_t + \epsilon_{y,t}, \tag{4.31}$$

where $\mathsf{E}[\epsilon_{y,t} \mid x_t] = 0 \Rightarrow \mathsf{cov}(\epsilon_{y,t}, x_t) = 0$ and β can be estimated by OLS also in the stationary case of $|\alpha_1| < 1$.

4.5.3 Model-based vs. data-based expectations

Apparently, it is often forgotten that the 'classical' regression formulation in (4.31) is consistent with the view that behaviour is driven by expectations, albeit not by model-based or rational expectations with unknown parameters that need to be estimated (unless they reside like memes in agents' minds). To establish the expectations interpretation of (4.31), replace (4.30) by

$$y_t^p = \beta x_{t+1}^e$$

and assume that agents solve $\Delta x_{t+1}^e = 0$ to obtain x_{t+1}^e. Substitution of $x_{t+1}^e = x_t$, and using (4.17) for y_t^p gives (4.31).

$\Delta x_{t+1}^e = 0$, is an example of a univariate prediction rule without any parameters but which is instead based directly on data properties, hence they are referred to as data-based expectations; see Hendry (1995*b*: ch. 6.2.3). Realistically, agents might choose to use data-based predictors because of the cost of information collection and processing associated with model-based predictors. It is true that agents who rely on $\Delta x_{t+1}^e = 0$ use a mis-specified model of the x-process in (4.25), and thus their forecasts will not attain the minimum mean square forecast error.[9] Hence, in a stationary world there are gains from estimating α_1 in (4.25). However, in practice there is no guarantee that the parameters of the x-process stay constant over the forecast horizon, and in this non-stationary state of the world a model-based forecast cannot be ranked as better than the forecast derived from the simple rule $\Delta x_{t+1}^e = 0$. In fact, depending on the dating of the regime shift relative to the 'production' of the forecast, the data-based forecast will be better than the model-based forecast in terms of bias.

In order to see this, we introduce a growth term in (4.25), that is,

$$x_t = \alpha_0 + \alpha_1 x_{t-1} + \epsilon_{x,t}, \quad \mathsf{E}[\epsilon_{x,t} \mid x_{t-1}] = 0 \tag{4.32}$$

and assume that there is a shift in α_0 (to α_0^*) in period $T + 1$.

---

[9] This is the well-known theorem that the conditional mean of a correctly specified model attains the minimum mean squared forecast error; see Granger and Newbold (1986: ch. 4), Brockwell and Davies (1991: ch. 5.1), or Clements and Hendry (1998: ch. 2.7).

We consider two agents, A and B, who forecast x_{T+1}. Agent A collects data for a period $t = 1, 2, 3, \ldots, T$ and is able to discover the true values of $\{\alpha_0, \alpha_1\}$ over that period. However, because of the unpredictable shift $\alpha_0 \to \alpha_0^*$ in period $T + 1$, A's forecast error will be

$$e_{A,T+1} = \alpha_0^* - \alpha_0 + \epsilon_{x,T+1}. \tag{4.33}$$

Agent B, using the data-based forecast $x_{T+1} = x_T$, will experience a forecast error

$$e_{B,T+1} = \alpha_0^* - (1 - \alpha_1)x_T + \epsilon_{x,T+1},$$

which can be expressed as

$$e_{B,T+1} = \alpha_0^* - \alpha_0 + (1 - \alpha_1)(x_s^0 - x_T) + \epsilon_{x,T+1}, \tag{4.34}$$

where x_s^0 denotes the (unconditional) mean of x_T (i.e. for the pre-shift intercept ϕ_0), $x_s^0 = \alpha_0/(1 - \alpha_1)$. Comparison of (4.33) and (4.34) shows that the only difference between the two forecast errors is the term $(1 - \alpha_1)(x_T - x_s^0)$ in (4.34). Thus, both forecasts are damaged by a regime shift that occurs after the forecast is made. The conditional means and variances of the two errors are

$$\mathsf{E}[e_{A,T+1} \mid T] = \alpha_0^* - \alpha_0, \tag{4.35}$$
$$\mathsf{E}[e_{B,T+1} \mid T] = \alpha_0^* - \alpha_0 + (1 - \alpha_1)(x_s^0 - x_T), \tag{4.36}$$
$$\mathsf{Var}[e_{A,T+1} \mid T] = \mathsf{Var}[e_{B,T+1} \mid T], \tag{4.37}$$

establishing that in this example of a *post-forecast* regime-shift, there is no ranking of the two forecasting methods in terms of the first two moments of the forecast error. The conditional forecast error variances are identical, and the bias of the model-based forecast are not necessarily smaller than the bias of the naive data-based predictor: assume, for example, that $\alpha_0^* > \alpha_0$—if at the same time $x_T < x_s^0$, the data-based bias can still be the smaller of the two. Moreover, unconditionally, the two predictors have the same bias and variance:

$$\mathsf{E}[e_{A,T+1}] = \mathsf{E}[e_{B,T+1}] = \alpha_0^* - \alpha_0, \tag{4.38}$$
$$\mathsf{Var}[e_{A,T+1}] = \mathsf{Var}[e_{B,T+1}]. \tag{4.39}$$

Next consider the forecasts made for period $T + 2$, conditional on $T + 1$, as an example of a *pre-forecast* regime shift ($\alpha_0 \to \alpha_0^*$ in period $T + 1$). Unless A discovers the shift in α_0 and successfully intercept-corrects the forecast, his error-bias will once again be

$$\mathsf{E}[e_{A,T+2} \mid T + 1] = [\alpha_0^* - \alpha_0]. \tag{4.40}$$

The bias of agent B's forecast error on the other hand becomes

$$\mathsf{E}[e_{B,T+2} \mid T + 1] = (1 - \alpha_1)(x_s^* - x_T), \tag{4.41}$$

where x_s^* denotes the post-regime shift unconditional mean of x, that is, $x_s^* = \alpha_0^*/(1 - \alpha_1)$. Clearly, the bias of the data-based predictor can easily be smaller than the bias of the model-based prediction error (but the opposite can of course also be the case). However,

$$\mathsf{E}[e_{\mathrm{A},T+2}] = [\alpha_0^* - \alpha_0],$$
$$\mathsf{E}[e_{\mathrm{B},T+2}] = 0,$$

and the unconditional forecast errors are always smallest for the data-based prediction in this case of *pre-forecast* regime shift.

The analysis generalises to the case of a unit root in the x-process, in fact it is seen directly from the above that the data-based forecast errors have even better properties for the case of $\alpha_1 = 1$, for example, $\mathsf{E}[e_{\mathrm{B},T+2} \mid T+1] = 0$ in (4.41). More generally, if x_t is $\mathsf{I}(d)$, then solving $\Delta^d x_{t+1}^e = 0$ to obtain x_{t+1}^e will result in forecast with the same robustness with respect to regime shifts as illustrated in our example; see Hendry (1995*a*, ch.6.2.3). This class of predictors belongs to forecasting models that are cast in terms of differences of the original data, that is, differenced vector autoregressions, denoted dVARs. They have a tradition in macroeconomics that goes back at least to the 1970s, then in the form of Box–Jenkins time-series analysis and ARIMA models. A common thread running through many published evaluations of forecasts, is that the naive time-series forecasts are often superior to the forecasts of the macroeconometric models under scrutiny (see, for example, Granger and Newbold 1986, ch. 9.4). Why dVARs tend to do so well in forecast competitions is now understood more fully, thanks to the work of, for example, Clements and Hendry (1996, 1998, 1999*a*). In brief, the explanation is exactly along the lines of our comparison of 'naive' and 'sophisticated' expectation formation above: the dVAR provides robust forecasts of non-stationary time-series that are subject to intermittent regime shifts. To beat them, the user of an econometric model must regularly take recourse to intercept corrections and other judgemental corrections (see Section 4.6). These issues are also discussed in further detail in Chapter 11.

4.5.4 Testing the Lucas critique

While it is logically possible that conventional Phillips curves are 'really' Lucas supply functions in reverse, that claim can be tested for specific models. Finding that the Phillips curve is stable over sample periods that included regime shifts and changes in the correlation structures is sufficient for refuting inversion. Likewise, the Lucas critique is a *possibility theorem*, not a truism (see Ericsson and Irons 1995), and its assumptions have testable implications. For example, the Lucas critique implies (1) that $\hat{\beta}_{\mathrm{OLS}}$ is non constant as α_1 changes (inside the unit circle), and (2) that determinants of α_1 (if identifiable in practice) should affect $\hat{\beta}_{\mathrm{OLS}}$ if included in the conditional model of y_t. Conversely, the Lucas critique is inconsistent with the *joint* finding of a stable conditional relationship *and* a regime shift occurring in the process which drives the explanatory variable; see Ericsson and Hendry (1999). Based on this logic methods of testing the Lucas critique have been developed: see, for example, Hendry (1988), Engle and Hendry (1993), and Favero and Hendry (1992).

Surveys of the empirical evidence for the Lucas critique are found in Ericsson and Irons (1995) and Stanley (2000). Though very different in methodology, the two studies conclude in a similar fashion, namely that there is little firm evidence supporting the empirical applicability of the Lucas critique. In Section 4.6 we review the applicability of the Lucas critique to the Norwegian Phillips curve. As an alternative to rational expectations, we note as a possibility that agents form expectations on the basis of observed properties of the data itself. Interestingly, there is a close relationship between data-based forecasting rules that agents may pick up, and the time-series models that have been successful in macroeconomic forecasting.

4.6   An empirical open economy Phillips curve system

In this section, we first specify and then evaluate an open economy Phillips curve for the Norwegian manufacturing sector. We use an annual data set for the period 1965–98, which is used again in later sections where competing models are estimated. In the choice of explanatory variables and of data transformations, we build on existing studies of the Phillips curve in Norway, cf. Stølen (1990, 1993). The variables are in log scale (unless otherwise stated) and are defined as follows:

wc_t = hourly wage cost in manufacturing;
q_t = index of producer prices (value added deflator);
p_t = the official consumer price index (CPI);
a_t = average labour productivity;
tu_t = rate of total unemployment (i.e. unemployment includes participants in active labour market programmes);
rpr_t = the replacement ratio;
h_t = the length of the 'normal' working day in manufacturing;
$t1_t$ = the manufacturing industry payroll tax-rate (not log).

Equation (4.42) shows the estimation results of a manufacturing sector Phillips curve which is as general as the number of observations allows. Arguably the use of OLS estimation may be defended by invoking the main-course theory (remembering that we model wages of an exposed industry), but the main reason here is plain simplicity, and we return to the estimation of the Phillips curve by system methods below.

The model is a straightforward application of the theoretical Phillips curve in (4.1): we include two lags in Δq_t and Δa_t, and, as discussed earlier, it is a necessary concession to realism to also include a lag polynomial of the consumer price inflation rate, Δp_t.[10] We use only one lag of the unemployment rate, since previous work on this data set gives no indication of any need to include a second lag of this variable.

$$
\begin{aligned}
\Delta wc_t - \Delta p_{t-1} = \ & -0.0287 & +0.133\Delta p_t & -0.716\Delta p_{t-1} & -0.287\Delta p_{t-2} \\
& (0.0192) & (0.182) & (0.169) & (0.163) \\[4pt]
& +0.0988\Delta a_t & +0.204\Delta a_{t-1} & -0.00168\Delta a_{t-2} \\
& (0.159) & (0.153) & (0.136) \\[4pt]
& +0.189\Delta q_t & +0.317\Delta q_{t-1} & +0.177\Delta q_{t-2} & -0.0156\ tu_t \\
& (0.0867) & (0.0901) & (0.0832) & (0.0128) \\[4pt]
& -0.00558\ tu_{t-1} & +0.796\Delta t1_t & +0.0464\ rpr_{t-1} \\
& (0.0162) & (0.531) & (0.0448) \\[4pt]
& -0.467\Delta h_t & +0.0293\ i1967_t & -0.0624IP_t & \hspace{2cm}(4.42)\\
& (0.269) & (0.0201) & (0.0146)
\end{aligned}
$$

OLS, $T = 34$ (1965–98)

$\hat{\sigma} = 0.01302$	$R^2 = 0.92$	RSS $= 0.002882$
$F_{Null} = 9.558[0.00]$	$F_{AR(1-2)} = 1.01[0.386]$	
$F_{ARCH(1-1)} = 0.115[0.700]$	$\chi^2_{normality} = 4.431[0.109]$	
$F_{Chow(1982)} = 2.512[0.4630]$	$F_{Chow(1995)} = 0.116[0.949]$	

The last five explanatory variables in (4.42) represent two categories; these are, first, the theoretically motivated variables: the change in the payroll tax rate ($\Delta t1_t$) and a measure of the generosity of the unemployment insurance system (the replacement ratio, rpr_{t-1}); and second, variables that capture the impact of changes in the institutional aspects of wage-setting in Norway. As indicated by its name, $i1967_t$ is an impulse dummy and is 1 in 1967 and zero elsewhere. It covers the potential impact of changes in legislation and indirect taxation in connection with the build up of the national insurance system in the late 1960s. Δh_t captures the short-run impact of income compensation in connection with the reforms in the length of the working week in 1964, 1968, and 1987 (see Nymoen 1989*b*). Finally, IP_t is a composite dummy representing a wage- and price-freeze in 1979 and centralised bargaining in 1988 and 1989: it is 1 in 1979 and 0.5 in 1980, 1 again in 1988 and 0.5 in 1989—zero elsewhere. The exact 'weighting' scheme is imported from Bårdsen and Nymoen (2003).[11]

---

[10] Below, and in the following, square brackets, [..], contain *p*-values whereas standard errors are stated in parentheses, (..).

[11] The dummy variable IP_t is designed to capture the effects of the wage-freeze in 1979 and the wage-laws of 1988 and 1989. Similar dummies for incomes policy appear with significant

The left-hand side variable in (4.42) is $\Delta wc_t - \Delta p_{t-1}$, since our earlier experience with this data set (see, for example, Bårdsen and Nymoen 2003, and Section 6.9.2), shows that the lagged rate of inflation is an important predictor of this year's nominal wage increase. Note, however, that the transformation on the left-hand side does not represent a restriction in (4.42) since Δp_{t-1} is also present on the right-hand side of the equation.

The general model (4.42) contains coefficient estimates together with conventionally computed standard errors (in brackets). Below the equation we report estimation statistics (T, number of observations; the residual sum of squares, RSS; the residual standard error $\hat{\sigma}$, R^2, and F_{Null} the probability of observing an F value as large or larger as the one we observe, given the null of 'no relationship'), and a set of mis-specification tests for the general unrestricted model (GUM): F-distributed tests of residual autocorrelation ($F_{AR(1-2)}$), heteroskedasticity ($F_{HET_{x^2}}$), autoregressive conditional heteroskedasticity ($F_{ARCH(1-1)}$) and the Doornik and Hansen (1994) Chi-square test of residual non-normality ($\chi^2_{normality}$). The last two diagnostics reported are two tests of parameter constancy based on Chow (1960). The first is a mid-sample split ($F_{Chow(1982)}$) and the second is an end-of-sample split ($F_{Chow(1995)}$). For each diagnostic test, the numbers in square brackets are p-values for the respective null hypotheses; they show that none of the tests are significant.

Automatised general to specific model selection using PcGets (see Hendry and Krolzig 2001), resulted in the Phillips curve in (4.43).

$$\widehat{\Delta wc_t - \Delta p_{t-1}} = -\underset{(0.0139)}{0.0683} \; -\underset{(0.105)}{0.743\Delta p_{t-1}} + \underset{(0.0851)}{0.203\Delta q_t}$$

$$+ \underset{(0.0851)}{0.29\Delta q_{t-1}} - \underset{(0.00431)}{0.0316 \; tu_t} - \underset{(0.0103)}{0.0647 IP_t} \qquad (4.43)$$

$$\text{OLS, } T = 34 \; (1965\text{--}98)$$

RSS $= 0.005608$	$\hat{\sigma} = 0.01415$	$R^2 = 0.84$
$F_{GUM} = 1.462[0.23]$	$F_{AR(1-2)} = 3.49[0.05]$	$F_{HET_{x^2}} = 0.732[0.69]$
$F_{ARCH(1-1)} = 0.157[0.90]$	$\chi^2_{normality} = 4.907[0.09]$	
$F_{Chow(1982)} = 0.575[0.85]$	$F_{Chow(1995)} = 0.394[0.76]$	

Whereas the GUM in (4.42) contains 16 explanatory variables, the final model (4.43) keeps only 5: the lagged rate of inflation, the current and lagged changes in the product price index, the rate of unemployment, and the composite incomes policy dummy. The test of the joint significance of the 11 restrictions is reported as F_{GUM} below the equation, with a p-value of 0.23, showing that the increase in residual standard error from 1.3% to 1.4% is statistically insignificant. The diagnostic tests confirm that the reduction process is valid, that is, only the test of 2. Order autocorrelation is marginally significant at the 5% level.

coefficients in earlier studies on both annual and quarterly data (see, for example, Johansen 1995a).

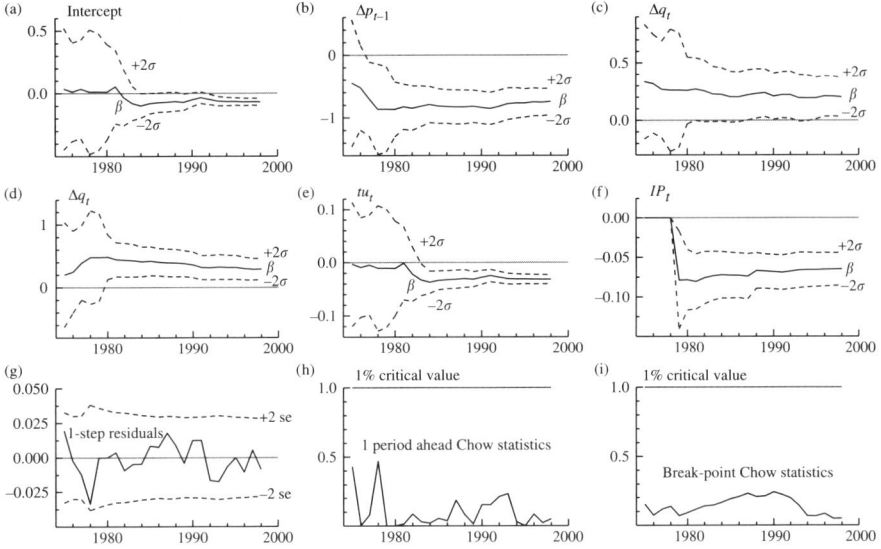

Figure 4.2. Recursive stability of final open economy wage Phillips curve model in equation (4.43)

As discussed earlier, a key parameter of interest in the Phillips curve model is the equilibrium rate of unemployment, that is, u^{phil} in (4.10). Using the coefficient estimates in (4.43), and setting the growth rate of prices (g_f) and productivity growth equal to their sample means of 0.06 and 0.027, we obtain $\hat{u}^{\mathrm{phil}} = 0.0305$, which is nearly identical to the sample mean of the rate of unemployment (0.0313).

In this section and throughout the book, figures often appear as panels of graphs, with each graph in a panel labelled sequentially by a suffix a,b,c,..., row by row. In Figure 4.2, the graphs numbered (a)–(f) show the recursively estimated coefficients in equation (4.43), together with ±2 estimated standard errors over the period 1976–98 (denoted β and $\pm2\sigma$ in the graphs). The last row with graphs in Figure 4.2 shows the sequence of 1-step residuals (with ±2 residual standard errors denoted $\pm2\,\mathrm{se}$), the 1-step Chow statistics and lastly the sequence of 'break-point' Chow statistics. Overall, the graphs show a considerable degree of stability over the period 1976–98. However, *Constant* (a) and the unemployment elasticity (e) are both imprecisely estimated on samples that end before 1986, and there is also instability in the coefficient estimates (for *Constant*, there is a shift in sign from 1981 to 1982). These results will affect the natural rate estimate, since u^{phil} depends on the ratio between *Constant* and the unemployment elasticity, cf. equation (4.5).

The period from 1984 to 1998 was a turbulent period for the Norwegian economy, and the manufacturing industry in particular. The rate of unemployment fell from 4.3% in 1984 to 2.6% in 1987, but already in 1989 it had risen to 5.4% and reached an 8.2% peak in 1989, before falling back to 3% in

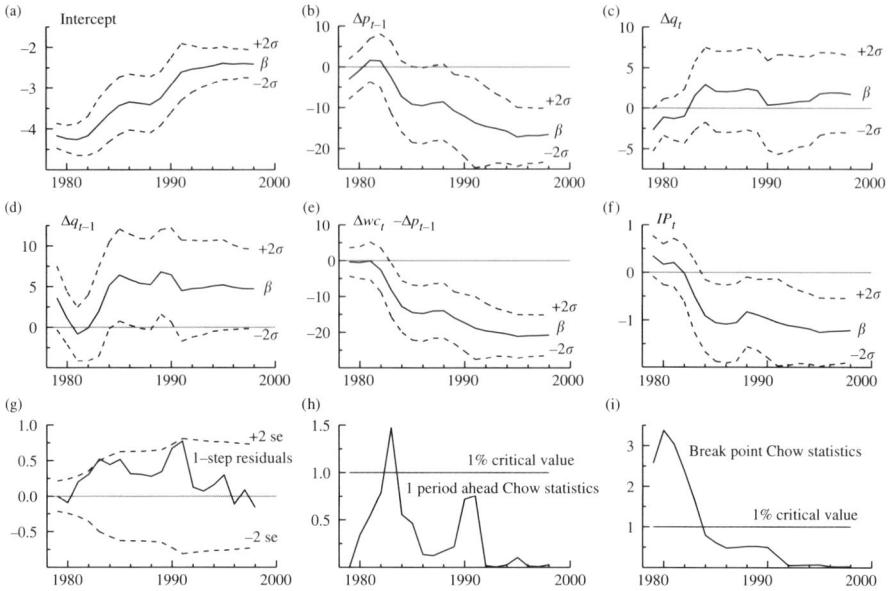

Figure 4.3. Recursive instability of the inverted Phillips curve model
(Lucas supply curve) in equation (4.43)

1998.[12] An aspect of this was a marked fall in manufacturing profitability in
the late 1980s. Institutions also changed (see Barkbu *et al.* 2003), as Norway
(like Sweden) embarked upon less coordinated wage settlements in the begin-
ning of the 1980s. The decentralisation was reversed during the late 1980s. The
revitalisation of coordination in Norway has continued in the 1990s. However,
according to (4.43), the abundance of changes have had only limited impact on
wage-setting, that is, the effect is limited to two shifts in the intercept in 1988
and 1989 as IP_t then takes the value of 1 and 0.05 as explained above. The
stability of the slope coefficients in Figure 4.2 over (say) the period 1984–98
therefore invalidates a Lucas supply curve interpretation of the estimated relationship in equation (4.43). On the contrary, given the stability of (4.43) and the
list of recorded changes, we are led to predict that the inverted regression will
be unstable over the 1980s and 1990s. Figure 4.3 confirms this interpretation
of the evidence.

   Given the non-invertibility of the Phillips curve, we can investigate more
closely the stability of the implied estimate for the equilibrium rate of unemployment. We simplify the Phillips curve (4.43) further by imposing dynamic
homogeneity ($F(1, 28) = 4.71[0.04]$), since under that restriction u^{phil} is

---

[12] The numbers refer to the 'total' rate of unemployment, that is, including persons on
active labour market programmes.

independent of the nominal growth rate (g_f). Non-linear estimation of the Phillips curve (4.43), under the extra restriction that the elasticities of the three price growth rates sum to zero, gives

$$\Delta wc_t - \Delta p_{t-1} - 0.027 = -0.668415\Delta p_{t-1} + 0.301663\Delta q_t + 0.289924\Delta q_{t-1}$$
$$\quad\quad (0.1077) \quad\quad\quad (0.07761) \quad\quad (-)$$
$$-0.0266204\,(tu_t - \log(0.033)) \; - 0.072\;IP_t$$
$$(0.003936) \quad\quad\quad (0.00376) \quad (0.01047)$$

$$(4.44)$$

NLS, $T = 34$ (1965–98)

RSS = 0.00655087875	$\hat{\sigma} = 0.0152957$
$F_{AR(1-2)} = 3.5071[0.0448]$	$F_{HET x^2} = 0.18178[0.9907]$
$F_{ARCH(1-1)} = 0.021262[0.8852]$	$\chi^2_{normality} = 0.85344[0.6526]$

The left-hand side has been adjusted for mean productivity growth (0.027) and the unemployment term has the interpretation: $(tu_t - u^{phil})$. Thus, the full sample estimate obtained of u^{phil} obtained from non-linear estimation is 0.033 with a significant 't-value' of 8.8. A short sample, like, for example, 1965–75 gives a very high, but also uncertain, u^{phil} estimate. This is as one would expect from Figure 4.2(a) and (e). However, once 1982 is included in the sample the estimates stabilise, and Figure 4.4 shows the sequence of u^{phil} estimates for the remaining samples, together with ±2 estimated standard errors and the actual unemployment rate for comparison. The figure shows that the estimated equilibrium rate of unemployment is relatively stable, and that it appears to be quite well determined. The years 1990 and 1991 are exceptions, where $\hat{u}^{phil}$

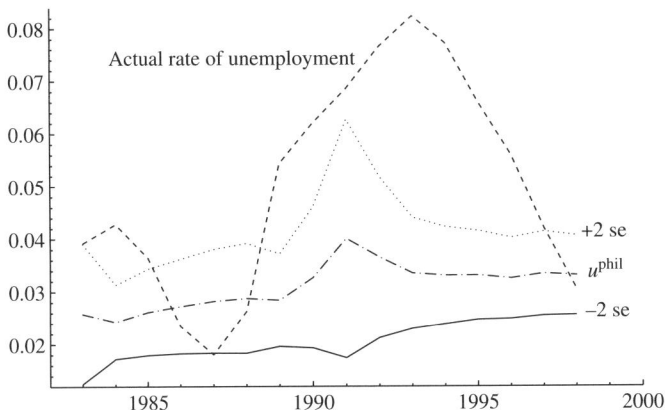

Figure 4.4. Sequence of estimated wage Phillips curve NAIRUs
(with ±2 estimated standard errors), and the actual rate of unemployment.
Wald-type confidence regions

Table 4.1

Confidence intervals for the Norwegian wage Phillips curve NAIRU

	NAIRU	95% confidence interval for NAIRU	
		Wald	Fieller and LR
Full sample: 1965–98	0.0330	[0.0253 ; 0.0407]	[0.0258 ; 0.0460]
Sub sample: 1965–91	0.0440	[0.0169 ; 0.0631]	[0.0255 ; 0.2600]
Sub sample: 1965–87	0.0282	[0.0182 ; 0.0383]	[0.0210 ; 0.0775]

Note: The Fieller method is applied to equation (4.43), with homogeneity imposed. The confidence intervals derived from the Wald and LR statistics are based on equation (4.44).

increases to 0.033 and 0.040 from 0.028 in 1989. However, compared to confidence interval for 1989, the estimated NAIRU increased significantly in 1991, which represents an internal inconsistency since one of the assumptions of this model is that u^{phil} is a time invariant parameter.

However, any judgement about the significance of jumps and drift in the estimated NAIRU assumes that the confidence regions in Figure 4.4 are approximately correct. As explained in Section 4.4, the confidence intervals are based on the Wald principle and may give a misleading impression of the uncertainty of the estimated NAIRU. In Table 4.1 we therefore compare the Wald interval with the Fieller (and Likelihood Ratio) confidence interval. Over the full sample the difference is not large, although the Wald method appears to underestimate the interval by 0.5%.

The two sub-samples end in 1987 (before the rise in unemployment), and in 1991 (when the rise is fully represented in the sample). On the 1965–87 sample, the Wald method underestimates the width of the interval by more than 3%; the upper limit being most affected. Hypothetically therefore, a decision maker who in 1987 was equipped with the Wald interval, might be excused for not considering the possibility of a rise in the NAIRU to 4% over the next couple of years. The Fieller method shows that such a development was in fact not unlikely. Over the sample that ends in 1991, the Wald method underestimates the uncertainty of the NAIRU even more dramatically; the Fieller method gives an interval from 2.6% to 26%.

A final point of interest in Figure 4.4 is how few times the actual rate of unemployment crosses the line for the estimated equilibrium rate. This suggests very sluggish adjustment of actual unemployment to the purportedly constant equilibrium rate. In order to investigate the dynamics formally, we graft the Phillips curve equation (4.43) into a system that also contains the rate of unemployment as an endogenous variable, that is, an empirical counterpart to equation (4.2) in the theory of the main-course Phillips curve. As noted, the endogeneity of the rate of unemployment is just an integral part of the Phillips curve framework as the wage Phillips curve itself, since without the 'unemployment equation' in place one cannot show that the equilibrium rate of unemployment obtained from the Phillips curve corresponds to a steady state of the system.

In the following, we model a three equation system similar to the theoretical setup in equations (4.7)–(4.9). The model explains the manufacturing sector wage, consumer price inflation and the rate of unemployment, conditional on incomes policy, average productivity and product price. In order to model Δp_t and tu_t we also need a larger set of explanatory variables, namely the GDP growth rate $(\Delta y_{\text{gdp},t})$, and an import price index (pi_t). In particular, the inclusion of $\Delta y_{\text{gdp},t}$ in the conditioning information set is important for consistency with our initial assumption about no unit roots in tu_t. It is shown by Nymoen (2002) that (1) conventional Dickey–Fuller tests do not reject the null of a unit root in the rate of unemployment, but (2) regressing tu_t on $\Delta y_{\text{gdp},t-1}$ (which in turn is not Granger caused by tu_t) turns that around, and establishes that tu_t is without a unit-root and non-stationary due to structural changes outside the labour market.

The first equation in Table 4.2 shows the Phillips curve (4.43), this time with full information maximum likelihood (FIML) coefficient estimates. There are only minor changes from the OLS results. The second equation models the change in the rate of unemployment, and corresponds to equation (4.2) in the theoretical model in Section 4.2.[13] The coefficient of the lagged unemployment rate is -0.147, and the t-value of -4.41 confirms that tu_t can be regarded as a I(0) series on the present information set, which includes $\Delta y_{\text{gdp},t}$ and its lag as important conditioning variables (i.e. z_t in (4.2)). In terms of economic theory, $\Delta y_{\text{gdp},t}$ represents an Okun's law type relationship. The elasticity of the lagged wage share is positive which corresponds to the sign restriction $b_{u2} > 0$ in equation (4.2). However, the estimate 0.65 is not significantly different from zero, so it is arguable whether equilibrium correction is strong enough to validate identification between the estimated u^{phil} and the true steady-state unemployment rate. Moreover, the stability issue cannot be settled from inspection of the first two equations alone, since the third equation shows that the rate of CPI inflation is a function of both u_{t-1} and $wc_{t-1} - q_{t-1} - a_{t-1}$. However, the characteristic roots of the companion matrix of the system

$$
\begin{array}{ccc}
0.1381 & 0 & 0.1381 \\
0.9404 & 0.1335 & 0.9498 \\
0.9404 & -0.1335 & 0.9498
\end{array}
$$

show that the model *is* dynamically stable (i.e. has a unique stationary solution for given initial conditions). That said, the large magnitude of the complex root implies that adjustment speeds are low. Thus, after a shock to the system,

---

[13] Residual standard deviations and model diagnostics are reported at the end of the table. Superscript v indicates that we report vector versions of the single equation mis-specification tests encountered above, see equation (4.42). The overidentification χ^2 is the test of the model in Table 4.2 against its unrestricted reduced form, see Anderson and Rubin (1949, 1950), Koopmans *et al.* (1950), and Sargan (1988, pp. 125 ff.).

Table 4.2

FIML results for a Norwegian Phillips curve model

$$\Delta wc_t - \Delta p_{t-1} = \underset{(0.0146)}{-0.0627} \underset{(0.104)}{- 0.7449\Delta p_{t-1}} + \underset{(0.0826)}{0.3367\Delta q_{t-1}}$$

$$\underset{(0.00994)}{- 0.06265 IP_t} + \underset{(0.0832)}{0.234\Delta q_t} - \underset{(0.00449)}{0.02874 tu_t}$$

$$\Delta tu_t = \underset{(0.0302)}{-0.1547 tu_{t-1}} - \underset{(1.47)}{7.216\Delta y_{\text{gdp},t}} - \underset{(0.333)}{1.055\Delta y_{\text{gdp},t-1}}$$

$$+ \underset{(0.333)}{1.055(wc - q - a)_{t-1}} + \underset{(0.139)}{0.366 i1989_t} - \underset{(0.443)}{2.188\Delta^2 pi_t}$$

$$\Delta p_t = \underset{(0.0203)}{0.06023} + \underset{(0.0992)}{0.2038\Delta p_{t-1}} - \underset{(0.00366)}{0.009452 tu_{t-1}} + \underset{(0.0564)}{0.2096(wc - q - a)_{t-1}}$$

$$+ \underset{(0.0313)}{0.2275\Delta_2 pi_t} - \underset{(0.0116)}{0.05303 i1979_t} + \underset{(0.0104)}{0.04903\ i1970_t}$$

$$wc_t - q_t - a_t \cong wc_{t-1} - q_{t-1} - a_{t-1} + \Delta wc_t - \Delta a_t - \Delta q_t;$$
$$tu \equiv tu_{t-1} + \Delta tu_t;$$

Note: The sample is 1964 to 1998, $T = 35$ observations

$\hat{\sigma}_{\Delta w} = 0.014586$

$\hat{\sigma}_{\Delta tu} = 0.134979$

$\hat{\sigma}_{\Delta p} = 0.0116689$

$F^v_{\text{AR}(1-2)}(18, 59) = 1.0260[0.4464]$

$\chi^{2,v}_{\text{normality}}(6) = 3.9186[0.6877]$

$\chi^2_{\text{overidentification}}(36) = 65.533[0.002]$

the rate of unemployment will take a long time before it eventually returns to the natural rate, thus confirming Figure 4.4.

Figure 4.5 offers visual inspection of some of the dynamic properties of the model. The first four graphs show the actual values of Δp_t, tu_t, Δwc_t, and the wage share $wc_t - q_t - a_t$ together with the results from dynamic simulation. As could be expected, the fits for the two growth rates are quite acceptable. However, the 'near unit root' property of the system manifests itself in the graphs for the level of the unemployment rate and for the wage share. In both cases there are several consecutive years of under- or overprediction. The last two displays contain the cumulated dynamic multipliers of tu and the wage share resulting from a 0.01 point increase in the unemployment rate. As one might expect from the characteristic roots, the stability property is hard to gauge from the two responses. For practical purposes, it is as if the level of unemployment and the wage share 'never' return to their initial values. Thus, in the model in Table 4.2, the equilibrium correction is extremely weak.

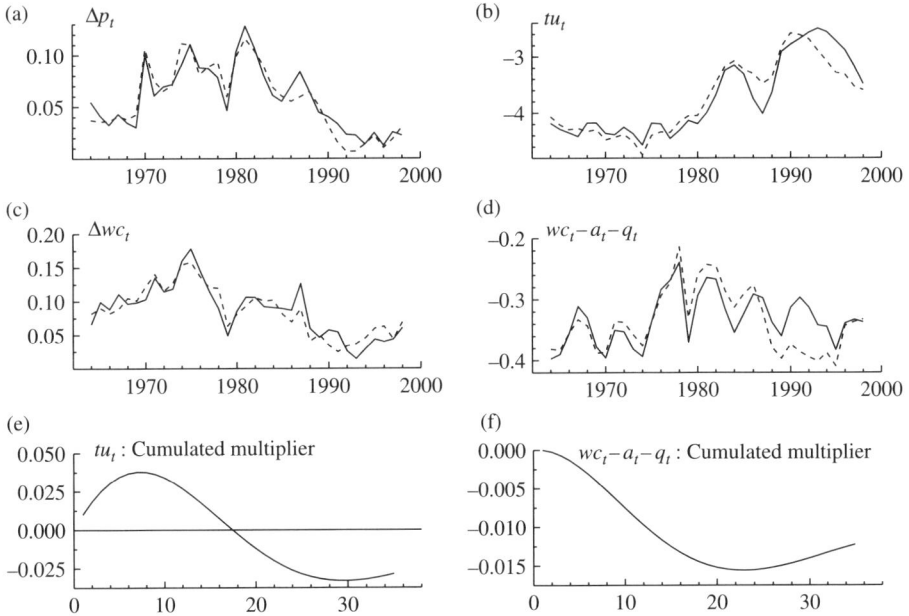

Figure 4.5. Dynamic simulation of the Phillips curve model in Table 4.2.
Panels (a–d) Actual and simulated values (dotted line). Panels (e–f):
multipliers of a one point increase in the rate of unemployment

As discussed at the end of Section 4.2, the belief in the empirical basis of the
Phillips curve natural rate of unemployment was damaged by the remorseless
rise in European unemployment in the 1980s, and the ensuing discovery of great
instability of the estimated natural rates. Thus, Solow (1986), commenting on
the large within country variation between different three-year sub-periods in
OECD estimates of the natural rate, concludes that

A natural rate that hops around from one triennium to another under the influence
of unspecified forces, *including past unemployment,* is not 'natural' at all. (Solow
1986, p. 33)

In that perspective, the variations in the Norwegian natural rate estimates
in Figure 4.4 are quite modest, and may pass as relatively acceptable as a
first-order approximation of the attainable level of unemployment. However,
the econometric system showed that equilibrium correction is very weak. After
a shock to the system, the rate of unemployment is predicted to drift away
from the natural rate for a very long period of time. Hence, the natural rate
thesis of asymptotical stability is not validated.

There are several responses to this result. First, one might try to patch
up the estimated unemployment equation, and try to find ways to recover

a stronger relationship between the real wage and the unemployment rate. In the following we focus instead on the other end of the problem, namely the Phillips curve itself. In Section 6.9.2 we show that when the Phillips curve framework is replaced with a wage model that allows equilibrium correction to any given rate of unemployment, rather than to the 'natural rate' only, all the inconsistencies are resolved. However, that kind of wage equation is first anchored in the economic theory of Chapters 5 and 6.

4.6.1 Summary

The Phillips curve ranges as the dominant approach to wage and price modelling in macroeconomics. In the United States, in particular, it retains its role as the operational framework for both inflation forecasting and for estimating of the NAIRU. In this chapter we have shown that the Phillips curve is consistent with cointegration between prices, wages and productivity, and a stationary rate of unemployment, and hence there is a common ground between the Phillips curve and the Norwegian model of inflation of the previous chapter. However, unlike the Norwegian model, the Phillips curve framework specifies a single equilibrating mechanism which supports cointegration—in the simplest case with fixed and exogenous labour supply, the equilibrium correction is due to a downward sloping labour demand schedule. The specificity of the equilibrating mechanism of the Phillips curve is not always recognised. In the context of macroeconomic models with a large number of equations, it has the somewhat paradoxical implication that the stationary value of the rate of unemployment can be estimated from a single equation.

We have also argued that the Phillips curve framework is consistent with a stable autoregressive process for the rate of unemployment, subject only to a few regime shifts that can be identified with structural breaks in the operation of labour markets. The development of European unemployment rates since the early 1980s is difficult to fit into this framework, and model builders started to look for alternative models. Interestingly, already in 1984 one review of the United Kingdom macroeconomic models concluded that 'developments in wage equations have led to the virtual demise of the Phillips curve as the standard wage relationship in macro models'.[14] These developments are the themes of the following two chapters.

[14] See Wallis *et al.* (1984, p. 134).

5

Wage bargaining and price-setting

In this chapter we go a step forward to compare both the main-course model and the Phillips curve by introducing the Layard–Nickell wage-curve model of incomplete competition. It marks a step forward in that it combines formal models of wage bargaining and models of monopolistic price-setting. Thus, compared to Aukrust's model, the hypothesised wage and price cointegrating vectors are better founded in economic theory, and specific candidates for explanatory variables flow naturally from the way the bargaining model is formulated. We will show that there are cases of substantive interest where the identification problem pointed out by Manning (1993) are resolved, and we will show applications with empirically stable and interpretable wage and price curves.

5.1 Introduction

In the course of the 1980s interesting developments took place in macroeconomics. First, the macroeconomic implications of imperfect competition with price-setting firms were developed in several papers and books; see, for example, Bruno (1979), Bruno and Sachs (1984), Blanchard and Kiyotaki (1987), and Blanchard and Fisher (1989: ch. 8). Second, the economic theory of labour unions, pioneered by Dunlop (1944), was extended and formalised in a game theoretic framework; see, for example, Nickell and Andrews (1983), Hoel and Nymoen (1988). Models of European unemployment, that incorporated elements from both these developments, appeared in Layard and Nickell (1986), Carlin and Soskice (1990), Layard *et al.* (1991), and Lindbeck (1993). The new standard model of European unemployment is incontestably linked to Layard and Nickell and their co-authors. However, we follow established practice and refer to the framework as the Incomplete Competition Model (ICM), or, interchangeably, as the *wage curve* framework (as opposed to the Phillips curve model of the previous chapter). *Incomplete competition* is particularly apt since the model's defining characteristic is the explicit assumption of imperfect competition in both product and labour markets, see; for example, Carlin and Soskice (1990).[1] The ICM was quickly incorporated into the supply side of macroeconometric models (see Wallis 1993, 1995), and purged European econometric models of the Phillips curve, at least until the arrival of the New Keynesian Phillips curve late in the 1990s (see Chapter 7 in this book).

Since the theory is cast in terms of levels variables, the ICM stands closer to the main-course model than the Phillips curve tradition. On the other hand, both the wage curve and the Phillips curve presume that it is the rate of unemployment that reconciles the conflict between wage earners and firms. Both models take the view that the equilibrium or steady-state rate of unemployment is determined by a limited number of factors that reflect structural aspects such as production technology, union preferences, and institutional factors (characteristics of the bargaining system, the unemployment insurance system). Thus, in both families of theories demand management and monetary policy have only a short-term effect on the rate of unemployment. In the (hypothetical) situation when all shocks are switched off, the rate of unemployment returns to a unique structural equilibrium rate, that is, the natural rate or the NAIRU. Thus, the ICM is unmistakably a model of the natural rate both in its motivation and in its implications: 'In the long run, unemployment is determined entirely by long-run supply factors and equals the NAIRU' (Layard *et al.* 1994, p. 23).

5.2   Wage bargaining and monopolistic competition

There is a number of specialised models of 'non-competitive' wage-setting; see, for example, Layard *et al.* (1991: ch. 7). Our aim in this section is to represent the common features of these approaches in a theoretical model of wage bargaining and monopolistic competition, building on Rødseth (2000: ch. 5.9) and Nymoen and Rødseth (2003). We start with the assumption of a large number of firms, each facing downward-sloping demand functions. The firms are price setters and equate marginal revenue to marginal costs. With labour being the only variable factor of production (and constant returns to scale) we obtain the following price-setting relationship:

$$Q_i = \frac{El_Q Y}{El_Q Y - 1} \frac{W_i}{A_i},$$

[1] Nevertheless, the ICM acronym may be confusing—in particular if it is taken to imply that the alternative model (the Phillips curve) contains perfect competition.

where $A_i = Y_i/N_i$ is average labour productivity, Y_i is output, and N_i is labour input. $El_Q Y > 1$ denotes the absolute value of the elasticity of demand facing each firm i with respect to the firm's own price. In general $El_Q Y$ is a function of relative prices, which provides a rationale for inclusion of, for example, the real exchange rate in aggregate price equations. However, it is a common simplification to assume that the elasticity is independent of other firms' prices and is identical for all firms. With constant returns technology aggregation is no problem, but for simplicity we assume that average labour productivity is the same for all firms and that the aggregate price equation is given by

$$Q = \frac{El_Q Y}{El_Q Y - 1} \frac{W}{A}. \tag{5.1}$$

The expression for real profits (π) is therefore

$$\pi = Y - \frac{W}{Q} N = \left(1 - \frac{W}{Q} \frac{1}{A}\right) Y.$$

We assume that the wage W is settled by maximising the Nash product:

$$(\nu - \nu_0)^{\mho} \pi^{1-\mho}, \tag{5.2}$$

where ν denotes union utility and ν_0 denotes the fall-back utility or reference utility. The corresponding break-point utility for the firms has already been set to zero in (5.2), but for unions the utility during a conflict (e.g. strike or work-to-rule) is non-zero because of compensation from strike funds. Finally $\mho$ represents the relative bargaining power of unions.

Union utility depends on the consumer real wage of an employed worker and the aggregate rate of unemployment, thus $\nu(W/P, U, Z_\nu)$ where P denotes the consumer price index (CPI).[2] The partial derivative with respect to wages is positive, and negative with respect to unemployment ($\nu_W' > 0$ and $\nu_U' \leq 0$). Z_ν represents other factors in union preferences. The fall-back or reference utility of the union depends on the overall real-wage level and the rate of unemployment, hence $\nu_0 = \nu_0(\bar{W}/P, U)$ where $\bar{W}$ is the average level of nominal wages which is one of the factors determining the size of strike funds. If the aggregate rate of unemployment is high, strike funds may run low in which case the partial derivative of ν_0 with respect to U is negative ($\nu_{0U}' < 0$). However, there are other factors working in the other direction, for example, that the probability of entering a labour market programme, which gives laid-off workers higher utility than open unemployment, is positively related to U. Thus, the sign of ν_{0U}' is difficult to determine from theory alone. However, we assume in the following that $\nu_U' - \nu_{0U}' < 0$.

---

[2] We abstract from income taxes.

With these specifications of utility and break-points, the Nash product, denoted $\mathcal{N}$, can be written as

$$\mathcal{N} = \left\{ \nu \left(\frac{W}{P}, U, Z_\nu \right) - \nu_0 \left(\frac{\bar{W}}{P} \right) \right\}^\upsilon \left\{ \left(1 - \frac{W}{Q} \frac{1}{A} \right) Y \right\}^{1-\upsilon}$$

or

$$\mathcal{N} = \left\{ \nu \left(\frac{W_q}{P_q}, U, Z_\nu \right) - \nu_0 \left(\frac{\bar{W}}{P} \right) \right\}^\upsilon \left\{ \left(1 - W_q \frac{1}{A} \right) Y \right\}^{1-\upsilon},$$

where $W_q = W/Q$ is the producer real wage and $P_q = P/Q$ is the wedge between the consumer and producer real wage. The first-order condition for a maximum is given by $\mathcal{N}_{W_q} = 0$ or

$$\upsilon \frac{\nu'_W (W_q/P_q, U, Z_\nu)}{\nu(W_q/P_q, U, Z_\nu) - \nu_0(\bar{W}/P, U)} = (1 - \upsilon) \frac{1/A}{(1 - W_q/A)}. \qquad (5.3)$$

In a symmetric equilibrium, $W = \bar{W}$, leading to $W_q/P_q = \bar{W}/P$ in equation (5.3), and the aggregate bargained real wage W_q^b is defined implicitly as

$$W_q^b = F(P_q, A, \upsilon, U). \qquad (5.4)$$

A log linearisation of (5.4), with subscript t for time period added, gives

$$w_{q,t}^b = m_{b,t} + \omega p_{q,t} - \varpi u_t, \qquad 0 \le \omega \le 1, \quad \varpi \ge 0. \qquad (5.5)$$

$m_{b,t}$ in (5.5) depends on A, υ, and Z_ν, and any one of these factors can of course change over time.

As noted above, the term $p_{q,t} = (p - q)_t$ is referred to as the *wedge* between the consumer real wage and the producer real wage. The role of the wedge as a source of wage pressure is contested in the literature. In part, this is because theory fails to produce general implications about the wedge coefficient ω—it can be shown to depend on the exact specification of the utility functions ν and ν_0 (see, for example, Rødseth 2000: ch. 8.5 for an exposition). We follow custom and restrict the elasticity ω of the wedge to be non-negative. The role of the wedge may also depend on the level of aggregation of the analysis. In the traded goods sector ('exposed' in the terminology of the main-course model of Chapter 3) it may be reasonable to assume that ability to pay and profitability are the main long-term determinants of wages, hence $\omega = 0$. However, in the sheltered sector, negotiated wages may be linked to the general domestic price level. Depending on the relative size of the two sectors, the implied weight on the consumer price may then become relatively large in an aggregate wage equation.

Equation (5.5) is a general proposition about the bargaining outcome and its determinants, and can serve as a starting point for describing wage formation in any sector or level of aggregation of the economy. In the rest of this section we view equation (5.5) as a model of the aggregate wage in the economy, which gives the most direct route to the predicted equilibrium outcome for real wages and for the rate of unemployment. However, in Section 5.4 we consider another frequently made interpretation, namely that equation (5.5) applies to the manufacturing sector.

The impact of the rate of unemployment on the bargained wage is given by the elasticity ϖ, which is a key parameter of interest in the wage curve literature. ϖ may vary between countries according to different wage-setting systems. For example, a high degree of coordination, especially on the employer side, and centralisation of bargaining is expected to induce more responsiveness to unemployment (a higher ϖ) than uncoordinated systems that give little incentives to solidarity in bargaining. At least this is the view expressed by authors who build on multi-country regressions; see, for example, Alogoskoufis and Manning (1988) and Layard *et al.* (1991: ch. 9). However, this view is not always shared by economists with detailed knowledge of, for example, the Swedish system of centralised bargaining (see Lindbeck 1993: ch. 8).

Figure 5.1 also motivates why the magnitude of ϖ plays such an important role in the wage curve literature. The horizontal line in the figure is consistent with the equation for price-setting in (5.1), under the assumption that productivity is independent of unemployment ('normal cost pricing'). The two downward sloping lines labelled 'low' and 'high' (wage responsiveness), represent different states of wage-setting, namely 'low' and 'high' ϖ. Point (i) in the figure represents a situation in which firm's wage-setting and the bargaining outcome are consistent in both countries—we can think of this as a low unemployment equilibrium. Next, assume that the two economies are hit by a supply-side shock, that shifts the firm-side real wage down to the dotted line. The Layard–Nickell model implies that the economy with the least real-wage responsiveness ϖ will experience the highest rise in the rate of unemployment, (ii) in the figure, while the economy with more flexible real wages ends up in point (iii) in the figure.

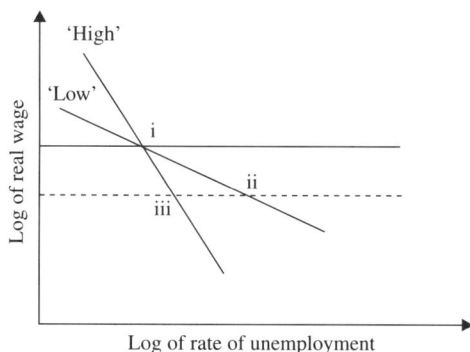

Figure 5.1. Role of the degree of wage responsiveness to unemployment

A slight generalisation of the price-setting equation (5.1) is to let the price markup on average cost depend on demand relative to capacity. If we in addition invoke an Okun's law relationship to replace capacity utilisation with the rate of unemployment, the real wage consistent with firms' price-setting, w_q^f, can be written in terms of log of the variables as

$$w_{q,t}^f = m_{f,t} + \vartheta u_t, \qquad \vartheta \geq 0. \tag{5.6}$$

$m_{f,t}$ depends on the determinants of the product demand elasticity $El_Q Y$ and average labour productivity a_t.

5.3   The wage curve NAIRU

Without making further assumptions, and for a given rate of unemployment, there is no reason why $w_{q,t}^b$ in (5.5) should be equal to $w_{q,t}^f$ in (5.6). However, there are really two additional doctrines of the Layard–Nickell model. First, that no equilibrium with a constant rate of inflation is possible without the condition $w_{q,t}^b = w_{q,t}^f$. Second, the adjustment of the rate of unemployment is the singular equilibrating mechanism that brings about the necessary equalisation of the competing claims.

The heuristic explanation usually given is that excessive real wage claims on the part of the workers, that is, $w_{q,t}^b > w_{q,t}^f$, result in increasing inflation (e.g. $\Delta^2 p_t > 0$), while $w_{q,t}^b < w_{q,t}^*$ goes together with falling inflation ($\Delta^2 p_t < 0$). The only way of maintaining a steady state with constant inflation is by securing that the condition $w_{q,t}^b = w_{q,t}^f$ holds, and the function of unemployment is to reconcile the claims, see Layard *et al.* (1994: ch. 3).

Equations (5.5), (5.6), and $w_q^b = w_q^f$ can be solved for the equilibrium real wage (w_q), and for the rate of unemployment that reconciles the real wage claims of the two sides of the bargain, the wage curve NAIRU, denoted $\bar{u}^w$:

$$\bar{u}_t^w = \left(\frac{m_b - m_f}{(\vartheta + \varpi)} + \frac{\omega}{(\vartheta + \varpi)} p_{q,t} \right), \qquad \omega \geq 0. \tag{5.7}$$

Thus, point (i) in Figure 5.1 is an example of $w_q = w_q^b = w_q^f$ and $u_t = u^w$, albeit for the case of normal cost pricing, that is, $\vartheta = 0$. Likewise, the analysis of a supply-side shock in the figure is easily confirmed by taking the derivative of u^w with respect to m_f.

In the case of $\omega = 0$, the expression for the wage curve NAIRU simplifies to

$$\bar{u}^w = \frac{m_b - m_f}{(\vartheta + \varpi)}, \qquad \text{if } \omega = 0, \tag{5.8}$$

meaning that the equilibrium rate of unemployment depends only on such factors that affect wage- and price-settings, that is, supply-side factors. This is the same type of result that we have seen for the Phillips curve under the condition of dynamic homogeneity, see Section 4.2.

The definitional equation for the log of the CPI, p_t is

$$p_t = \phi q_t + (1 - \phi)pi_t, \qquad (5.9)$$

where pi denotes the log of the price of imports in domestic currency, and we abstract from the indirect tax rate. Using (5.9), the wedge p_q in equation (5.7) can be expressed as

$$p_{q,t} = (1 - \phi)pi_{q,t},$$

where $pi_q \equiv pi - q$, denotes the real exchange rate. Thus it is seen that, for the case of $\omega > 0$, the model can alternatively be used to determine a real exchange rate that equates the two real wage claims for a given level of unemployment; see Carlin and Soskice (1990: ch. 11.2), Layard *et al.* (1991: ch. 8.5), and Wright (1992). In other words, with $\omega > 0$, the wage curve natural $\bar{u}^w$ is more of an intermediate equilibrium which is not completely supply-side determined, but depends on demand-side factors through the real exchange rate. To obtain the long-run equilibrium, an extra constraint of balanced current account is needed.[3]

Earlier in this section we have seen that theory gives limited guidance to whether the real-wage wedge affects the bargained wage or not. The empirical evidence is also inconclusive; see, for example, the survey by Bean (1994). However, when it comes to short-run effects of the wedge, or to components of the wedge such as consumer price growth, there is little room for doubt: dynamic wedge variables have to be taken into account. In Chapter 6 we present a model that includes these dynamic effects in full.

At this stage, it is nevertheless worthwhile to foreshadow one result, namely that the 'no wedge' condition, $\omega = 0$, is *not* sufficient to ensure that $\bar{u}^w$ in equation (5.7) corresponds to an asymptotically stable stationary solution of a dynamic model of wage- and price-setting. Other and additional parameter restrictions are required. This suggests that something quite important is lost by the ICM's focus on the static price and wage relationships, and in Chapter 6 we therefore graft these long-run relationships into a dynamic theory framework. As a first step in that direction, we next investigate the econometric specification of the wage curve model, building on the idea that the theoretical wage- and price-setting schedules may correspond to cointegrating relationships between observable variables.

5.4 Cointegration and identification

In Chapter 3, we made the following assumptions about the time-series properties of the variables we introduced: nominal and real wages and productivity

---

[3] Rødseth (2000: ch. 8.5) contains a model with a richer representation of the demand side than in the model by Layard *et al.* (1991). Rødseth shows that the long-run equilibrium must satisfy both a zero private saving condition and the balanced current account condition.

are I(1), while, possibly after removal of deterministic shifts, the rate of unemployment is without a unit root. A main concern is clearly how the theoretical wage curve model can be reconciled with these properties of the data. In other words: how should the long-run wage equation be specified to attain a true cointegrating relationship for real wages, and to avoid the pitfall of spurious regressions?

As we have seen, according to the bargaining theory, the term $m_{b,t}$ in (5.5) depends on average productivity, A_t.[4] Having assumed $u_t \sim$ I(0), and keeping in mind the possibility that $\omega = 0$, it is seen that it follows directly from cointegration that productivity *has* to be an important variable in the relationship. In other words, a positive elasticity $El_A W_q$ is required to balance the I(1) trend in the product real wage on the left-hand side of the expression.

Thus, the general long-run wage equation implied by the wage bargaining approach becomes

$$w_{q,t}^b = m_b + \iota a_t + \omega p_{q,t} - \varpi\, u_t, \qquad 0 < \iota \le 1, \quad 0 \le \omega \le 1, \quad \varpi \ge 0, \quad (5.10)$$

where $w_{q,t}^b \equiv w_t^b - q_t$ denotes the 'bargained real wage' as before. The intercept m_b is redefined without the productivity term, which is now singled out as an I(1) variable on the left-hand side of the expression, and with the other determinants assumed to be constant. Finally, defining

$$ecm_{b,t} = w_{q,t} - w_{q,t}^b \sim \text{I}(0)$$

allows us to write the hypothesised cointegrating wage equation as

$$w_{q,t} = m_b + \iota a_t + \omega p_{q,t} - \varpi\, u_t + ecm_{b,t}. \qquad (5.11)$$

Some writers prefer to include the reservation wage (the wage equivalent of being unemployed) in (5.10). For example, from Blanchard and Katz (1999) (but using our own notation to express their ideas):

$$w_{q,t}^b = m_b + \iota' a_t + (1 - \iota')w_t^r + \omega p_{q,t} - \varpi\, u_t, \qquad 0 < \iota' \le 1, \qquad (5.12)$$

where w_t^r denotes the reservation wage. However, since real wages are integrated, any meaningful operational measure of w_t^r must logically cointegrate with $w_{q,t}$ directly. In fact, Blanchard and Katz hypothesise that w_t^r is a linear function of the real wage and the level of productivity.[5] Using that (second) cointegrating relationship to substitute out w_t^r from (5.12) implies a relationship which is observationally equivalent to (5.11).

The cointegration relationship stemming from price-setting is anchored in equation (5.6). In the same way as for wage-setting, it becomes important in

---

[4] Recall that we expressed the Nash-product as

$$\mho \frac{\nu_W'(W_q/P_q,U,Z_\nu)}{\nu(W_q/P_q,U,Z_\nu) - \nu_0(W_q/P_q,U)} = (1 - \mho)\frac{1/A}{(1 - W_q/A)},$$

in (5.3).

[5] See their equation (4), which uses the lagged real wage, which cointegrates with current real wage, on the right-hand side.

applied work to represent the productivity term explicitly in the relationship. We therefore rewrite the long-term price-setting schedule as

$$w_{q,t}^{f} = m_f + a_t + \vartheta u_t, \tag{5.13}$$

where the composite term m_f in (5.6) has been replaced by $m_f + a_t$. Introducing $ecm_{f,t} = w_{q,t} - w_{q,t}^{f} \sim I(0)$, the second implied cointegration relationship becomes

$$w_{q,t} = m_f + a_t + \vartheta u_t + ecm_{f,t}. \tag{5.14}$$

While the two cointegrating relationships are not identified in general, identifying restrictions can be shown to apply in specific situations that occur frequently in applied work. From our own experience with modelling both disaggregate and aggregate data, the following three 'identification schemes' have proven themselves useful:

One cointegrating vector. In many applications, especially on sectorial data, formal tests of cointegration support only one cointegration relationship, thus either one of $ecm_{b,t}$ and $ecm_{f,t}$ is I(1), instead of both being I(0). In this case it is usually possible to identify the single cointegrating equation economically by restricting the coefficients, and by testing the weak exogeneity of one or more of the variables in the system.

No wedge. Second, and still thinking in terms of a sectorial wage–price system: assume that the price markup is not constant as assumed above, but is a function of the relative price (via the price elasticity ϵ). In this case, the price equation (5.14) is augmented by the real exchange rate $p_t - pi_t$. If we furthermore assume that $\omega = 0$ (no wedge in wage formation) and $\vartheta = 0$ (normal cost pricing), identification of both long-run schedules is logically possible.

Aggregate price–wage model. The third cointegrating identification scheme is suited for the case of aggregated wages and prices. The long-run model is

$$w_t = m_b + (1 - \omega)q_t + \iota a_t + \omega p_t - \varpi u_t + ecm_{b,t}, \tag{5.15}$$
$$q_t = -m_f + w_t - a_t - \vartheta u_t - ecm_{f,t}, \tag{5.16}$$
$$p_t = \phi q_t + (1 - \phi)pi_t,$$

solving out for producer prices q_t then gives a model in wages w_t and consumer prices p_t only,

$$w_t = m_b + \frac{1 - \omega(1 - \phi)}{\phi} p_t + \iota a_t - \varpi u_t$$
$$- \frac{(1 - \omega)(1 - \phi)}{\phi} pi_t + ecm_{b,t} \tag{5.17}$$
$$p_t = -\phi m_f + \phi(w_t - a_t) - \phi \vartheta u_t + (1 - \phi)pi_t - \phi ecm_{f,t}, \tag{5.18}$$

that implicitly implies non-linear cross-equation restrictions in terms of ϕ.

By simply viewing (5.17) and (5.18) as a pair of simultaneous equations, it is clear that the system is unidentified in general. However, if the high level of aggregation means that ω can be set to unity (while retaining cointegration), and there is normal cost pricing in the aggregated price relationship, identification is again possible. Thus $\omega = 1$ and $\vartheta = 0$ represent one set of necessary (order) restrictions for identification in this case:

$$w_t = m_b + p_t + \iota a_t - \varpi u_t + ecm_{b,t}, \tag{5.19}$$

$$p_t = -\phi m_f + \phi(w_t - a_t) + (1 - \phi)pi_t - \phi ecm_{f,t}. \tag{5.20}$$

We next give examples of how the first and third schemes can be used to identify cointegrating relationships in Norwegian manufacturing and in a model of aggregate United Kingdom wages and prices.

5.5   Cointegration and Norwegian manufacturing wages

We analyse the annual data set for Norwegian manufacturing that was used to estimate a main-course Phillips curve in Section 4.6. We estimate a vector autoregressive model (VAR), check for mis-specification and then for cointegration, and discuss identification. Several of the variables were defined in Section 4.6.

The endogenous variables in the VAR are all in log scale and are denoted as follows: wc_t (wage cost per hour), q_t (producer price index), a_t (average labour productivity), tu_t (the total rate of unemployment, that is, including labour market programmes), rpr_t (the replacement ratio), and we (the real-wage wedge in manufacturing). The operational measure of the wedge is defined as

$$we_t = p_t - q_t + t1_t + t2_t \equiv p_{q,t} + t1_t + t2_t,$$

where $t1$ and $t2$ denote payroll and average income tax rates respectively. The annual sample period is 1964–98, so there are only 36 observations of the six variables. We estimate a first-order VAR, extended by four conditioning variables:

two dummies ($i1967_t$ and IP_t);
the lagged inflation rate, Δcpi_{t-1};
the change in normal working hours, Δh_t,

all of which were discussed in Section 4.6. Table 5.1 contains the residual diagnostics for the VAR. To save space we have used * to denote a statistic which is significant at the 10% level, and ** to denote significance at the 5% level. There are only two significant mis-specification tests and both indicate heteroskedasticity in the residuals of the replacement ratio.[6]

---

[6] The statistics reported in the table are explained in Section 4.6, Table 4.2, and in connection with equation (4.43).

Table 5.1

Diagnostics for a first-order conditional VAR for
Norwegian manufacturing 1964–98

	wc	q	a	tu	we	rpr	VAR
$F_{AR(1-2)}(2,22)$	0.56	0.41	0.59	1.29	1.27	2.36	
$\chi^2_{normality}(2)$	0.60	1.42	0.42	0.38	0.26	2.87	
$F_{HETx^2}(12,11)$	0.45	0.28	0.54	0.42	1.47	8.40*	
$F_{ARCH(1-1)}(1,22)$	0.16	1.12	0.44	0.10	2.51	13.4**	
$F^v_{AR(1-2)}(72,43)$							1.518
$\chi^{2,v}_{normality}(12)$							6.03
$\chi^{2,v}_{HETx^2}(252)$							269.65

Table 5.2

Cointegration analysis, Norwegian manufacturing
wages 1964–98

$$
\begin{bmatrix}
r & 1 & 2 & 3 & 4 & 5 & 6 \\
\text{Eigenvalue} & 0.92 & 0.59 & 0.54 & 0.29 & 0.16 & 0.01 \\
\text{Max} & 72.49^{**} & 25.64 & 22.5 & 10.12 & 4.98 & 0.31 \\
\text{Tr} & 136^{**} & 63.56 & 37.92 & 15.42 & 5.29 & 0.31
\end{bmatrix}
$$

The results of the cointegration analysis are shown in Table 5.2 which contains the eigenvalues and associated maximum eigenvalue (Max) and trace (Tr) statistics, which test the hypothesis of $(r-1)$ vs. r cointegration vectors, and r vs. less than $(r-1)$ cointegrating vectors, respectively. These eigenvalue tests are corrected for degree of freedom (see Doornik and Hendry 1997b), and give formal evidence for one cointegrating relationship, namely

$$wc_t = m_{wc} + 0.93q_t + 1.20a_t - 0.0764tu_t + 0.0318we_t + 0.11614rpr_t + ecm_{b,t}, \tag{5.21}$$

when we normalise on wc, and let $ecm_{b,t}$ denote the I(0) equilibrium correction term.

Equation (5.21) is unique, *qua* cointegrating relationship, but it can either represent a wage equation or a long-run price-setting schedule. Both interpretations are consistent with finding long-run price homogeneity and a unit long-run elasticity of labour productivity. The joint test of these two restrictions gives $\chi^2(2) = 4.91[0.09]$, and a restricted cointegrating vector becomes

$$wc_t - q_t - a_t = m_{wc} - 0.069tu_t + 0.075we_{q,t} + 0.1644rpr_t + ecm_{b,t}.$$

The real-wage wedge can be omitted from the relationship, and thus imposing $\omega = 0$, we obtain the final estimated cointegration relationship as:

$$wc_t - q_t - a_t = m_{wc} - 0.065tu_t + 0.184rpr_t + ecm_{b,t} \tag{5.22}$$

and the test statistics for all three restrictions $\chi^2(3) = 5.6267[0.1313]$. Equation (5.22) is the empirical counterpart to (5.11), with $\omega = 0$ and $m_b = m_{wc} + 0.184 rpr$.

In simplified form, the six variable $I(0)$ system can be written as:

$$
\begin{pmatrix} \Delta wc_t \\ \Delta q_t \\ \Delta a_t \\ \Delta tu_t \\ \Delta we_t \\ \Delta rpr_t \end{pmatrix} = \begin{pmatrix} -0.476(0.05) \\ -0.017(0.121) \\ -0.074(0.086) \\ 0.800(0.787) \\ -0.309(0.168) \\ -0.006(0.177) \end{pmatrix} \cdot ecm_{b,t-1} + \text{additional terms,} \qquad (5.23)
$$

which shows that there is significant evidence of equilibrium correction in wage-setting.

Interestingly, the real-wage wedge we_t also appears to be endogenous. However, since we_t does not enter into the cointegration relationship, its endogeneity poses no problems for identification. A set of sufficient restrictions that establishes (5.22) as a long-run wage equation is given by the weak exogeneity of q_t, a_t, u_t, and rpr_t with respect to the parameters of the cointegrating relationship (5.22). The test of the 4 restrictions gives $\chi^2(4) = 2.598[0.6272]$, establishing that (5.22) has been identified as a long-run wage equation.

Of particular interest is the significance (or otherwise) of the adjustment coefficients of the product price index q_t and average productivity a_t, since the answer to that question relates to whether the causality thesis ($H4_{mc}$) of Aukrust's main-course model in Section 3.2.1 applies to the Norwegian manufacturing sector. Again, from (5.23) there is clear indication that the $ecm_{b,t-1}$-coefficients of Δq_t and Δa_t are insignificant, and a test of their joint insignificance gives $\chi^2(2) = 0.8315[0.6598]$. Thus, we not only find that the cointegration equation takes the form of the extended main-course equation discussed in Chapter 3, but also that deviations from the long-run relationship seem to be corrected through wage adjustments and not through prices and productivity.[7]

Visual inspection of the strength of cointegration is offered by Figure 5.2, where panel (a) shows the sequence of (largest) eigenvalues over the period 1980–98. Although the canonical correlation drops somewhat during the 1980s, it settles at a value close to 0.92 for the rest of the sample. Panels (b) and (c) show that the elasticities of the rate of unemployment and of the replacement

---

[7] This result is the opposite of Rødseth and Holden (1990, p. 253), who found that deviation from the main course is corrected by Δmc_t defined as $\Delta a_t + \Delta q_t$. However, that result is influenced by invalid conditioning, since their equation for Δmc_t has not only ecm_{t-1}, but also Δwc_t on the right-hand side. Applying their procedure to our data gives their results: for the sample period 1966–98, ecm_{t-1} obtains a 't-value' of 2.94 and a (positive) coefficient of 0.71. However, when Δwc_t is dropped from the right-hand side of the equation (thus providing the relevant framework for testing) the 't-value' of ecm_{t-1} for Δa_t falls to 0.85.

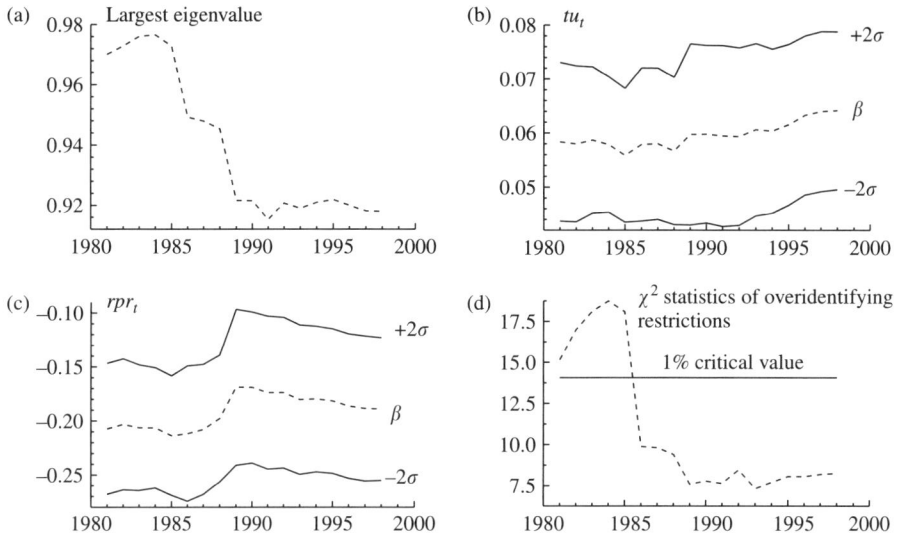

Figure 5.2. Norwegian manufacturing wages, recursive cointegration results
1981–98: (a) Sequence of highest eigenvalue; (b) and (c) coefficients of
identified equation; (d) sequence of χ^2 test of 7 overidentifying restrictions.

ratio are stable, and significant when compared to the ± 2 estimated standard
errors.[8] Over the period 1964–98, the joint test of all the 7 restrictions yields
$\chi^2(7) = 8.2489[0.3112]$. Figure 5.2 shows that we would have reached the same
conclusion about no rejection on samples that end in 1986 and later.

The findings are interpretable in the light of the theories already discussed.
First, equation (5.22) conforms to an extended main-course proposition that we
discussed in Chapter 3: the wage share is stationary around a constant mean,
conditional on the rate of unemployment and the replacement ratio. However,
it is also consistent with the wage curve of Section 5.2. The elasticity of the
rate of unemployment is 0.065 which is somewhat lower than the 0.1 elasticity
which has come to be regarded as an empirical law following the comprehensive empirical documentation in Blanchflower and Oswald (1994). Finally, the
exogeneity tests support the main-course model assumption about exogenous
productivity and product price trends, and that wages are correcting deviation
from the main course. The analysis also resolves the inconsistency that ham-
pered the empirical Phillips curve system in Section 4.6, namely that there was
little sign of an equilibrium correction which is necessary to keep the wage on
the main course. In the cointegration model, wages are adjusting towards the
main course, and the point where the Phillips curve goes wrong is exactly by

---

[8] Note that these estimates are conditioned by the restrictions on the loadings matrix
explained in the text and that the the signs of the coefficients are reversed in the graphs.

insisting that we should look to unemployment for provision of the equilibrating mechanism. In Chapter 6 we develop the theoretical implication of this type of dynamics further. Specifically, in Section 6.9.2, we incorporate the long-run wage curve in (5.22) into a dynamic model of manufacturing wages and the rate of unemployment in Norway.

5.6 Aggregate wages and prices: UK quarterly data

Bårdsen *et al.* (1998) present results of aggregate wage and price determination in the United Kingdom, that can be used to illustrate the third identification scheme above. In the quarterly data set for the United Kingdom the wage variable w_t is average actual earnings. The price variable p_t is the retail price index, excluding mortgage interest payments and the Community Charge. In this analysis, mainland productivity a_t, import prices pi_t, and the unemployment rate u_t are initially treated as endogenous variables in the VAR, and the validity of restrictions of weak exogeneity is tested. The variables that are treated as non-modelled without testing are employers' taxes $t1_t$, indirect taxes $t3_t$, and a measure of the output gap gap_t, approximated by mainland GDP-cycles estimated by the Hodrick–Prescott (HP) filter. Finally, two dummies are included to take account of income policy events.

The equilibrium relationships presented by Bårdsen *et al.* (1998) are shown in Table 5.3 (to simplify the table, the constants appearing in equations (5.15)–(5.20) are omitted along with the residuals $ecm_{b,t}$ and $ecm_{f,t}$). The first panel simply records the two long-run relationships (5.17) and (5.18), with the noted changes. Panel 2 records the unidentified cointegrating vectors, using the Johansen procedure (residual diagnostics are given at the bottom of the table). Panel 3 reports the estimated relationships after imposing weak exogeneity restrictions for u_t, a_t, and pi_t. The estimated β coefficients do not change much, and the reported test statistic $\chi^2(6) = 10.02[0.12]$ does not reject the exogeneity restrictions. Panel 4 then applies the restrictions discussed in Section 5.4—$\omega = 1$ and $\vartheta = 0$—hence the two estimated equations correspond to the theoretical model (5.19) and (5.20). The impact of the identification procedure on the estimated β coefficients is clearly visible. Panel 5 shows the final wage and price equations reported by Bårdsen *et al.* (1998), that is, their equation (14a) and (14b). The recursive estimates of the cointegration coefficients (note the sign change in the graphs) together with confidence intervals and the sequence of tests of the overidentifying restrictions are shown in Figure 5.3.

The identifying restrictions are statistically acceptable on almost any sample size, and the coefficients of the two identified relationships are stable over the same period. Bårdsen *et al.* (1998) perform an analysis of aggregate Norwegian

Table 5.3
Cointegrating wage- and price-setting schedules in the United Kingdom

Panel 1: The theoretical equilibrium

$$w_t = \frac{1 - \omega(1 - \phi)}{\phi}p_t + \iota a_t - \delta_1 t1_t - \varpi u_t - \frac{(1 - \omega)(1 - \phi)}{\phi}pi_t - \frac{\delta_3(1 - \omega)}{\phi}t3_t$$

$$p_t = \phi(w_t + t1_t - a_t) - \phi\vartheta u_t + (1 - \phi)pi_t + \delta_3 t3_t$$

Panel 2: No restrictions

$$w = 1.072p + 1.105a - 0.005u - 0.101pi - 0.892t1 - 0.395t3$$
$$p = 0.235w + 0.356a - 0.215u + 0.627pi - 0.775t1 + 3.689t3$$

Panel 3: Weak exogeneity

$$w = 1.103p + 1.059a - 0.005u - 0.139pi - 0.936t1 - 0.421t3$$
$$p = 0.249w + 0.325a - 0.212u + 0.535pi - 0.933t1 + 3.796t3$$
$$\chi^2(6) = 10.02[0.12]$$

Panel 4: Non-linear cross equation restrictions, weak exogeneity

$$w = 0.99p + 1.00a - 0.05u - 0.01pi - 1.32t1 - 0.05t3$$
$$\qquad (0.10) \quad (0.01) \quad (0.03) \quad (0.31)$$
$$p = 0.89w - 0.89a + 0.11pi + 0.89t1 + 0.61t3$$
$$\qquad (0.02) \qquad\qquad (0.15)$$
$$\chi^2(10) = 15.45[0.12]$$

Panel 5: Simplified linear restrictions, weak exogeneity

$$w = p + a + 0.065u - t1$$
$$\qquad (0.013)$$
$$p = 0.89w - 0.89a + 0.11pi + 0.89t1 + 0.62t3$$
$$\qquad (0.017) \qquad\qquad (0.17)$$
$$\chi^2(13) = 20.08[0.09]$$

Diagnostic tests for the unrestricted conditional subsystem

$$F^v_{AR(1-5)} = 0.95[0.61]$$
$$\chi^{2,v}_{normality}(10) = 19.844[0.03]$$
$$F^v_{HETx^2}(360, 152) = 0.37[1.00]$$

Note: The sample is 1976(3) to 1993(1), 67 observations.

wages and prices, and show that the results are very similar for the two economies.

5.7   Summary

The Layard–Nickell wage-curve model of incomplete competition marks a step forward compared to both the Norwegian model and the Phillips curve, in that it combines formal models of wage bargaining and models of monopolistic price-setting. Thus, compared to Aukrust's model, the hypothesised wage and

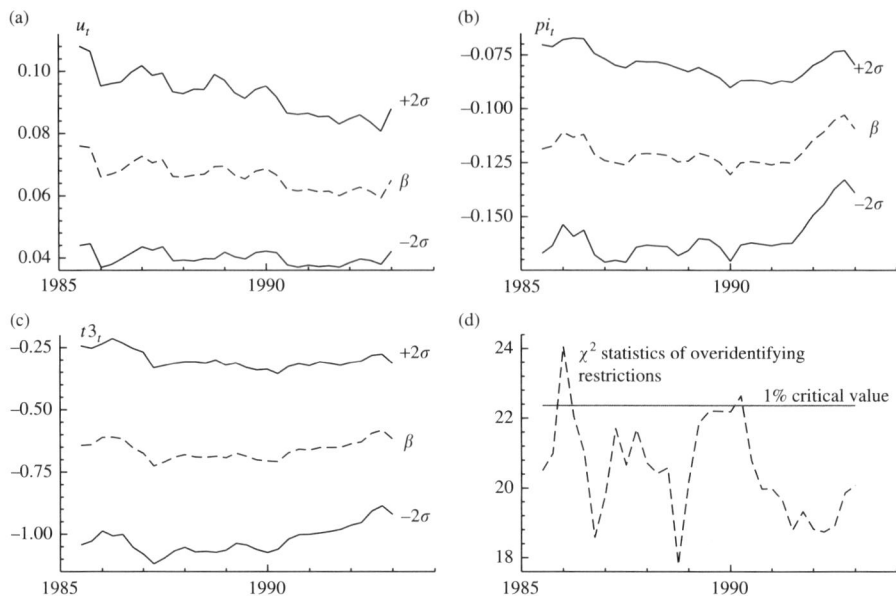

Figure 5.3. United Kingdom quarterly aggregate wages and prices, recursive cointegration results: (a)–(c) coefficients of identified equations from Panel 5 in Table 5.3; (d) sequence of χ^2 test of 7 overidentifying restrictions

price cointegrating vectors are better founded in economic theory, and specific candidates for explanatory variables flow naturally from the way the bargaining model is formulated. We have shown that there are cases of substantive interest where the identification problem pointed out by Manning (1993) are resolved, and have shown applications with empirically stable and interpretable wage and price curves.

As a model of equilibrium unemployment, the framework is incomplete since only the cointegrating part of the dynamic system is considered. To evaluate the natural rate implication of the theory, which after all is much of the rationale for the whole framework, a broader setting is required. That also defines the theme of the next chapter.

6

Wage–price dynamics

This chapter discusses the modelling of the wage–price subsystem of the economy. We show that under relatively mild assumptions about price- and wage-setting behaviour, there exists a conditional steady state (for inflation and real wages) for any given long-run mean of the rate of unemployment. The view that asymptotic stability of inflation 'requires' that the rate of unemployment simultaneously converges to a NAIRU (which only depends on the properties of the wage and price equations) will be refuted both logically and empirically.

6.1 Introduction

The open economy Phillips curve and the Incomplete Competition Model (ICM) appear to be positioned at opposite ends of a scale, with a simple dynamic model at the one end, and an economically more advanced but essentially static system at the other. In this section, we present a model of wage and price dynamics that contains the Phillips curve and the wage curve as special cases, building on the analyses in Kolsrud and Nymoen (1998) and Bårdsen and Nymoen (2003).

Section 6.2 presents the basic set of equations, and defines the concepts of static and dynamic homogeneity and their relationships to nominal rigidity and absence of neutrality to nominal shocks. Section 6.3 defines the asymptotically stable solution of the system, and Section 6.4 discusses some important economic implications of the conditional wage–price model as well as special cases that are of substantive interest (e.g. the no wedge case, and a small open economy interpretation, akin to the Norwegian model of inflation). The comparison with the ICM is drawn in Section 6.5, while Section 6.6 covers the Phillips curve case. Section 6.7 then gives an overview of the existing evidence in support of the restrictions that defines the two natural rate models. Since we find that the evidence of the respective NAIRU restrictions are at best flimsy, we expect that endeavours to estimate a time varying NAIRU run the danger of misrepresenting time varying coefficients of, for example, wage equations as changes in structural features of the economy, and Section 6.8 substantiates that claim for the four main Nordic countries.

In brief, the natural rate thesis that stability of inflation is tantamount to having the rate of unemployment converging to a natural rate, is refuted both theoretically and empirically in this chapter. Section 6.9 therefore sketches a model of inflation and unemployment dynamics that is consistent with the evidence and presents estimates of the system for the Norwegian data set that is used throughout.

6.2   Nominal rigidity and equilibrium correction

The understanding that conflict is an important aspect to take into account when modelling inflation in industrialised economies goes back at least to Rowthorn (1977), and has appeared frequently in models of the wage–price spiral; see, for example, Blanchard (1987).[1] In Rowthorn's formulation, a distinction is drawn between the negotiated profit share and the target profit share. If these shares are identical, there is no conflict between the two levels of decision-making (wage bargaining and firm's unilateral pricing policy), and no inflation impetus.[2] But if they are different, there is conflict and inflation results as firms adjust prices unilaterally to adjust to their target. In the model presented later, not only prices but also wages are allowed to change between two (central) bargaining rounds. This adds realism to the model, since even in countries like Norway and Sweden, with strong traditions for centralised wage settlements, wage increases that are locally determined regularly end up with accounting for a significant share of the total annual wage growth (i.e. so-called wage drift, see Rødseth and Holden 1990 and Holden 1990).

The model is also closely related to Sargan (1964, 1980), in that the difference equations are written in equilibrium correction form, with nominal wage and price changes reacting to past disequilibria in wage formation and in price-setting. In correspondence with the previous chapter, we assume that wages, prices, and productivity are I(1) variables, and that equations (5.11) and (5.14) are two cointegrating relationships. Cointegration implies equilibrium

---

[1] Norwegian economists know such models as 'Haavelmo's conflict model of inflation', see Qvigstad (1975).

[2] Haavelmo formulated his model, perhaps less deliberately, in terms of two separate target real wage rates for workers and firms (corresponding to w^b and w^f of Chapter 5), but the implications for inflation are the same as in Rowthorn's model.

correction, so we specify the following equations for wage and price growth:

$$\Delta w_t = \theta_w(w^b_{q,t-1} - w_{q,t-1}) + \psi_{wp}\Delta p_t + \psi_{wq}\Delta q_t - \varphi u_{t-1} + c_w + \varepsilon_{w,t},$$
$$0 \le \psi_{wp} + \psi_{wq} \le 1, \quad \varphi \ge 0, \quad \theta_w \ge 0 \qquad (6.1)$$

and

$$\Delta q_t = -\theta_q(w^f_{q,t-1} - w_{q,t-1}) + \psi_{qw}\Delta w_t + \psi_{qi}\Delta pi_t - \varsigma u_{t-1} + c_q + \varepsilon_{q,t},$$
$$0 \le \psi_{qw} + \psi_{qi} \le 1, \quad 0 \le \theta_q, \quad \varsigma \ge 0, \qquad (6.2)$$

where $\varepsilon_{w,t}$ and $\varepsilon_{q,t}$ are assumed to be uncorrelated white noise processes. The expressions for $w^b_{q,t-1}$ and $w^f_{q,t-1}$ were established in Chapter 5, and they are repeated here for convenience:

$$w^b_{q,t} = m_b + \iota a_t + \omega p_{q,t} - \varpi u_t, \qquad 0 < \iota \le 1, \quad 0 \le \omega \le 1, \quad \varpi \ge 0,$$
$$w^f_{q,t} = m_f + a_t + \vartheta u_t, \qquad\qquad \vartheta \ge 0,$$

that is, (5.10) for wage bargaining, and (5.13) based on modified normal cost pricing. In Rowthorn's terminology, the negotiated profit share is $(1 - w^b_{q,t} - a_t)$, while the target profit share is $(1 - w^f_{q,t} - a_t)$.

For the wage side of the inflation process, equations (5.10) and (6.1) yield

$$\Delta w_t = k_w + \psi_{wp}\Delta p_t + \psi_{wq}\Delta q_t - \mu_w u_{t-1}$$
$$+ \theta_w \omega p_{q,t-1} - \theta_w w_{q,t-1} + \theta_w \iota a_{t-1} + \varepsilon_{w,t}, \qquad (6.3)$$

where $k_w = (c_w + \theta_w m_b)$. In equation (6.3), the coefficient of the rate of unemployment μ_w, is defined by

$$\mu_w = \theta_w \varpi \quad \text{(when } \theta_w > 0\text{)} \quad \text{or} \quad \mu_w = \varphi \quad \text{(when } \theta_w = 0\text{)}, \qquad (6.4)$$

which may seem cumbersome at first sight, but is required to secure internal consistency: note that if the nominal wage rate is adjusting towards the long-run wage curve, $\theta_w < 0$, logic requires that the value of φ in (6.1) is zero, since u_{t-1} is already contained in the equation, with coefficient $\theta_w \varpi$. Conversely, if $\theta_w = 0$, it is nevertheless possible that there is a wage Phillips curve relationship, hence $\mu_w = \varphi \ge 0$ in this case. In equation (6.3), long-run price homogeneity is ensured by the two lagged level terms—the wedge $p_{q,t-1} \equiv (p - q)_{t-1}$ and the real wage $w_{q,t-1} \equiv (w - q)_{t-1}$.

For producer prices, equations (5.13) and (6.2) yield a dynamic equation of the cost markup type:

$$\Delta q_t = k_q + \psi_{qw}\Delta w_t + \psi_{qi}\Delta pi_t - \mu_q u_{t-1} + \theta_q w_{q,t-1} - \theta_q a_{t-1} + \varepsilon_{q,t}, \qquad (6.5)$$

where $k_q = (c_q - \theta_q m_f)$ and

$$\mu_q = \theta_q \vartheta \quad \text{or} \quad \mu_q = \varsigma. \qquad (6.6)$$

The definition of μ_q reflects exactly the same considerations as explained above for wage-setting.

In terms of economic interpretation (6.3) and (6.5) are consistent with wage and price staggering and lack of synchronisation among firms' price-setting (see, for example, Andersen 1994, ch. 7). An underlying assumption is that firms preset nominal prices prior to the period and then within the period meet the demand forthcoming at this price (which exceeds marginal costs, as in Chapter 5). Clearly, the long-run price homogeneity embedded in (6.3) is joined by long-run homogeneity with respect to wage costs in (6.5). Thus we have overall long-term nominal homogeneity as a direct consequence of specifying the cointegrating relationships in terms of relative prices.[3]

In static models, nominal homogeneity is synonymous with neutrality of output to changes in nominal variables since relative prices are unaffected (see Andersen 1994). This property does not carry over to the dynamic wage and price system, since relative prices (e.g. $w_{q,t}$) will be affected for several periods following a shift in, for example, the price of imports. In general, the model implies nominal rigidity along with long-term nominal homogeneity. Thus, care must also be taken when writing down the conditions that eventually remove short-run nominal rigidity from the system. Specifically, the conditions for 'dynamic homogeneity', that is, $\psi_{wp} + \psi_{wq} = 1$ and $\psi_{qw} + \psi_{qi} = 1$, *do not* eliminate nominal rigidity as an implied property; see Section 6.4.2. As will become clear, there is a one-to-one relationship between nominal neutrality and the natural rate property, and a set of sufficient conditions are given in Section 6.5. First however, Section 6.3 defines the asymptotically stable solution of the system with long-term homogeneity (but without neutrality) and Section 6.4 discusses some important implications.

6.3   Stability and steady state

We want to investigate the dynamics of the wage–price system consisting of equations (6.3), (6.5), and the definitional equation

$$p_t = \phi q_t + (1 - \phi)pi_t, \qquad 0 < \phi < 1. \tag{6.7}$$

Following Kolsrud and Nymoen (1998), the model can be rewritten in terms of the product real wage $w_{q,t}$, and the real exchange rate

$$pi_{q,t} = pi_t - q_t. \tag{6.8}$$

In order to close the model, we make two additional assumptions:

1. u_t follows a separate ARMA process with mean u_{ss}.
2. pi_t and a_t are random walks with drift.

---

[3] See Kolsrud and Nymoen (1998) for an explicit parameterisation with nominal variables with long-run homogeneity imposed.

The NAIRU thesis states that the rate of unemployment *has* to be at an appropriate equilibrium level if the rate of inflation is to be stable. Assumption 1 is made to investigate whether that thesis holds true for the present model: with no feedback from $w_{q,t}$ and/or $pi_{q,t}$ on the rate of unemployment there is no way that u_t can serve as an equilibrating mechanism. If a steady state exists in spite of this, the NAIRU thesis is rejected, even though the model incorporates both the ICM and the Phillips curve as special cases.

Obviously, in a more comprehensive model of inflation we will relax assumption 1 and treat u_t as an endogenous variable, in the same manner as in the Phillips curve case of Chapter 4. In the context of the imperfect competition model, that step is postponed until Section 6.9. Assumption 2 eliminates stabilising adjustments that might take place via the nominal exchange rate and/or in productivity. In empirical work this amounts to the question of whether it is valid to condition upon import prices (in domestic currency) and/or productivity. Section 5.5 gives an empirical example of such valid conditioning, since we found weak exogeneity of productivity with respect to the identified long-run wage curve.

The reduced form equation for the product real wage $w_{q,t}$ is

$$w_{q,t} = \delta_t + \xi \Delta pi_t + \kappa w_{q,t-1} + \lambda pi_{q,t-1} - \eta u_{t-1} + \epsilon_{w_q,t},$$
$$0 \le \xi \le 1, \quad 0 \le \kappa \le 1, \quad 0 \le \lambda, \tag{6.9}$$

with disturbance term $\epsilon_{w_q,t}$, a linear combination of $\varepsilon_{w,t}$ and $\varepsilon_{q,t}$, and coefficients which amalgamate the parameters of the structural equations:

$$\delta_t = [(c_w + \theta_w m_b + \theta_w \iota a_{t-1})(1 - \psi_{qw})$$
$$- (c_q - \theta_q m_f - \theta_q a_{t-1})(1 - \psi_{wq} - \psi_{wp}\phi)]/\chi,$$
$$\xi = [\psi_{wp}(1 - \psi_{qw})(1 - \phi) - \psi_{qi}(1 - \psi_{wq} - \psi_{wp}\phi)]/\chi,$$
$$\lambda = \theta_w \omega (1 - \psi_{qw})(1 - \phi)/\chi, \tag{6.10}$$
$$\kappa = 1 - [\theta_w(1 - \psi_{qw}) + \theta_q(1 - \psi_{wq} - \psi_{wp}\phi)]/\chi,$$
$$\eta = [\mu_w(1 - \psi_{qw}) - \mu_q(1 - \psi_{wq} - \psi_{wp}\phi)]/\chi.$$

The denominator of the expressions in (6.10) is given by

$$\chi = (1 - \psi_{qw}(\psi_{wq} + \psi_{wp}\phi)). \tag{6.11}$$

The corresponding reduced form equation for the real exchange rate $pi_{q,t}$ can be written as

$$pi_{q,t} = -d_t + e \Delta pi_t - k w_{q,t-1} + l pi_{q,t-1} + n u_{t-1} + \epsilon_{pi_q,t},$$
$$0 \le e \le 1, \quad l \le 1, \quad 0 \le n, \tag{6.12}$$

where the parameters are given by

$$
\begin{aligned}
d_t &= [(c_q - \theta_q m_f - \theta_q a_{t-1}) + (c_w + \theta_w m_b + \theta_w \iota a_{t-1}) \psi_{qw}] / \chi, \\
e &= 1 - [\psi_{qw} \psi_{wp} (1 - \phi) - \psi_{qi}] / \chi, \\
l &= 1 - [\psi_{qw} \theta_w \omega (1 - \phi) / \chi], \\
k &= (\theta_q - \psi_{qw} \theta_w) / \chi, \\
n &= (\mu_w \psi_{qw} + \mu_q) / \chi.
\end{aligned}
\tag{6.13}
$$

Equations (6.9) and (6.12) constitute a system of first-order difference equations that determines the real wage $w_{q,t}$ and the real exchange rate $pi_{q,t}$ at each point in time. As usual in dynamic economics we consider the deterministic system, corresponding to a hypothetical situation in which all shocks $\varepsilon_{w,t}$ and $\varepsilon_{q,t}$ (and thus $\epsilon_{w_q,t}$ and $\epsilon_{pi_q,t}$) are set equal to zero. Once we have obtained the solutions for $w_{q,t}$ and $pi_{q,t}$, the time paths for Δw_t, Δp_t, and Δq_t can be found by backward substitution.

The roots of the characteristic equation of the system are given by

$$
r = \frac{1}{2} \left[(\kappa + l) \pm \sqrt{(\kappa - l)^2 - 4k\lambda} \right],
\tag{6.14}
$$

hence the system has a unit root whenever $k\lambda = 0$ *and* either $\kappa = 1$ or $l = 1$. Using (6.10) and (6.13), we conclude that the wage–price system has both its roots inside the unit circle *unless* one or more of the following conditions hold:

$$
\theta_w \omega = 0, \tag{6.15}
$$
$$
\theta_w = \theta_q = 0, \tag{6.16}
$$
$$
\psi_{qw} (1 - \psi_{qw}) = \theta_q = 0. \tag{6.17}
$$

Based on (6.15)–(6.17), we can formulate a set of sufficient conditions for stable roots, namely

$$
\theta_w > 0 \quad \text{and} \quad \theta_q > 0 \quad \text{and} \quad \omega > 0 \quad \text{and} \quad \psi_{qw} < 1. \tag{6.18}
$$

The first two conditions represent equilibrium correction of wages and prices with respect to deviations from the wage curve and the long-run price-setting schedule. The third condition states that there is a long-run wedge effect in wage-setting. Finally, a particular form of dynamic response is precluded by the fourth condition: for stability, a one point increase in the rate of wage growth must lead to less than one point increase in the rate of price growth. Note that $\psi_{qw} = 1$ is different from (and more restrictive than) dynamic homogeneity in general, which would entail $\psi_{qw} + \psi_{qi} = 1$ and $\psi_{wp} + \psi_{wq} = 1$. Dynamic homogeneity, in this usual sense, is consistent with a stable steady state.

6.4 The stable solution of the conditional wage–price system

If the sufficient conditions in (6.18) hold, we obtain a dynamic equilibrium—the 'tug of war' between workers and firms reaches a stalemate. The system is stable in the sense that, if all stochastic shocks are switched-off, $pi_{q,t} \rightarrow pi_{q,\text{ss}}(t)$ and $w_{q,t} \rightarrow w_{q,\text{ss}}(t)$, where $pi_{q,\text{ss}}(t)$ and $w_{q,\text{ss}}(t)$ denote the deterministic steady-state growth paths of the real exchange rate and the product real wage. The steady-state growth paths are independent of the historically determined initial conditions $pi_{q,0}$ and $w_{q,0}$ but depend on the steady-state growth rate of import prices (g_{pi}), of the mean of u_t denoted u_{ss}, and of the expected time path of productivity:

$$w_{q,\text{ss}}(t) = \xi^0 g_{pi} + \eta^0 u_{\text{ss}} + g_a(t-1) - \delta^0, \tag{6.19}$$

$$pi_{q,\text{ss}}(t) = e^0 g_{pi} + n^0 u_{\text{ss}} + \frac{1-\iota}{\omega(1-\phi)} g_a(t-1) - d^0, \tag{6.20}$$

where g_a is the drift parameter of productivity.[4] The coefficients of the two steady-state paths in (6.19) and (6.20) are given by (6.21):

$$
\begin{aligned}
\xi^0 &= (1 - \psi_{qw} - \psi_{qi})/\theta_q, \\
\eta^0 &= \mu_q/\theta_q, \\
\delta^0 &= (c_q - \theta_q m_f)/\theta_q + \text{coeff} \times g_a, \\
e^0 &= [\theta_q(1 - \psi_{wq} - \psi_{wp}) + \theta_w(1 - \psi_{qw})]/\theta_w\theta_q\omega(1-\phi), \\
n^0 &= (\theta_q\mu_w + \theta_w\mu_q)/\theta_w\theta_q\omega(1-\phi), \\
d^0 &= [\theta_q(c_w + \theta_w m_b) + \theta_w(c_q - \theta_q m_f)]/\theta_w\theta_q\omega(1-\phi) + \text{coeff} \times g_a.
\end{aligned}
\tag{6.21}
$$

One interesting aspect of equations (6.19) and (6.20) is that they represent formalisations and generalisations of the main-course theory of Chapter 3. In the current model, domestic firms adjust their prices in response to the evolution of domestic costs and foreign prices, they do not simply take world prices as given. In other words, the one-way causation of Aukrust's model has been replaced by a wage–price spiral. The impact of this generalisation is clearly seen in (6.19) which states that the trend growth of productivity $g_a(t-1)$ traces out a main course, not for the nominal wage level as in Figure 3.1, but for the real wage level. It is also consistent with Aukrust's ideas that the steady state of the wage share: $ws_{\text{ss}}(t) \equiv w_{q,s}(t) - a_{\text{ss}}(t)$, is without trend, that is,

$$ws_{\text{ss}} = \xi^0 g_{pi} + \eta^0 u_{\text{ss}} - \delta^0 \tag{6.22}$$

but that it can change due to, for example, a deterministic shift in the long-run mean of the rate of unemployment.

---

[4] Implicitly, the initial value a_0 of productivity is set to zero.

According to (6.20), the real exchange rate in general also depends on the productivity trend. Thus, if $\iota < 1$ in the long-run wage equation (5.10), the model predicts continuing depreciation in real terms. Conversely, if $\iota = 1$ the steady-state path of the real exchange rate is without a deterministic trend. Note that Sections 5.5 and 5.6 showed results for two data sets, where $\iota = 1$ appeared to be a valid parameter restriction.

Along the steady-state growth path, with $\Delta u_{ss} = 0$, the two rates of change of real wages and the real exchange rate are given by:

$$\Delta w_{q,ss} = \Delta w_{ss} - \Delta q_{ss} = g_a,$$
$$\Delta pi_{q,ss} = \Delta pi_{ss} - \Delta q_{ss} = \frac{1 - \iota}{\omega(1 - \phi)} g_a.$$

Using these two equations, together with (6.7)

$$\Delta p_{ss} = \phi \Delta q_{ss} + (1 - \phi) g_{pi},$$

we obtain

$$\Delta w_{ss} = g_{pi} + g_a, \tag{6.23}$$

$$\Delta q_{ss} = g_{pi} - \frac{1 - \iota}{\omega(1 - \phi)} g_a, \tag{6.24}$$

$$\Delta p_{ss} = g_{pi} - \frac{\phi(1 - \iota)}{\omega(1 - \phi)} g_a. \tag{6.25}$$

It is interesting to note that equation (6.23) is fully consistent with the Norwegian model of inflation of Section 3.2.2. However, the existence of a steady state was merely postulated in that section. The present analysis improves on that, since the steady state is derived from set of difference equations that includes wage bargaining theory and equilibrium correction dynamics. Equations (6.24) and (6.25) show that the general solution implies a wedge between domestic and foreign inflation. However, in the case of $\iota = 1$ (wage earners benefit fully in the long term from productivity gains), we obtain the standard open economy result that the steady-state rate of inflation is equal to the rate of inflation abroad.

What does the model tell us about the status of the NAIRU? A succinct summary of the thesis is given by Layard *et al.* (1994):

'Only if the real wage (W/P) desired by wage-setters is the same as that desired by price-setters will inflation be stable. *And, the variable that brings about this consistency is the level of unemployment.*'[5]

Compare this to the equilibrium consisting of $u_t = u_{ss}$, and $w_{q,ss}$ and $pi_{q,ss}$ given by (6.19) and (6.20): clearly, inflation is stable, since (6.23)–(6.25) is implied, even though u_{ss} is determined 'from outside', and is not determined

---

[5] Layard *et al.* (1994, p. 18), authors' italics.

by the wage- and price-setting equations of the model. Hence, the (emphasised) second sentence in the quotation is not supported by the steady state. In other words, it is not necessary that u_{ss} corresponds to $\bar{u}^w$ in equation (5.7) in Chapter 5 for inflation to be stable. This contradiction of the quotation occurs in spite of the model's closeness to the ICM, that is, their wage- and price-setting schedules appear crucially in our model as cointegration relationships.

In Sections 6.5 and 6.6, we return to the NAIRU issue. We show there that both the wage curve and Phillips curve versions of the NAIRU are special cases of the model formulated above. But first, we need to discuss several important special cases of wage–price dynamics.

6.4.1 Cointegration, long-run multipliers, and the steady state

There is a correspondence between the elasticities in the equations that describe the steady-state growth paths and the elasticities in the cointegrating relationships (5.11) and (5.14). However, care must be taken when mapping from one representation to the other. For example, since much applied work pays more attention to wage-setting (the bargaining model) than to price-setting, it is often implied that the coefficient of unemployment in the estimated cointegrating wage equation also measures how much the steady-state growth path of real wages changes as a result of a permanent shift in the rate of unemployment. In other words, the elasticity in the cointegrating equation is interpreted as the long-run multiplier of real wages with respect to the rate of unemployment. However, from (6.19) the general result (from the stable case) is that the long-run multiplier of the producer real wage w_q is

$$\frac{\partial w_q}{\partial u} = \eta^0 = \vartheta \geq 0,$$

that is, the elasticity of unemployment in the long-term price-setting equation (5.13), not the one in the wage curve (5.11).

Moreover, long-run multipliers are not invariant to the choice of deflator. Thus, if we instead consider the long-run multiplier of the consumer real wage $w - p$, we obtain

$$\frac{\partial(w - p)}{\partial u} = \left[\vartheta\left(1 - \frac{1}{\omega}\right) - \varpi\right] \leq 0.$$

Comparing the multipliers for the two definitions of the real wage, it is evident that it is only the multiplier of the consumer real wage curve that has the conventional negative sign. However, also $\partial(w - p)/\partial u$ is a function of the elasticities from both cointegrating relationships (price *and* wage).

The one-to-one correspondence between the long-run multiplier and the unemployment elasticity in the 'wage curve' (5.11) requires additional assumptions. Consider, for example, the case of $\omega = 1$ and $\vartheta = 0$, that is, only costs of living (not product prices) play a role in wage bargaining, and domestic firms practice normal cost pricing. As argued in Section 5.4, this corresponds to the case of aggregate wage–price dynamics, and we obtain $\partial w_q/\partial u = 0$ and $\partial(w-p)/\partial u = -\varpi$. Thus, the long-run multiplier of the consumer real wage is identical to the elasticity of unemployment in the wage curve in this case.

6.4.2   Nominal rigidity despite dynamic homogeneity

At first sight, one might suspect that the result that u_{ss} is undetermined by the wage- and price-setting equations has to do with dynamic inhomogeneity, or 'monetary illusion'. For example, this is the case for the Phillips curve model where the steady-state rate of unemployment corresponds to the natural rate whenever the long-run Phillips curve is vertical, which in turn requires that dynamic homogeneity is fulfilled. Matters are different in the model in this section, though. As explained above, the property of dynamic homogeneity requires that we impose $\psi_{qw} + \psi_{qi} = 1$ in the equation representing price formation, and $\psi_{wq} + \psi_{wp} = 1$ in the dynamic wage curve. It is seen directly from (6.18) that the model is asymptotically stable even when made subject to these two restrictions. Thus the equilibrium conditioned on a level of unemployment u_{ss} determined outside the system, does not require dynamic inhomogeneity. Put differently, the two restrictions, $\psi_{qw} + \psi_{qi} = 1$ and $\psi_{wq} + \psi_{wp} = 1$ (dynamic homogeneity) do not remove nominal rigidity from the system.

The stable solution even applies to the case of $\psi_{qw} = 1$ ($\psi_{qi} = 0$), in which case the coefficients of the reduced form equation for $w_{q,t}$ reduce to

$$\delta_t = -(c_q - \theta_q m_f - \theta_q a_{t-1}),$$
$$\xi = 0,$$
$$\lambda = 0, \qquad\qquad\qquad\qquad\qquad (6.26)$$
$$\kappa = 1 - \theta_q,$$
$$\eta = -\mu_q,$$

while the coefficients (6.13) of the reduced form equation (6.12) for pi_q become

$$d = \left[(c_q - \theta_q m_f - \theta_q a_{t-1}) + (c_w + \theta_w m_b + \iota a_{t-1})\right]/\psi_{wp}(1-\phi),$$
$$e = 0,$$
$$l = 1 - \theta_w \omega/\psi_{wp}, \qquad\qquad\qquad\qquad (6.27)$$
$$k = (\theta_q - \theta_w)/(\psi_{wp}(1-\phi)),$$
$$n = (\mu_w + \mu_q)/(\psi_{wp}(1-\phi)).$$

Since $\lambda = 0$ in (6.26), there is no effect of the real exchange rate in the reduced-form equation for real wages, hence the solution for real wages can be obtained from equation (6.9) alone. Note also how all coefficients of the real wage equation (6.9) depend only on parameters from the firms' price-setting, whereas the competitiveness equation (6.12) still amalgamates parameters from both sides of the wage bargain, as is seen from the coefficients in (6.27).

The steady state is given by (6.19) and (6.20) as before. The expressions for η^0 and n^0 are unchanged, but $\xi^0 = e^0 = 0$ as a result of dynamic homogeneity, hence we obtain the expected result that the steady-state real exchange rate and the real wage are both unaffected by the rate of international inflation.

6.4.3   An important unstable solution: the 'no wedge' case

Real-wage resistance is an inherent aspect of the stable solution, as $\theta_w \omega \neq 0$ is one of the conditions for the stability of the wage–price system, cf. equation (6.15). However, as we have discussed earlier, the existence or otherwise of wedge effects remains unsettled, both theoretically and empirically, and it is of interest to investigate the behaviour of the system in the absence of real wage resistance, that is, $\theta_w \omega = 0$ due to $\omega = 0$.

Inspection of (6.9) and (6.12) shows that in this case, the system partitions into a stable real wage equation

$$w_{q,t} = \delta_t + \xi \Delta pi_t + \kappa w_{q,t-1} - \eta u_{t-1}, \tag{6.28}$$

and an unstable equation for the real exchange rate

$$\Delta pi_{q,t} = -d_t + e \Delta pi_t - k w_{q,t-1} + n u_{t-1}. \tag{6.29}$$

Thus, in the same way as in the stable case of $\omega > 0$, the real wage follows a stationary autoregressive process around the productivity trend which is included in δ_t. However, from (6.29), the real exchange rate is seen to follow a unit root process, albeit with $w_{q,t-1}$, u_{t-1}, and a (suppressed) disturbance term as I(0) variables on the right-hand side.[6]

The steady-state real-wage path is given by:

$$w_{q,\mathrm{ss}}(t) = \frac{\delta_t}{(1-\kappa)} + \frac{\xi}{(1-\kappa)} g_{pi} - \frac{\eta}{(1-\kappa)} u_{\mathrm{ss}}. \tag{6.30}$$

Unlike the real wage given by (6.19), the coefficients of the long-run real wage in (6.30) contain parameters from both sides of the bargain, not only price-setting. The expression for the long-run multiplier with respect to the unemployment rate, $\partial w_{q,\mathrm{ss}}/\partial u_{\mathrm{ss}}$, shows interesting differences from the stable case in Section 6.4:

$$\frac{\partial w_{q,\mathrm{ss}}}{\partial u_{\mathrm{ss}}} = -\frac{[\varpi \theta_w (1 - \psi_{qw}) - \theta_q \vartheta (1 - \psi_{wq} - \psi_{wp}\phi)]}{[\theta_w (1 - \psi_{qw}) + \theta_q (1 - \psi_{wq} - \psi_{wp}\phi)]}.$$

The multiplier is now a weighted sum of the two coefficients ϖ (wage curve) and ϑ (price-setting). With normal cost pricing $\vartheta = 0$, the long-run multiplier is seen to be negative.

---

[6] Of course, if there is a long-run effect of competitiveness on prices, that is (5.6) is extended by a competitiveness term, $\omega = 0$ is not sufficient to produce an unstable solution.

The long-run elasticity of $w_{q,\mathrm{ss}}$ with respect to productivity becomes

$$\frac{\partial w_{q,\mathrm{ss}}}{\partial a} = \frac{\iota + (\theta_q/\theta_w)(1 - \psi_{wq} - \psi_{wp}\phi)/(1 - \psi_{qw})}{1 + (\theta_q/\theta_w)(1 - \psi_{wq} - \psi_{wp}\phi)/(1 - \psi_{qw})}.$$

Hence, in the case of $\iota = 1$ in the cointegrating wage equation, the long-run multiplier implies that the product real-wage will increase by 1% as a result of a 1% permanent increase in productivity. Thus, the steady-state wage share is again without a deterministic trend.

The steady-state rate of inflation in the no-wedge case is obtained by substituting the solution for the real wage (6.30) back into the two equilibrium correction equations (6.3), imposing $\omega = 0$, and (6.5), and then using the definition of consumer prices in (5.9). The resulting steady-state rate of inflation can be shown to depend on the unemployment rate and on import price growth, that is, $\Delta p \neq \Delta pi$ in the equilibrium associated with the 'no wedge' case ($\omega = 0$). Instead, the derived long-run Phillips curve is downward-sloping provided that $\eta > 0$.

Finally we note that, unlike the static wage curve of Chapter 5, the 'no wedge' restriction ($\omega = 0$) *in itself* does not imply a supply-side determined equilibrium rate of unemployment.[7] The restrictions that are sufficient for the model to imply a purely supply-side determined equilibrium rate of unemployment is considered in Section 6.5.

6.4.4   A main-course interpretation

In Chapter 3 we saw that an important assumption of Aukrust's main-course model is that the wage-share is I(0), and that causation is one way: it is only the exposed sector wage that corrects deviations from the equilibrium wage share. Moreover, as maintained throughout this chapter, the reconstructed Aukrust model had productivity and the product price as exogenous I(1) processes.

The following two equations, representing wage-setting in the exposed sector, bring these ideas into our current model:

$$w_{q,t}^b = m_b + a_t - \varpi\, u_t, \qquad 0 < \iota \le 1, \quad \varpi \ge 0, \tag{6.31}$$

and

$$\Delta w_t = \theta_w(w_{q,t-1}^b - w_{q,t-1}) + \psi_{wp}\Delta p_t + \psi_{wq}\Delta q_t + c_w + \varepsilon_{w,t},$$
$$0 \le \psi_{wp} + \psi_{wq} \le 1, \quad \theta_w \ge 0. \tag{6.32}$$

In Section 3.2 we referred to (6.31) as the extended main-course hypothesis. It is derived from (5.10) by setting $\omega = 0$, since in Aukrust's theory, there is no role for long-run wedge effects, and in the long run there is full pass-through from productivity on wages, $\iota = 1$. Equation (6.32) represents wage dynamics

---

[7] See, for example, Layard *et al.* (1991, p. 391).

in the exposed industry and is derived from (6.1) by setting $\varphi = 0$ (since by assumption $\theta_w > 0$). Note that we include the rate of change in the consumer price index (CPI) Δp_t, as an example of a factor that can cause wages to deviate temporarily from the main course (i.e. domestic demand pressure, or a rise in indirect taxation that lead to a sharp rise in the domestic costs of living).

The remaining assumptions of the main-course theory, namely that the sheltered industries are wage followers, and that prices are marked up on normal costs, can be represented by using a more elaborate definition of the CPI than in (6.7), namely

$$\Delta p_t = \phi_1 \Delta q_t + \phi_2 (\Delta w_t - \Delta a_t) + (1 - \phi_1 - \phi_2) \Delta p i_t, \qquad (6.33)$$

where ϕ_1 and ϕ_2 are the weights of the products of the two domestic industries in the log of the CPI. The term $\phi_2 \Delta w_t$ amalgamates two assumptions: followership in the sheltered sector's wage formation and normal cost pricing.

The three equations (6.31)–(6.33) imply a stable difference equation for the product real wage in the exposed industry. Equations (6.31) and (6.32) give

$$\Delta w_t = k_w + \psi_{wp} \Delta p_t + \psi_{wq} \Delta q_t - \theta_w [w_{q,t-1} - a_{t-1} + \varpi u_{t-1}] + \varepsilon_{w,t}, \quad (6.34)$$

and when (6.33) is used to substitute Δp_t, the equilibrium correction equation for $w_{q,t}$ can be written as

$$\Delta w_{q,t} = \tilde{k}_w + (\tilde{\psi}_{wq} - 1)\Delta q_t + \tilde{\psi}_{wpi} \Delta p i_t - \tilde{\psi}_{wpa} \Delta a_t$$
$$- \tilde{\theta}_w [w_{q,t-1} - a_{t-1} + \varpi u_{t-1}] + \tilde{\varepsilon}_{w,t}, \qquad (6.35)$$

with coefficients

$$\tilde{k}_w = k_w / (1 - \psi_{wp}\phi_1),$$
$$\tilde{\psi}_{wq} = (\psi_{wq} + \psi_{wp}\phi_1)/(1 - \psi_{wp}\phi_1),$$
$$\tilde{\psi}_{wpi} = \psi_{wp}\phi_2 / (1 - \psi_{wp}\phi_1),$$
$$\tilde{\psi}_{wpi} = \psi_{wp}(1 - \phi_1 - \phi_2)/(1 - \psi_{wp}\phi_1),$$
$$\tilde{\theta}_w = \theta_w / (1 - \psi_{wp}\phi_1),$$

and disturbance $\tilde{\varepsilon}_{w,t} = \varepsilon_{w,t}/(1 - \psi_{wp}\phi_1)$.

In the same manner as before, we define the steady state as a hypothetical situation where all shocks have been switched off. From equation (6.35), and assuming dynamic homogeneity for simplicity, the steady-state growth path becomes

$$w_{q,\text{ss}}(t) = \tilde{k}_{w,\text{ss}} - \varpi u_{\text{ss}} + g_a(t - 1) + a_0, \qquad (6.36)$$

where $\tilde{k}_{w,\text{ss}} = \{\tilde{k}_w + (\psi_{wp}\phi_2 - 1)g_a\}/\tilde{\theta}_w$. This steady-state solution contains the same productivity trend as the unrestricted steady-state equation (6.19), but there is a notable difference in that the long-run multiplier is $-\varpi$, the slope coefficient of the wage curve.

In Section 6.9.2 we estimate an empirical model for the Norwegian manufacturing industry which corresponds closely to equations (6.31)–(6.33).

6.5   Comparison with the wage-curve NAIRU

In Chapter 5 we saw that the model with bargained wages and price-setting firms defined a certain level of unemployment denoted $\bar{u}^w$ at which the conflicting real wage claims were reconciled. Moreover, if there is no wedge term in wage-setting, theory implies that $\bar{u}^w$ depends only of factors in wage- and price-setting.

Recall first that the two long-term relationships are

wage-setting $\quad w_{q,t} = m_b + \iota a_t + \omega p_{q,t} - \varpi\, u_t + ecm_{b,t}, \qquad \mathsf{E}[ecm_{b,t}] = 0,$

price-setting $\quad w_{q,t} = m_f + a_t + \vartheta u_t + ecm_{f,t}, \qquad\qquad \mathsf{E}[ecm_{f,t}] = 0.$

The counterpart to $\bar{u}^w$ is derived by taking the (unconditional) expectation on both sides of (5.11) and (5.14) and solving for the rate of unemployment:

$$\bar{u}_t^w = \frac{m_b - m_f}{(\vartheta + \varpi)} + \frac{\iota - 1}{(\vartheta + \varpi)}\mathsf{E}[a_t] + \frac{\omega}{(\vartheta + \varpi)}\mathsf{E}[p_{q,t}], \qquad \omega \geq 0. \qquad (6.37)$$

We add a time subscript to $\bar{u}^w$, since the mean of productivity, a non-stationary variable, enters on the right-hand side of the expression. Remember that a_t in wage-setting is essential for the framework to accommodate the integration properties of the wage and price data. However, in the case of $\iota = 1$ (full pass through of productivity on real wages) and $\omega = 0$ (no wedge) the expression of the wage curve NAIRU simplifies to

$$\bar{u}^w = \frac{m_b - m_f}{(\vartheta + \varpi)}, \qquad\qquad (6.38)$$

which corresponds to the fundamental supply-side determined NAIRU of the static incomplete competition model (see equation (5.8)).

In Section 6.4 we established the general result that $u_t \to u_{ss} \neq \bar{u}^w$, which contradicts ICM though we build on the same long-run wage and price equations. The difference is that we model the implied equilibrium correction behaviour of wages and prices. Thus, there is in general no correspondence between the wage curve NAIRU and the steady state of the wage–price system (the correspondence principle of Samuelson (1941) appears to be violated).

Interestingly, elimination of 'money illusion', by imposing $\psi_{wp} + \psi_{wq} = 1$ (workers) and $\psi_{qw} + \psi_{qi} = 1$ (firms), is not enough to establish dynamic correspondence between $\bar{u}^w$ and u_{ss}, see Section 6.4.2. Instead, to formulate a dynamic model that captures the heuristic dynamics of the static wage curve model, we invoke the following set of restrictions:

(1) Eliminate the wedge in the long-run wage equation, $\omega = 0$, but maintain $\theta_w > 0$.
(2) Impose short-run homogeneity of the particular form $\psi_{qw} = \psi_{wq} = 1$, and hence $\psi_{wp} = \psi_{qi} = 0$.

The implication of (1) and (2) is that (6.3) and (6.5) are two *conflicting equations* of the product real wage $w_{q,t}$. Essentially, all nominal rigidity is eliminated from the model. The assumption of an exogenously determined rate of unemployment can no longer be reconciled with dynamic stability.[8] Instead, we argue (heuristically) that unemployment *has* to converge to the level necessary to reconcile the 'battle of markups' incarnated in two conflicting real wage equations. Formally, the system that determines the time paths of $w_{q,t}$ and u_t becomes

$$\Delta w_{q,t} = k_w - \theta_w \varpi u_{t-1} - \theta_w w_{q,t-1} + \theta_w \iota a_{t-1} + \varepsilon_{w,t}, \qquad (6.39)$$

$$\Delta w_{q,t} = -k_q + \theta_q \vartheta u_{t-1} - \theta_q w_{q,t-1} + \theta_q a_{t-1} + \varepsilon_{q,t}. \qquad (6.40)$$

Consistency with cointegration implies that θ_q and/or θ_w are strictly positive, and the roots of (6.39) and (6.40) are therefore within the unit circle. Hence, in a situation where all shocks are switched off, $u_t \to \bar{u}^w$:

$$u^w = \bar{u}^w + \frac{\theta_q - \theta_w}{(\vartheta + \varpi)} g_a, \qquad (6.41)$$

where the second term on the right-hand side reflects that the model (6.39)–(6.40) is a dynamic generalisation of the conventional static ICM.

Figure 6.1 illustrates the different equilibria. The upward sloping line represents firms' price-setting and the downward sloping line represents wage-setting (they define a phase diagram). According to the wage curve model, the only possible equilibrium is where the two line cross, hence the NAIRU $\bar{u}^w$ is also the dynamic equilibrium. It is not surprising to find that the natural rate property is equivalent to having a wage–price system that is free of any form of nominal rigidity, but the restrictions needed to secure nominal neutrality are seldom acknowledged: neither long-term nor dynamic homogeneity are sufficient, instead the full set of restrictions in conditions (1) and (2) is required. There is no logical or practical reason which forces these restrictions on the dynamic wage–price system, and without them, a rate of unemployment like u_{ss} is fully consistent with a steady-state rate growth of the real wage, and a stationary wage share, cf. Section 6.4.

---

[8] The roots of the system (where u_t is exogenous) are $r_1 = 1 - \theta_q$ and $r_2 = 1$.

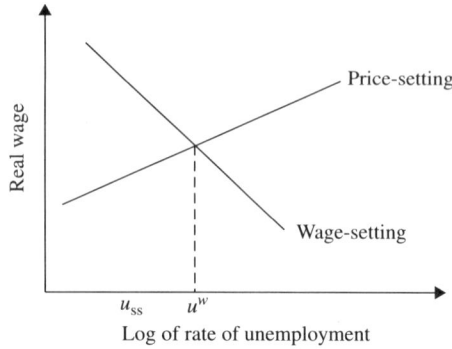

Figure 6.1. Real wage and unemployment determination.
Static and dynamic equilibrium

On the other hand, there is nothing that says that (1) and (2) cannot hold, and econometric specification and testing of wage–price systems should investigate that possibility.

6.6 Comparison with the wage Phillips curve NAIRU

In the case of no equilibrium correction in nominal wage-setting, $\theta_w = 0$, equation (6.1) simplifies to

$$\Delta w_t = c_w + \psi_{wp}\Delta p_t + \psi_{wq}\Delta q_t - \varphi u_{t-1} + \varepsilon_{w,t}, \qquad (6.42)$$

which is consistent with the short-run Phillips curve in equation (4.1) of Chapter 4. From the stability analysis of Section 6.3, $\theta_w = 0$ implies $\lambda = 0$ and $\kappa = 0$ in (6.9) and (6.12), and the solution of the system is qualitatively identical to the 'no wedge' case: the real wage is stable around the productivity trend, whereas the real exchange rate is unstable because of the unit root. Thus there is a paradox in the sense that despite the open economy Phillips curve in (6.42), there is no implied equilibrium rate of unemployment (u^{phil}) of the form found in equation (4.10) in Chapter 4. However, it is clear that the Phillips curve system involves an important extra assumption: foreign prices were assumed to be taken as given by domestic producers, which in the present model translates into $\theta_q = \psi_{qw} = 0$. Thus, restricting both wage- and price-setting by imposing

$$\theta_w = \theta_q = \psi_{qw} = 0,$$

is seen to imply two unit roots, and the system is now cast in terms of the two difference variables $\Delta w_{q,t}$ and $\Delta pi_{q,t}$. Consequently, neither the real wage level nor the real exchange rate are dynamically stable (even subtracting the productivity trend). Heuristically, in order to re-establish a stable steady state for the real wage, the assumption of a separate stationary model for u_t must be replaced by something like equation (4.2) in Chapter 4, that is, a separate equation for the rate of unemployment.[9]

6.7 Do estimated wage–price models support the NAIRU view of equilibrium unemployment?

The analysis of this chapter has shown that there is no logical reason why dynamic stability of real wages and inflation should imply or 'require' a supply-side determined NAIRU. Conversely, by claiming that a derived (and estimated) NAIRU from an incomplete system of equations corresponds to the dynamic equilibrium level of unemployment in the economy, one invokes restrictions on the (unspecified) wage–price dynamics that may or may not hold empirically.

As we have seen, there are necessary conditions for correspondence that can be tested from wage equations alone. This is fortunate, since a range of studies estimate wage models of the ICM type. Often the aim of the studies have been to estimate the NAIRU, or at least to isolate its determinants. They represent a body of research evidence that can be re-interpreted using our framework. While not claiming to be complete, Section 6.7.1 aims to summarise the evidence found in several econometric studies of (single-equation) wage models. Section 6.7.2 then discusses in more detail the NAIRU implications of a wage–price system estimated for United Kingdom aggregate data.

6.7.1 Empirical wage equations

Empirical models of Nordic manufacturing wage formation are reviewed and updated in Nymoen and Rødseth (2003). Their results for Denmark, Finland, Norway, and Sweden strongly reject the Phillips curve specification. The evidence against the Phillips curve hypothesis, $\theta_w = 0$, is not confined to the Nordic countries; see, for example, Grubb (1986) and Drèze and Bean (1990) who analyse manufacturing wages for a number of European economies.

---

[9] Note that an identical line of reasoning starts from setting $\theta_q = 0$ and leads to a price Phillips curve NAIRU. This seems to give rise to an issue about logical (and empirical) indeterminacy of the NAIRU, but influential papers like Gordon (1997) are not concerned with this, reporting instead different NAIRU estimates for different operational measures of inflation.

Turning to the bargaining model, the main idea is that the NAIRU can be derived from the long-run real wage and price equations. If there is no wedge term in the wage equation, the NAIRU is independent of the real exchange rate. However, the above analysis shows that only subject to specific restrictions does the wage curve NAIRU correspond to the steady state of the system. The Nordic study by Nymoen and Rødseth (2003), while supporting that $\omega = 0$, implies strong rejection of the NAIRU restrictions on the dynamics. Results for other European countries give the same impression: for example, 6 out of 10 country-studies surveyed by Drèze and Bean (1990) do not imply a wage curve NAIRU, since they are not genuine *product* real wage equations: either there is a wedge effect in the levels part of the equation ($\omega > 0$), or the authors fail to impose $\psi_{wq} = 1$, $\psi_{wp} = 0$.[10]

For the United Kingdom, there are several individual studies to choose from, some of which include a significant wedge effect, that is, $\omega > 0$ (see, for example, Carruth and Oswald 1989 and Cromb 1993). In a comprehensive econometric study of United Kingdom inflation, Rowlatt (1992) is able to impose dynamic homogeneity, $\psi_{wp} + \psi_{wq} = 1$ in wage formation, but the NAIRU restriction $\psi_{wq} = 1$ is not supported by the data.[11] The work of Davies and Schøtt-Jensen (1994) contains similar evidence for several EU countries. For the majority of the data sets, consumer price growth is found to be important alongside producer prices, and as we have shown this is sufficient to question the logical validity of the claims made in the same study, namely that a steady-state unemployment equilibrium is implied by the estimated real-wage equations.

OECD (1997*b*, table 1.A.1) contains detailed wage equation results for 21 countries. For 14 countries the reported specification is of the wage-curve type but the necessary restrictions derived above on the short-run dynamics are rejected. Phillips curve specifications are reported for the other seven countries, notably for the United States, which corroborate evidence in other studies; see Blanchard and Katz (1997) for a discussion.

This brief overview confirms the impression that the evidence from European data supports a wage curve rather than a Phillips curve specification. However, in the light of the model framework of this section, the estimated wage curves do not support the identification of the implied NAIRUs with the equilibrium level of unemployment.

---

[10] From Drèze and Bean (1990, table 1.4), and the country papers in Drèze and Bean (1990) we extract that the equations for Austria, Britain, and (at least for practical purposes) Germany are 'true' product real-wage equations. The equation for France is of the Phillips-curve type. For the other countries we have, using our own notation: Belgium and the Netherlands: consumer real-wage equations, that is, $\psi_{wp} = 1$, $\psi_{wq} = 0$, and $\omega = 1$. Denmark: $\omega = 1$, $\psi_{wp} = 0.24$, $\psi_{wq} = 0.76$. Italy: $\omega = 0$, $\psi_{wp} = 0.2(1 - \phi)$, $\psi_{wq} = 0.8(1 - \phi)$. United States: $\omega = 0.45(1 - \phi)$, $\psi_{wq} = 1$, $\psi_{wp} = 0$. Spain: $\omega = 0.85 \cdot 0.15$, $\theta_w = 1$, $\psi_{wp} = \omega$, $\psi_{wq} = 1 - \omega$ (the equation for Spain is static).

[11] See Rowlatt (1992: ch. 3.6).

6.7.2 Aggregate wage–price dynamics in the United Kingdom

In Section 5.6 we showed that, using aggregate wage and price data for the period 1976(3)–1993(1), the following long-term wage and price equations were identified (see Table 5.3).

$$(1) \quad w = p + a - t1 - 0.065u + \text{constant};$$ (6.43)

$$(2) \quad p = 0.89(w + t1 - a) + 0.11pi + 0.6t3 + \text{constant}.$$ (6.44)

Next, consider the model in Table 6.1 which is estimated by full information maximum likelihod (FIML). Equations (6.43) and (6.44) are incorporated into the dynamic model as equilibrium-correction terms, and their importance is clearly shown. In addition to the equilibrium-correction term, wages are driven by growth in consumer prices over the last two periods and by productivity gains. With an elasticity estimate of 0.66 and a standard error of 0.039, short-run homogeneity is clearly rejected.

The negative coefficient estimated for the change in the indirect tax-rate $(\Delta t3_t)$ is surprising at first sight. However, according to equation (6.44), consumer prices respond when the tax rate is increased which in turn is passed on to wages. Hence, the *net* effect of a discretionary change in the indirect tax rate on wages is estimated to be effectively zero in the short run and positive in the intermediate and long run. The effect of an increase in the payroll tax rate is to reduce earnings, both in the short- and long-run.

According to the second equation in Table 6.1, prices respond sharply (by 0.96%) to a 1 percentage change in wage costs. Hence short-run homogeneity is likely to hold for prices. In addition to wage increases and equilibrium-correction behaviour, price inflation is seen to depend on the output gap, as captured by the variable *gap*.

Finally, note that the two dummy variables for incomes policy, *BONUS* and *IP4*, are significant in both equations, albeit with different signs. Their impact in the first equation is evidence of incomes policy raising wages, and their reversed signs in the price equation indicate that these effects were not completely anticipated by price-setters.

The diagnostics reported at the bottom of Table 6.1 give evidence of a well-determined model. In particular, the insignificance of the overidentification χ^2 statistic, shows that the model encompasses the implied unrestricted reduced form—see Bårdsen *et al.* (1998) for evidence of recursive stability.

The significant equilibrium correction terms are consistent with previous cointegration results, and are clear evidence against a Phillips curve NAIRU, that is, $\theta_q > 0$ and $\theta_w > 0$ in the theory model. As for the wage curve NAIRU, note that the model formulation implies $\omega = 1$ in the theory model, rather than $\omega = 0$ which is one necessary requirement for correspondence between u^w

Table 6.1
The model for the United Kingdom

The wage equation

$$\widehat{\Delta w_t} = \underset{(0.075)}{0.187}\,\Delta w_{t-1} + \underset{(0.039)}{0.332}\,(\Delta_2 p_t + \Delta a_t) - \underset{(0.100)}{0.341}\,\Delta^2 t1_t$$

$$- \underset{(0.064)}{0.162}\,\Delta_2 t3_t - \underset{(0.023)}{0.156}\,(w_{t-2} - p_{t-2} - a_{t-1} + t1_{t-2} + 0.065 u_{t-1})$$

$$+ \underset{(0.071)}{0.494} + \underset{(0.003)}{0.013}\,BONUS_t + \underset{(0.001)}{0.003}\,IP4_t$$

$$\hat{\sigma} = 0.45\%$$

The price equation

$$\widehat{\Delta p_t} = \underset{(0.149)}{0.963}\,\Delta w_t - \underset{(0.118)}{0.395}\,\Delta a_t + \underset{(0.059)}{0.153}\,\Delta(p+a)_{t-1}$$

$$- \underset{(0.019)}{0.044}\,\Delta u_{t-1} + \underset{(0.092)}{0.536}\,\Delta t3_t$$

$$- \underset{(0.047)}{0.480}\,[p_{t-1} - 0.89(w + t1 - a)_{t-2} - 0.11 pi_{t-2} - 0.6 t3_{t-1}]$$

$$+ \underset{(0.099)}{0.238}\,gap_{t-1} - \underset{(0.131)}{1.330} - \underset{(0.005)}{0.019}\,BONUS_t - \underset{(0.001)}{0.005}\,IP4_t$$

$$\hat{\sigma} = 0.71\%$$

Diagnostic tests

$$\chi^2_{\text{overidentification}}(16) = 24.38[0.08]$$
$$F^v_{\text{AR}(1-5)}(20, 94) = 0.97[0.50]$$
$$\chi^{2,v}_{\text{normality}}(4) = 3.50[0.48]$$
$$F^v_{\text{HET}x^2}(84, 81) = 0.63[0.98]$$

Note: The sample is 1976(3)–1993(1), 67 observations. Estimation is by FIML. Standard errors are in parentheses below the estimates. The symbol $\hat{\sigma}$ denotes the estimated percentage residual standard error. The *p*-values of the diagnostic tests are in brackets.

and u_{ss}. In addition, although the estimates suggest that dynamic homogeneity can be imposed in the price equation, a similar restriction is statistically rejected in the wage equation.

6.8 Econometric evaluation of Nordic structural employment estimates

While early models treated the NAIRU as a quasi fixed parameter, cf. the open economy Phillips curve NAIRU of Chapter 4, the ICM framework provides the intellectual background for inclusion of a wider range of supply-side and socioeconomic structural characteristics. Such factors vary over time and across countries. There has been a large output of research that looks for the true structural sources of fluctuations in the NAIRU. This includes the joint estimation of wage- and price-equations (see Nickell 1993 and Bean 1994 for surveys), but also reduced form estimation of unemployment equations with variables representing structural characteristics as explanatory variables. However, despite these efforts, the hypothesis of shifts in structural characteristics have failed to explain why the unemployment rates have risen permanently since the 1960s; see Cross (1995), Backhouse (2000), and Cassino and Thornton (2002).

An alternative approach to the estimation of the NAIRU is based on some sort of filtering technique, ranging from the simplest HP filter, to advanced methods that model the natural rate and trend output jointly in a 'stochastic parameter' framework estimated by the Kalman filter (see Apel and Jansson 1999 and Richardson *et al.* 2000). A common assumption of these studies is that the (stochastic) NAIRU follows a random walk, that is, its mean does not exist. As discussed in Section 4.3, in connection with the Phillips curve NAIRU, this may represent an internal inconsistency, at least if the NAIRU is to represent the mean rate of unemployment in a dynamically stable system. However, proponents of the time varying NAIRU approach could claim that they capture the essence of the natural rate dichotomy, since only supply-side shocks (not nominal or demand shocks) are allowed to affect the estimated NAIRU process.

In this section, we show that the idea of a time varying NAIRU can be evaluated with conventional econometric methods.[12] The basic insight is that the amount of variation in the NAIRU ought to match up with the amount of instability that one can identify in the underlying wage- and price-equations. Because of its practical importance and its simplicity, we focus on OECD's 'NAWRU' method.

6.8.1 The NAWRU

The NAWRU indicator has been used extensively by the OECD and others on several important issues, including policy evaluation and estimation of potential output and the structural budget balance; see Holden and Nymoen (2002) for a discussion. Elmeskov and MacFarland (1993) and Elmeskov (1994) define the non-accelerating wage rate of unemployment, NAWRU, in terms of a stylised wage-pressure equation

$$\Delta^2 w_t = -c_t(U_t - U_t^{\text{NAWRU}}), \qquad c_t > 0, \tag{6.45}$$

where U^{NAWRU} is the NAWRU level of unemployment. In words, it is assumed that wage inflation is affected in a linear way by the difference between the actual level of unemployment and the NAWRU. Equation (6.45) can either be

---

[12] The analysis follows Holden and Nymoen (2002).

seen as a vertical wage Phillips curve (dynamic homogeneity is imposed); or as representing the heuristic dynamics of the wage curve model. The linear functional form is not essential, but is used in exposition and in applications of the method.

Based on an assumption that U_t^{NAWRU} is unchanged between consecutive observations, (6.45) is used to calculate the parameter c_t, for each observation separately

$$c_t = -\Delta^3 w_t / \Delta U_t. \tag{6.46}$$

Substituting the observation dependent parameter values c_t back into (6.45) the NAWRU is calculated as:

$$U_t^{\mathrm{NAWRU}} = U_t - (\Delta U_t / \Delta^3 w_t)\Delta^2 w_t. \tag{6.47}$$

In all four Nordic countries, actual unemployment has risen since the early 1970s, first in Denmark, more recently in the other countries. The raw NAWRU estimates as given by equation (6.47) are very volatile (see Holden and Nymoen 2002, figure 2), and published NAWRUs are based on HP filtering of these raw NAWRU estimates. Figure 6.2 records the NAWRUs that are cited in policy analysis discussions—see OECD Economic Surveys for Norway and Sweden, OECD (1997a,b).

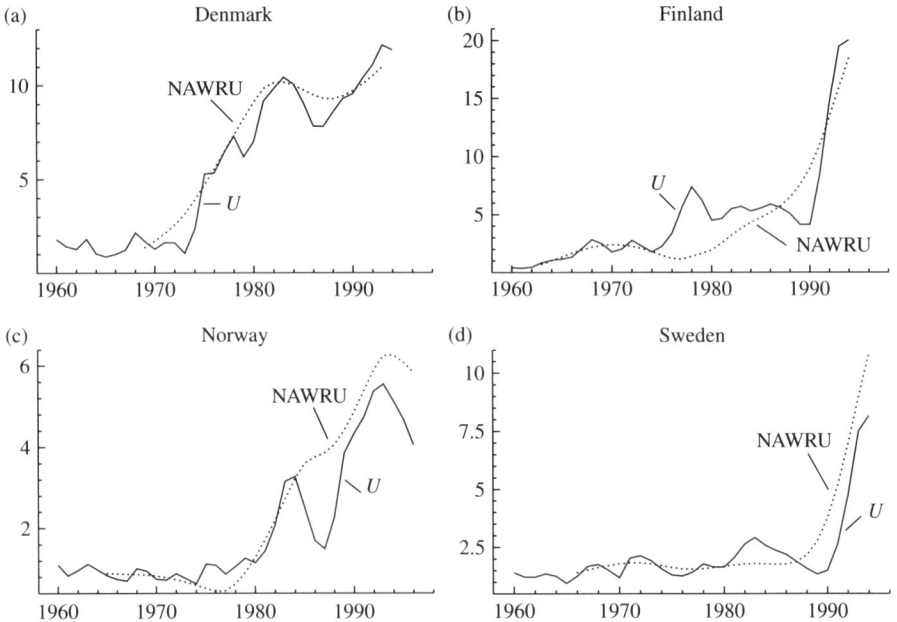

Figure 6.2. Actual rates of unemployment (U) and NAWRUs for
the four Nordic countries

For all countries, the NAWRU estimates indicate a corresponding increase in structural unemployment. Hence accepting this evidence at face value, one is led to the conclusion that the rise in unemployment is associated with a structural change in the labour market. However, Solow's 1986 critique of natural rates that 'hops around from one triennium to another under the influence of unspecified forces ... is not natural at all', clearly applies to NAWRUs.[13] Hence, in the following we investigate whether the dramatic changes in Figure 6.2 can be rationalised in a satisfactory way.

6.8.2 Do NAWRU fluctuations match up with structural changes in wage formation?

We have estimated equilibrium correction wage equations:

$$\Delta wc_t = \beta_0 - \beta_1(wc - q - a)_{t-1} - \beta_2 u_t + \beta_x' X_t + \varepsilon_{wt}, \qquad (6.48)$$

which are similar to, for example, Nymoen (1989*a*). The results are for the manufacturing sectors of each country, and draw on the analysis of Nymoen and Rødseth (2003). For Norway, the variables have been defined in earlier sections (see Section 4.6), and the data set contains the same variables for the other countries: wc = log of hourly wage cost in manufacturing; q = log of the index of value added prices; a = log of value added labour productivity; u = log of the rate of unemployment.[14] The terms $\beta_x' X_t$ should be viewed as composite, containing both growth rate variables, for example, the rates of change in the CPI, and variables that capture the impact of changes in policy or in the institutional set-up, as in equation (6.3) of the theoretical model. Finally, Δ is the difference operator and ε_{wt} is a disturbance term.

Table 6.2 shows that wage growth in *Norway* is found to depend negatively on the lagged wage share and of the level of open unemployment, and positively on the replacement ratio variable, rpr_{t-1}. The model is *dynamically homogeneous*, since the elasticities of the changes in the consumer and product price indices (Δp_t and Δq_t) sum to unity (a test of this restriction yields $F(1, 21) = 0.03$, which is insignificant). Another empirically valid restriction is that the elasticities of growth in product prices and productivity are equal. Thus wage-setting adjusts to changes in value added, irrespective of whether the change originates in price or in productivity. As discussed earlier (Section 4.6) the hours-variable (Δh_t) picks up the direct wage compensation in connection with reductions in the length of the working day.

---

[13] The full quotation is given in Section 4.6.

[14] Note that in the Norwegian Phillips curve of Section 4.6 and in Section 6.9.2, the log of the *total* unemployment rate was used. In the cross-country results reported here we chose to use open unemployment for all countries. However, as documented in Nymoen and Rødseth (2003), the choice has little influence on the estimation results.

Table 6.2

Nordic manufacturing wage equations

Norway

$$\Delta(\widehat{wc_t - p_{t-1}}) = \underset{(0.007)}{-0.0584} + \underset{(0.037)}{0.446} \left\{ 0.5\Delta_2(q+a)_t - \Delta p_{t-1} \right\} - \underset{(0.01)}{0.276}\,\Delta h_t$$
$$- \underset{(0.0023)}{0.0286}\,u_t + \underset{(0.017)}{0.109}\,\Delta lmp_t - \underset{(0.025)}{0.2183}\,(wc_{t-1} - q_{t-1} - a_{t-1})$$
$$+ \underset{(0.013)}{0.075}\,rpr_{t-1} + \underset{(0.007)}{0.039}\,i67_t - \underset{(0.005)}{0.054}\,IP_t$$

Method: OLS $\qquad$ $T = 31[1964\text{–}1994]$, $\qquad$ $R^2 = 0.98$, $\hat{\sigma} = 0.58\%$

$t_{\text{EqCM}} = -8.8$ $\qquad$ $\text{Stab}_\sigma(1) = 0.07\{0.5\}$ $\quad$ $\text{Stab}_{\beta,\sigma}(9) = 1.24\{2.54\}$

$\chi^2_{\text{normality}}(2) = 0.19[0.901]$ $\quad$ $F_{\text{AR}(1-1)} = 2.03[0.17]$ $\quad$ $F_{\text{HET}x^2} = 0.55[0.84]$

Sweden

$$\Delta(\widehat{w_t - p_{t-1}}) = \underset{(0.028)}{-0.157} + \underset{(0.066)}{0.360} \left\{ \Delta(q+a)_t - \Delta p_{t-1} \right\} - \underset{(0.338)}{0.849}\,\Delta h_{t-1}$$
$$- \underset{(0.007)}{0.042}\,u_{t-1} - \underset{(0.043)}{0.273}\,(wc_{t-1} - q_{t-1} - a_{t-1})$$

Method: OLS $\qquad$ $T = 30[1964\text{–}1994]$, $\qquad$ $R^2 = 0.854$, $\hat{\sigma} = 1.49\%$

$t_{\text{EqCM}} = -6.4$ $\qquad$ $\text{Stab}_\sigma(1) = 0.18\{0.5\}$ $\quad$ $\text{Stab}_{\beta,\sigma}(6) = 0.71\{1.7\}$

$\chi^2_{\text{normality}}(2) = 0.01[0.99]$ $\quad$ $F_{\text{AR}(1-1)} = 0.04[0.84]$ $\quad$ $F_{\text{HET}x^2} = 0.43[0.87]$

Finland

$$\Delta(\widehat{wc - p})_t = \underset{(0.017)}{0.110} + \underset{(0.015)}{0.111}\,rpr_t - \underset{(0.009)}{0.070}\,\Delta tu_t - \underset{(0.003)}{0.008}\,u_{t-1}$$
$$- \underset{(0.033)}{0.146}\,(wc_{t-1} - q_{t-2} - a_{t-2})$$

Method: OLS $\qquad$ $T = 33[1962\text{–}1994]$, $\qquad$ $R^2 = 0.809$, $\hat{\sigma} = 1.17\%$

$t_{\text{EqCM}} = -4.49$ $\qquad$ $\text{Stab}_\sigma(1) = 0.24\{0.5\}$ $\quad$ $\text{Stab}_{\beta,\sigma}(6) = 0.76\{1.7\}$

$\chi^2_{\text{normality}}(2) = 0.36[0.84]$ $\quad$ $F_{\text{AR}(1-1)} = 0.57[0.46]$ $\quad$ $F_{\text{HET}x^2} = 0.50[0.84]$

Denmark

$$\Delta(\widehat{wc - p})_t = \underset{(0.022)}{-0.032} - \underset{(0.231)}{0.644}\,\Delta_2 h_t + \underset{(0.097)}{0.428}\,\Delta(q+a-p)_t - \underset{(0.006)}{0.0322}\,u_{t-1}$$
$$- \underset{(0.087)}{0.336}\,(wc_t - q_t - a_{t-2}) + \underset{(0.058)}{0.150}\,rpr_{t-1}$$

Method: OLS $\qquad$ $T = 27[1968\text{–}1994]$, $\qquad$ $R^2 = 0.85$, $\hat{\sigma} = 1.51\%$

$t_{\text{EqCM}} = -3.88$ $\qquad$ $\text{Stab}_\sigma(1) = 0.29[0.5]$ $\quad$ $\text{Stab}_{(\beta,\sigma)}(7) = 0.86[1.9]$

$\chi^2_{\text{normality}}(2) = 2.15[0.34]$ $\quad$ $F_{\text{AR}(1-1)} = 3.53[0.08]$ $\quad$ $F_{\text{HET}x^2}(10,10) = 0.79[0.64]$

The estimated coefficient of the variable Δlmp_t indicates that the active use of programmes in order to contain open unemployment reduces wage pressure—lmp being the log of the share of open unemployment in total unemployment.[15] Finally, there are two dummy variables in the Norwegian equation, already explained in Section 4.6: IP_t and $i67_t$.

Below the equation we report the estimation method (ordinary least squares, OLS), the sample length T, the squared multiple correlation coefficient R^2, and the percentage residual standard error $\hat{\sigma}$. t_{EqCM} is the t-value of the coefficient of the lagged wage share and is used here as a direct test of the hypothesis of no cointegration; see Kremers *et al.* (1992). Compared to the relevant critical values in MacKinnon (1991, table 1) $t_{\mathsf{EqCM}} = -8.8$ gives formal support for cointegration between the wage-share, the rate of unemployment, and the replacement ratio. This conclusion is supported by the results of multivariate cointegration methods (see Bårdsen and Nymoen 2003).

Together with the standard tests of fit and of residual properties (defined in Section 4.6), we also report two of Hansen's (1992) statistics of parameter non-constancy: $\mathsf{Stab}_\sigma(1)$ tests the stability of the residual standard error (σ) individually. $\mathsf{Stab}_{\beta,\sigma}(10)$ tests the joint stability of σ and the set regression coefficients (β). The degrees of freedom are in parentheses, and, since the distributions are non-standard, the 5% critical values are reported in curly brackets. Neither of the statistics are significant, which indicates that the empirical wage equation is stable over the sample.

The equation for the other countries in Table 6.2 have several features in common with the Norwegian model: dynamic homogeneity, strong effects of consumer price growth, and of pay compensation for reductions of the length of the working week.

The *Swedish* equation contains only two levels variables, the rate of unemployment and the wage share. Unlike Norway, there is no effect of the replacement ratio; adding rpr_t and rpr_{t-1} to the equation yields $\mathsf{F}(2, 23) = 1.1$, with a p-value of 0.36, for the joint null hypothesis of both coefficients being equal to zero. The insignificance of $\mathsf{Stab}_\sigma(1)$ and $\mathsf{Stab}_{\sigma,\beta}(6)$ indicates that the equation is stable over the sample period. We also tested the impact of intervention dummies that have been designed to capture the potential effects of the following episodes of active incomes policy and exchange-rate regime changes—see Calmfors and Forslund (1991) and Forslund and Risager (1994) (i.e. a 'Post devaluation dummy': 1983–85; Incomes policy: 1974–76 and 1985; Devaluation/decentralised bargaining: 1983–90). None of the associated dummies came close to statistical significance when added to the Swedish equation in Table 6.2.

The *Danish* and *Finnish* equations contain three levels variables; the replacement ratio, the unemployment rate, and the lagged wage share. In the

---

[15] The appearance of this variable has to do with the use of the open rate of unemployment, rather than the total rate.

Finnish model, the estimated coefficient of the lagged rate of unemployment is seen to be economically rather insignificant, while the change in the rate of total unemployment ($\Delta t u_t$) has a much stronger effect. Both these features are consistent with previous findings; cf. Calmfors and Nymoen (1990) and Nymoen (1992).

The four wage equations are thus seen to be congruent with the available data evidence. We have also checked the robustness of the models, by testing the significance of potential 'omitted variables', for example, the levels and the changes in the average income tax rates, and a composite 'wedge' term, without finding any predictive power of these variables; see Holden and Nymoen (2002).

Figure 6.3 confirms the stability of the equations already suggested by the insignificance of the Stab_σ and $\mathsf{Stab}_{\sigma,\beta}$ statistics. The first column shows the 1-step residuals with ± 2 residual standard errors, ± 2se in the graphs. The second column contains the estimated elasticities of the wage share, with ± 2 estimated coefficient standard errors, denoted β and $\pm 2\sigma$ in the graphs. All graphs show a high degree of stability, which stands in contrast to the instability of the NAWRU estimates.

The stability of the empirical wage equations does not preclude a shift in the wage curve in the employment—real wage space, that is, if other explanatory

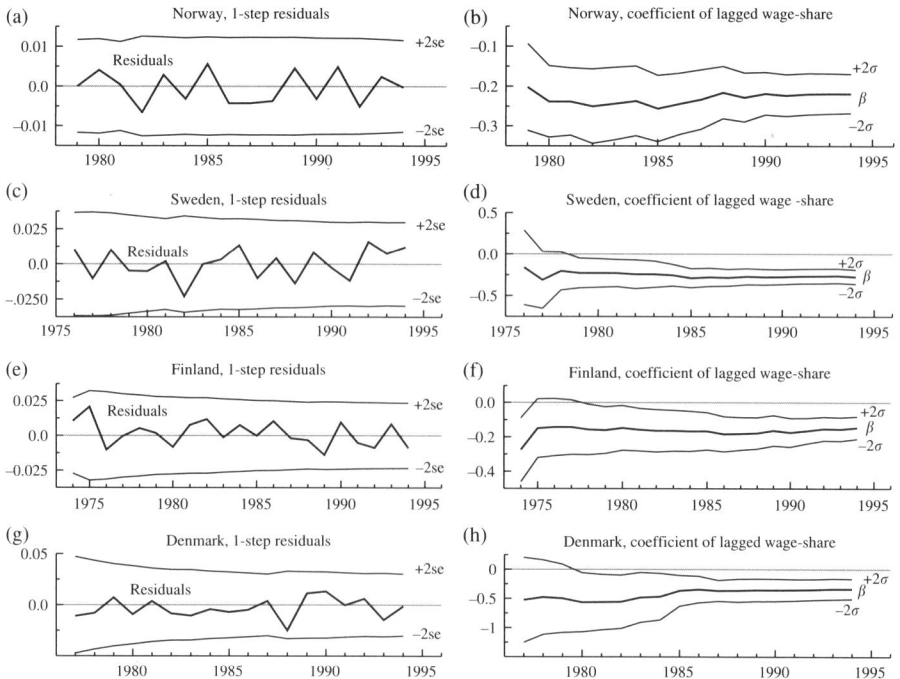

Figure 6.3. Recursive stability of Nordic wage equations

variables have changed. The question is whether changes in the explanatory variables of the wage equation amount to anything like the movement of the NAWRUs. To investigate this, we construct a new variable, the Average Wage-Share rate of Unemployment (AWSU). This variable is defined as the rate of unemployment that (according to our estimated wage equations) in each year would have resulted in a constant wage-share in that year, if the actual lagged wage share were equal to the sample mean.

To clarify the calculation and interpretation of AWSU, consider a 'representative' estimated wage equation

$$\Delta(wc_t - p_t) = \hat{\beta}_0 - \hat{\beta}_1(\overline{wc - q - a}) - \hat{\beta}_2 u_t + \hat{\beta}_3 \Delta(q + a - p)_t \quad (6.49)$$
$$+ \hat{\beta}'_x X_t,$$

where $(\overline{wc - q - a})$ is the sample mean of the wage share, and we recognise dynamic price homogeneity, a wage scope variable with estimated elasticity $\hat{\beta}_3$ and $\hat{\beta}'_x X_t$ which contains other, country-specific effects. Solving for u_t with $\Delta(wc - q - a)_t = 0$ imposed yields

$$u_t = \frac{\hat{\beta}_0}{\hat{\beta}_2} - \frac{\hat{\beta}_1}{\hat{\beta}_2}(\overline{wc - q - a}) + \frac{\hat{\beta}_3 - 1}{\hat{\beta}_2}\Delta(q - p + a)_t + \frac{\hat{\beta}'_x}{\hat{\beta}_2} X_t \quad (6.50)$$

and the exponential of the left-hand side of (6.50) is the AWSU. In the calculations of the AWSU, actual values are used for all the variables appearing in the estimated equations. Increased upward wage pressure (due to other factors than lower unemployment and lower lagged wage share) leads to a rise in the AWSU, because to keep the wage share constant the rate of unemployment must be higher.

The graphs of the AWSU for Denmark, Norway, and Sweden are displayed in Figure 6.4. Finland is omitted, because the very low estimated coefficient of lagged unemployment implies that the mapping of wage pressure into unemployment is of little informative value. In the case of Denmark, the increase in the replacement ratio in the late 1960s explains the high AWSU estimates of the 1970s. In the 1990s, a reversion of the replacement ratio, and high growth in value added per man-hours, explain why AWSU falls below the actual rate of unemployment. For Norway and Sweden the AWSUs show quite similar developments: periods when consumer price growth is rapid relative to growth in manufacturing value added per hour (the late 1970s and early 1980s), are marked by an increase in the AWSU. In the case of Norway, the replacement rate also contributes to the rise. However, the important overall conclusion to draw from the graphs is that there is little correlation between wage pressure (as measured by the AWSU) and unemployment; in particular the rise in unemployment in the early 1990s cannot be explained by a rise in wage pressure.

(a)

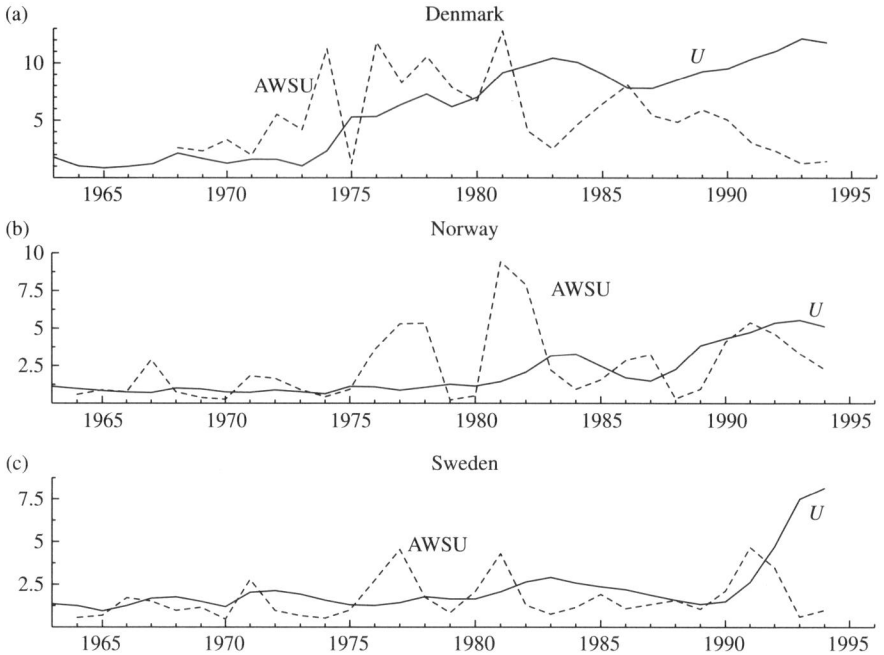

(b)

(c)

Figure 6.4. Unemployment and the Average Wage-Share rates of
Unemployment (AWSU; see explanation in text)

6.8.3 Summary of time varying NAIRUs in the Nordic countries

In sum, for all three countries, we obtain stable empirical wage equations over
the period 1964–94 (Denmark 1968–94). Nor do we detect changes in explanatory variables in the wage-setting that can explain the rise in unemployment
(as indicated by absence of an increasing trend in the AWSU indicator in
Figure 6.4). The instability of the NAWRU estimate appears to be an arte-
fact of a mis-specified underlying wage equation, and is not due to instability
in the wage-setting itself. Note also that the conclusion is not specific to the
NAWRU but extends to other methods of estimating a time varying NAIRU:
as long as the premise of these estimations are that any significant changes in
the NAIRU is due to changes in wage (or price) setting, they also have as a
common implication that the conditional wage equations in Table 6.2 should
be unstable. Since they are not, a class of models is seen to be inconsistent with
the evidence.

The results bring us back to the main question: should empirical macroeconomic modelling be based on the natural rate doctrine? The evidence presented

in this section more than suggests that there is a negative answer to this question. Instead we might conclude that if the equilibrium level of unemployment is going to be a strong attractor of actual unemployment, without displaying incredible jumps or unreasonably strong drift, the dichotomy between structural supply-side factors and demand-side influences has to be given up. In the next section, we outline a framework that goes beyond the natural rate model.

6.9 Beyond the natural rate doctrine: unemployment–inflation dynamics

In this section, we relax the assumption, made early in the section, of exogenously determined unemployment which, after all, was made for a specific purpose, namely for showing that under reasonable assumptions about price- and wage-setting, there exist a steady-state rate of inflation, and a steady-state growth rate for real wages for a given long-run mean of the rate of unemployment. Thus, the truism that a steady state requires that the rate of unemployment simultaneously converges to the NAIRU has been refuted. Moreover, we have investigated special cases where the natural doctrine represents the only logically possible equilibrium, and have discussed how the empirical relevance of those special cases can be asserted.

6.9.1 A complete system

Equations (6.51)–(6.57) are a distilled version of an interdependent system for real wages, the real exchange rate and unemployment that we expect to encounter in practical situations.

$$\Delta w_{q,t} = \delta_t + \xi \Delta pi_t + (\kappa - 1)w_{q,t-1} + \lambda pi_{q,t-1} - \eta u_{t-1} + \varepsilon_{w_q,t}, \quad (6.51)$$

$$\Delta pi_{q,t} = -d_t + e\,\Delta pi_t - k\,w_{q,t-1} + (l - 1)\,pi_{q,t-1} + nu_{t-1} + \varepsilon_{pi_q,t}, \quad (6.52)$$

$$\Delta u_t = \beta_{u0} - (1 - \beta_{u1})u_{t-1} + \beta_{u2}w_{q,t-1} + \beta_{u3}a_{t-1}$$
$$+ \beta_{u4}pi_{q,t-1} - \beta_{u5}z_{ut} + \varepsilon_{u,t}, \quad (6.53)$$

$$\Delta pi_t = g_{pi} + \varepsilon_{pi,t}, \quad (6.54)$$

$$\Delta q_t = \Delta pi_t - \Delta pi_{q,t}, \quad (6.55)$$

$$\Delta w_t = \Delta w_{q,t} + \Delta q_t, \quad (6.56)$$

$$\Delta p_t = b_{p1}(\Delta w_t - \Delta a_t) + b_{p2}\Delta pi_t + \varepsilon_{p,t}. \quad (6.57)$$

Equations (6.51) and (6.52) are identical to equations (6.9) and (6.12) of Section 6.3, where the coefficients were defined. Note that the two intercepts have time subscripts since they include the (exogenous) labour productivity a_t, cf. (6.10) and (6.13). The two equations are the reduced forms of the theoretical model that combines wage bargaining and monopolistic price-setting with equilibrium correction dynamics. Long-run dynamic homogeneity is incorporated, but the system is characterised by nominal rigidity. Moreover, as explained above, not even dynamic homogeneity in wage- and price-setting is in general sufficient to remove nominal rigidity as a system property.

A relationship equivalent to (6.53) was introduced already in Section 4.2, in order to close the open economy Phillips curve model. However there are two differences as a result of the more detailed modelling of wages and prices: first, since the real exchange rate is endogenous in the general model of wage price dynamics, we now include $pi_{q,t-1}$ with non-negative coefficient $(\beta_{u,4} \geq 0)$.[16] Second, since we maintain the assumption about stationarity of the rate of unemployment (in the absence of structural break), that is, $|\beta_{u1}| < 1$, we include a_{t-1} unrestricted, in order to balance the productivity effects on real wages and/or the real exchange rate. In the same way as in the section on the Phillips curve system, z_{ut} represents a vector consisting of I(0) stochastic variables, as well as deterministic explanatory variables.

Equation (6.54) restates the assumption of random walk behaviour of import prices made at the start of the section, and the following two equations are definitions that back out the nominal growth rates of the product price and nominal wage costs. The last equation of the system, (6.57), is a hybrid equation for the rate of inflation that has normal cost pricing in the non-tradeables sector built into it.[17]

The essential difference from the wage–price model of Section 6.2 is of course equation (6.53) for the rate of unemployment. Unless $\beta_{u2} = \beta_{u3} = 0$, the stability analysis of Section 6.4 no longer applies, and it becomes impractical to map the conditions for stable roots back to the parameters. However, for estimated versions of (6.51)–(6.57) the stability or otherwise is checked from the eigenvalues of the associated companion matrix (as demonstrated in the next paragraph). Subject to stationarity, the steady-state solution is easily obtained from (6.51)–(6.53) by setting $\Delta w_{q,t} = g_a$, $\Delta pi_{q,t} = 0$, $\Delta u_{ss} = 0$, and solving for $w_{q,ss}$, $pi_{q,ss}$, and u_{ss}. In general, all three steady-state variables become functions of the steady states of the variables in the vector $z_{u,t}$, the conditioning variables in the third unemployment equation, in particular

$$u_{ss} = f(z_{u,ss}).$$

Note that while the real wage is fundamentally influenced by productivity, $u_t \sim$ I(0) implies that equilibrium unemployment u_{ss} is unaffected by the level of productivity. Is this equilibrium rate of unemployment a 'natural rate'? If we think of the economic interpretation of (6.53) this seems unlikely: equation (6.53) is a reduced form consisting of labour supply, and the labour demand of private firms as well as of government employment. Thus, one can

---

[16] The other elasticities in (6.53) are also non-negative.

[17] This equation is similar to (4.9) in the Phillips curve chapter. The only difference is that we now let import prices represent imported inflation.

think of several factors in $z_{u,t}$ that stem from domestic demand, as well as from the foreign sector. At the end of the day, the justification of the specific terms included in $z_{u,ss}$ and evaluation of the relative strength of demand- and supply-side factors, must be made with reference to the institutional and historical characteristics of the data.

In the next section, we give an empirical example of (6.51)–(6.57), and Chapters 9 and 10 present operational macroeconomic models with a core wage–price model, and where (6.51) is replaced by a system of equations describing output, domestic demand, and financial markets.

6.9.2 Wage–price dynamics: Norwegian manufacturing

In this section, we return to the manufacturing data set of Section 4.6 (Phillips curve), and 5.5 (wage curve). In particular, we recapitulate the cointegration analysis of Section 5.5:

1. A long-run wage equation for the Norwegian manufacturing industry:

$$wc_t - q_t - a_t = -\underset{(0.081)}{0.065}\, tu_t + \underset{(0.036)}{0.184}\, rpr_t + ecm_{w,t}, \qquad (6.58)$$

   that is, equation (5.22). rpr_t is the log of the replacement ratio.
2. No wedge term in the wage curve cointegration relationship (i.e. $\omega = 0$).
3. Nominal wages equilibrium correct, $\theta_w > 0$.
4. Weak exogeneity of q_t, a_t, tu_t, and rpr_t with respect to the parameters of the cointegration relationship.

These results suggest a 'main-course' version of the system (6.51)–(6.57): as shown in Section 6.4.4, the no-wedge restriction together with one-way causality from product prices (q_t) and productivity (a_t) on to wages imply a dynamic wage equation of the form

$$\Delta w_t = k_w + \psi_{wp}\Delta p_t + \psi_{wq}\Delta q_t - \theta_w[w_{q,t-1} - a_{t-1} + \varpi u_{t-1}] + \varepsilon_{w,t}, \qquad (6.59)$$

(cf. equation (6.34)). The term in square brackets has its empirical counterpart in $ecm_{w,t}$.

Given items 1–3, our theory implies that the real exchange rate is dynamically unstable (even when we control for productivity). This has further implications for the unemployment equation in the system: since there are three I(1) variables on the right-hand side of (6.53), and two of them cointegrate $(w_{q,t}$ and $a_t)$, the principle of balanced equations implies that $\beta_{u4} = 0$. However, the exogeneity of the rate of unemployment (item 4) does not necessarily carry over from the analysis in Section 5.5, since z_{ut} in equation (6.53) includes I(0) conditioning variables. From the empirical Phillips curve system in Section 4.6, the main factor in z_{ut} is the GDP growth rate $(\Delta y_{gdp,t-1})$.

We first give the details of the econometric equilibrium-correction equation for wages, and then give FIML estimation of the complete system, using a slightly extended information set.

Equation (6.60) gives the result of a wage generalised unrestricted model (GUM) which uses $ecm_{w,t}$ defined in item 1 as a lagged regressor.

$$\widehat{\Delta w_t} = -\ \underset{(0.0349)}{0.183}\ -\ \underset{(0.0795)}{0.438}\ ecm_{t-1}\ +\ \underset{(0.387)}{0.136}\ \Delta t1_t\ +\ \underset{(0.116)}{0.0477}\Delta p_t$$

$$+\ \underset{(0.115)}{0.401}\ \Delta p_{t-1}\ +\ \underset{(0.114)}{0.0325}\Delta p_{t-2}\ +\underset{(0.102)}{0.0858}\Delta a_t\ +\ \underset{(0.0917)}{0.0179}\ \Delta a_{t-1}$$

$$-\ \underset{(0.0897)}{0.0141}\ \Delta a_{t-2}\ +\ \underset{(0.0632)}{0.299}\ \Delta q_t\ +\ \underset{(0.0818)}{0.0209}\ \Delta q_{t-1}\ -\underset{(0.0665)}{0.000985}\Delta q_{t-2}$$

$$-\ \underset{(0.185)}{0.738}\ \Delta h_t\ -\ \underset{(0.00843)}{0.0106}\ \Delta tu_t\ +\ \underset{(0.0128)}{0.0305}\ i1967_t\ -\ \underset{(0.00789)}{0.0538}\ IP_t$$

$$(6.60)$$

$$\text{OLS},\ T = 34(1965\text{--}98)$$

$\hat{\sigma} = 0.008934$	$R^2 = 0.9714$	$RSS = 0.001437.$
$F_{\text{Null}} = (16, 17) = 17.48[0.00]$	$F_{\text{AR}(1-2)} = 4.0021[0.039]$	
$F_{\text{ARCH}(1-1)} = 1.2595[0.2783]$	$\chi^2_{\text{normality}} = 1.983[0.371]$	
$F_{\text{Chow}(1982)} = 0.568[0.7963]$	$F_{\text{Chow}(1995)} = 0.248[0.861]$	

It is interesting to compare equation (6.60) with the Phillips curve GUM for the same data; cf. equation (4.42) of Section 4.6. In (6.60) we have omitted the second lag of the price and productivity growth rates, and the levels of tu_{t-1} and rpr_{t-1} are contained in $ecm_{w,t-1}$, but in other respects the two GUMs are identical. The residual standard error is down from 1.3% (Phillips curve) to 0.89% (wage curve). To a large extent the improved fit is due to the inclusion of ecm_{t-1}, reflecting that the Phillips curve restriction $\theta_w = 0$ is firmly rejected by the t-test.

The mis-specification tests show some indication of (negative) autoregressive residual autocorrelation, which may suggest overfitting of the GUM, and which no longer represents a problem in the final model shown in equation (6.61):

$$\widehat{\Delta w_t} = -\ \underset{(0.0143)}{0.197}\ -\ \underset{(0.0293)}{0.478}\ ecm_{w,t-1}\ +\ \underset{(0.0535)}{0.413}\ \Delta p_{t-1}\ +\ \underset{(0.0449)}{0.333}\ \Delta q_t$$

$$-\ \underset{(0.129)}{0.835}\ \Delta h_t\ +\ \underset{(0.00823)}{0.0291}\ i1967_t\ -\ \underset{(0.00561)}{0.0582}\ IP_t$$

$$(6.61)$$

$$\text{OLS},\ T = 34(1965\text{--}98)$$

$RSS = 0.001695$	$\hat{\sigma} = 0.007922$	$R^2 = 0.9663$
$F_{p\text{GUM}} = 0.9402$	$F_{\text{AR}(1-2)} = 0.857[0.44]$	$F_{\text{HET}x^2} = 0.818[0.626].$
$F_{\text{ARCH}(1-1)} = 2.627[0.118]$	$\chi^2_{\text{normality}} = 1.452[0.4838]$	
$F_{\text{Chow}(1982)} = 0.954$	$F_{\text{Chow}(1995)} = 0.329[0.8044]$	

The estimated residual standard error is lower than in the GUM, and by F_{pGUM}, the final model formally encompasses the GUM in equation (6.60). The model in (6.61) shows close correspondence with the theoretical (6.32) in Section 6.4.4, with $\hat{\theta}_w = 0.48$ $(t_{\hat{\theta}_w} = 16.3)$, and $\hat{\psi}_{wp} + \hat{\psi}_{wq} = 0.75$, which is significantly different from one $(F(1, 27) = 22.17[0.0001])$.

As already said, the highly significant equilibrium-correction term is evidence against the Phillips curve equation (4.43) in Section 4.6 as a congruent model of manufacturing industry wage growth. One objection to this conclusion is that the Phillips curve is ruled out from the outset in the current specification search, that is, since it is not nested in the equilibrium–correction models (EqCM–GUM). However, we can rectify that by first forming the union model of (6.61) and (4.43), and next do a specification search from that starting point. The results show that PcGets again picks equation (6.61), which thus encompasses also the wage Phillips curve of Section 4.6.

Figure 6.5 shows the stability of equation (6.61) over the period 1978–94. All graphs show a high degree of stability. The two regressors (Δp_{t-1} and Δq_t) that also appear in the Phillips curve specification in Section 4.6 have much

Table 6.3

FIML results for a model of Norwegian manufacturing wages, inflation, and total rate of unemployment

$$\Delta wc_t = -0.1846 - 0.4351 ecm_{w,t-1} + 0.5104 \Delta p_{t-1} + 0.2749\ \Delta q_t$$
$$(0.016)\quad (0.0352)\qquad\qquad (0.0606)\qquad\quad (0.0517)$$
$$-0.7122\,\Delta h_t + 0.03173\ i1967_t - 0.05531\ IP_t + 0.2043\,\Delta y_{\text{gdp},t-1}$$
$$(0.135)\qquad\ (0.00873)\qquad\quad (0.00633)\qquad (0.104)$$

$$\Delta tu_t = -0.2319\ tu_{t-1} - 8.363\,\Delta y_{\text{gdp},t-1} + 1.21\ ecm_{w,t-1}$$
$$(0.0459)\qquad\quad (1.52)\qquad\qquad (0.338)$$
$$+0.4679\,i1989_t - 2.025\ \Delta^2 pi_t$$
$$(0.148)\qquad (0.468)$$

$$\Delta p_t = 0.01185 + 0.1729\ \Delta w_t - 0.1729\Delta a_t - 0.1729\Delta q_{t-1} + 0.3778\ \Delta p_{t-1}$$
$$(0.00419)\ (0.0442)\qquad (-)\qquad\quad (-)\qquad\qquad (0.0864)$$
$$+0.2214\,\Delta_2 pi_t - 0.4682\,\Delta h_t + 0.04144\,i1970_t$$
$$(0.0325)\qquad (0.174)\qquad (0.0115)$$
$$ecm_{w,t} = ecm_{w,t-1} + \Delta wc_t - \Delta q_t - \Delta a_t + 0.065\Delta tu_t - 0.184\Delta rpr_t$$
$$tu_t = tu_{t-1} + \Delta tu_{t-1};$$

Note: The sample is 1964–98, $T = 35$ observations.

$$\hat{\sigma}_{\Delta w} = 0.00864946$$
$$\hat{\sigma}_{\Delta tu} = 0.130016$$
$$\hat{\sigma}_{\Delta p} = 0.0110348$$
$$F^v_{\text{AR}(1-2)}(18, 59) = 0.65894[0.84]$$
$$\chi^{2,v}_{\text{normality}}(6) = 4.5824[0.60]$$
$$\chi^2_{\text{overidentification}}(32) = 47.755[0.04].$$

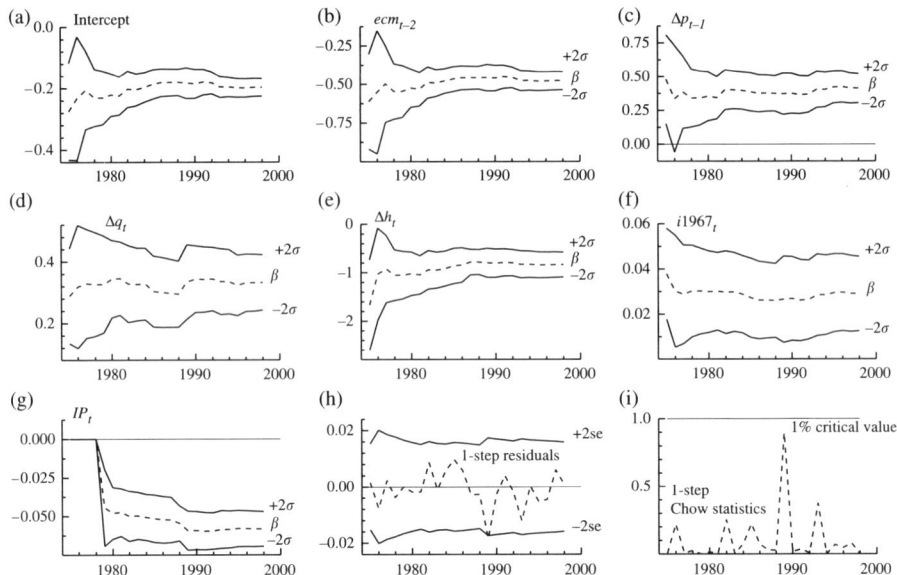

Figure 6.5. Recursive estimation of the final EqCM wage equation

narrower confidence bands in this figure than in Figure 4.2. In sum, the single-equation results are in line with earlier 'equilibrium correction' modelling of Norwegian manufacturing wages; see, for example, Nymoen (1989a). In particular, Johansen (1995a) who analyses annual data, contains results that are in agreement with our findings: he finds no evidence of a wedge effect but reports a strong wage response to consumer price growth as well as to changes in the product price.

Thus, the results imply that neither the Phillips curve, nor the wage-curve NAIRU, represent valid models of the unemployment steady-state in Norway. Instead, we expect that the unemployment equilibrium depends on forcing variables in the unemployment equation of the larger system (6.51)–(6.57). The estimated version of the model is shown in Table 6.3, with coefficients estimated by FIML.

It is interesting to compare this model to the Phillips curve system in Table 4.2 of Section 4.6. For that purpose we estimate the model on the sample 1964–98, although that means that compared to the single equation results for wages just described, one year is added at the start of the sample. Another, change from the single equation results is in Table 6.3: the wage equation in augmented by $\Delta y_{\text{gdp},t-1}$, that is, the lagged GDP growth rate. This variable was included in the information set because of its anticipated role the equation for unemployment. Finding it to be marginally significant also in the wage equation creates no inconsistencies, especially since it appears to be practically orthogonal to the explanatory variables that were included in the information set of the single equation PcGets modelling.

The second equation in Table 6.3 is similar to (6.53) in the empirical Phillips curve system estimated on this data set in Section 4.6. However, due to cointegration, the feedback from wages on unemployment is captured by $ecm_{w,t-1}$, thus there are cross-equation restrictions between the parameters in the wage and unemployment equations. The third equation in the table is consistent with the theoretical inflation equation (6.33) derived in Section 6.4.4.[18]

The model is completed by the two identities, first for $ecm_{w,t}$ which incorporates the cointegrating wage–curve relationship, and second, the identity for the rate of unemployment. The three non-trivial roots of the characteristic equation are

$$
\begin{array}{ccc}
0.6839 & 0 & 0.6839 \\
0.5969 & 0.1900 & 0.6264 \\
0.5969 & -0.1900 & 0.6264
\end{array}
$$

that is, a complex pair, and a real root at 0.68. Hence the system is dynamically stable, and compared to the Phillips curve version of the main-course model of Section 4.6 the adjustment speed is quicker.

Comparison of the two models is aided by comparing Figure 6.6 with Figure 4.5 of Section 4.6. For each of the four endogenous variables shown in Figure 6.6, the model solution ('simulated') is closer to the actual values than in the corresponding Figure 4.5. The two last panels of Figure 6.6 show the cumulated dynamic multiplier of a point increase in the rate of unemployment. The difference from Figure 4.5, where the steady state was not even 'in sight' within the 35 years simulation period, is striking. In Figure 6.6, 80% of the long-run effect is reached within four years, and the system is clearly stabilising in the course of a 10-year simulation period.

6.10 Summary

This chapter has discussed the modelling of the wage–price subsystem of the economy. We have shown that under relatively mild assumptions about price- and wage-setting behaviour, there exists a conditional steady-state (for inflation, and real wages) for any given long-run mean of the rate of unemployment. The view that asymptotic stability of inflation 'requires' that the rate of unemployment simultaneously converges to a NAIRU (which only depends on the

---

[18] The inflation rate depends on Δwc_t, a feature which is consistent with the result about an endogenous real-wage wedge in the cointegration analysis of Chapter 5, Section 5.5: $p_t - q_t$ was found to be endogenous, while the product price (q_t) was weakly exogenous.

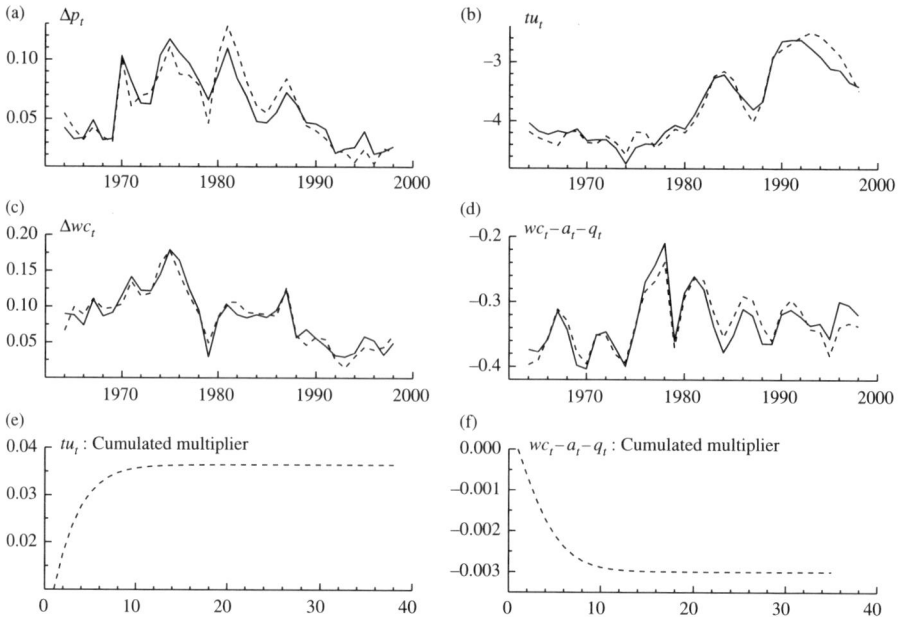

Figure 6.6. Dynamic simulation of the EqCM model in Table 6.3.
Panels (a–d): actual and simulated values (dotted line). Panels (e–f):
multipliers of a one point increase in the rate of unemployment

properties of the wage and price and equations) has been refuted both logically
and empirically. To avoid misinterpretations, it is worth restating that this
result in no way justifies a return to demand driven macroeconomic models.
Instead, as sketched in the earlier section, we favour models where unemployment is determined jointly with real wages and the real exchange rate, and
this implies that wage- and price-equations are grafted into a bigger system
of equations which also includes equations representing the dynamics in other
parts of the economy. This is also the approach we pursue in the following
chapters. As we have seen, the natural rate models in the macroeconomic literature (Phillips curve and ICM) are special cases of the model framework
emerging from this section.

The finding that long-run unemployment is left undetermined by the wage–
price sub-model is a strong rationale for building larger systems of equations,
even if the first objective and primary concern is the analysis of wages, prices,
and inflation. Another thesis of this section is that stylised wage–price models
run the danger of imposing too much in the form of nominal neutrality
(absence of nominal rigidity) prior to the empirical investigation. Conversely,
no inconsistencies or overdetermination arise from enlarging the wage–price-
setting equations with a separate equation of the rate of unemployment into

the system, where demand variables may enter. The enlarged model will have a steady state (given some conditions that can be tested). The equilibrium rate of unemployment implied by this type of model is not of the natural rate type, since factors (in real growth rate form) from the demand side may have lasting effects. On the other hand, 'money illusion' is not implied, since the variables conditioned upon when modelling the rate of unemployment are all defined in real terms.

7

The New Keynesian Phillips curve

Hitherto, we have considered models that have a unique backward solution, given a set of initial conditions. Even though individual variables may be dominated by unit roots, models defined in terms of differences and cointegration relationships are also asymptotically stable. Models with forward-looking expectations are not contained by this framework. Recently a coherent theory of price-setting with rational expectations has gained in popularity. In this chapter, we give an appraisal of the New Keynesian Phillips curve model (hereafter NPCM) as an empirical model of inflation. The favourable evidence for NPCMs on Euro-area data reported in earlier studies is illusive. The empirical support for the economic forcing variable is fragile, and little distinguishes the performance of the estimated NPCM from a pure time-series model of the inflation rate. The NPCM can be reinterpreted as a highly restricted (and therefore unlikely) equilibrium correction model. Using that framework, we construct tests based on variable addition and encompassing. The results show that economists should not accept the NPCM too readily, and that specific hypotheses about expectations terms are better handled as potential extensions of existing econometrically adequate models.

7.1 Introduction

The previous four chapters have analysed alternative models of wage–price setting in small open economies. A common underlying assumption has been that all processes are causal or future independent processes, that is, the roots of the characteristic polynomials are on (unit roots) or inside the unit circle. This means that the model can be solved uniquely from known initial conditions. In this chapter, we turn to rational expectations models—systems where expected future values of endogenous variables enter as explanatory variables, in one or more equations. Rational expectations models yield different types of solutions than causal models. In principle, a solution depends on (all) future values of the model's disturbances. However, if some of the characteristic roots have modulus less than unity while the others have modulus bigger than unity, saddle-path solutions may exist. Saddle-path solutions are not asymptotically stable but depend on very specific initial conditions. Assume that the system is initially in a stationary situation A. If a shock occurs that defines a new stationary situation B, there are no stable dynamic trajectories starting from A, due to the lack of asymptotic dynamic stability. The endogenous variables of a macroeconomic model can be classified as state or jump variables. The time derivatives of state variables are always finite. In contrast, and as the name suggests, jump variables can shift up or down to new levels quite instantaneously (exchange rates and other asset prices are common examples). Jump variables play a key role in saddle-path equilibria. Essentially, if a shock occurs in a stationary situation A, instability is avoided by one or more jump variables jumping instantaneously to establish a new set of initial conditions that set the dynamics on to the saddle path leading to the new stationary situation B. Models with saddle-path solutions are important in academic macroeconomics, as demonstrated by, for example, the monetary theory of the exchange rate and Dornbusch's (1976) overshooting model. Whether saddle-path equilibria have a role in econometric models of inflation is a separate issue, which we address by considering the New Keynesian Phillips curve.

The New Keynesian Phillips Curve Model (NPCM) is aspiring to become the new consensus theory of inflation in modern monetary economics. This position is due to its stringent theoretical derivation, as laid out in Clarida *et al.* (1999), Svensson (2000), and Woodford (2003: ch. 3). In addition, empirical evidence is accumulating rapidly. For example, the recent studies of Galí and Gertler (1999) and Galí *et al.* (2001), hereafter GG and GGL, claim to have found considerable empirical support for the NPCM—using European as well as United States data. Moreover, Batini *et al.* (2000) derives an open economy NPCM which they have on United Kingdom data with supportive results for the specification. In this chapter, we re-analyse the data used in two of these studies, namely GGL and the study by Batini *et al.* (2000). The results show that the empirical relevance of the NPCM on these data sets is very weak. We reach this surprising conclusion by applying encompassing tests, where the NPCM is tested against earlier econometric inflation models, as opposed to the corroborative approach of the NPCM papers. In addition we also examine the relevance of the NPCM for Norwegian inflation.[1]

The structure of the chapter is as follows. After defining the model in Section 7.2, we investigate the dynamic properties of the NPCM in Section 7.3. This entails not only the NPCM equation, but also specification of a process

---

[1] This chapter draws on Bårdsen *et al.* (2002*b*, 2004).

for the forcing variable. Given that a system of linear difference equations is the right framework for theoretical discussions about stability and the type of solution (forward or backward), it follows that the practice of deciding on these issues on the basis of single equation estimation is not robust to extensions of the information set. For example, a forward solution may suggest itself from estimation of the NPCM equation alone, while system estimation may show that the forcing variable is endogenous, giving rise to a different set of characteristic roots and potentially giving support to a backward solution.

Section 7.4 discusses estimation issues of the NPCM, using Euro-area data for illustration. After conducting a sensitivity analysis of estimates of the model under the assumption of correct specification, we apply several methods for testing and evaluating the specification in Section 7.5. We conclude that the specification is not robust. In particular, building on the insight from Section 7.3, we show that it is useful to extend the evaluation from the single equation NPCM to a system consisting of the rate of inflation and the forcing variable.

Another strategy of model evaluation is to consider competing theories, resulting in alternative model specifications. For example, there are several studies that have found support for incomplete competition models, giving rise to systems with cointegrating relationships between wages, prices, unemployment, and productivity, as well a certain ordering of causality. In Section 7.5.4 we show that these existing results can be used to test the encompassing implications of the NPCM. This approach is applied to the open economy version of the NPCM of Batini *et al.* (2000). Finally we add to the existing evidence by evaluating the NPCM on Norwegian data and testing the encompassing implications. Appendix A.2 provides the necessary background material on solution and estimation of rational expectations models.

7.2 The NPCM defined

Let p_t be the log of a price level index. The NPCM states that inflation, defined as $\Delta p_t \equiv p_t - p_{t-1}$, is explained by $E_t \Delta p_{t+1}$, expected inflation one period ahead conditional upon information available at time t, and excess demand or marginal costs x_t (e.g. output gap, the unemployment rate, or the wage share in logs):

$$\Delta p_t = b_{p_1} E_t \Delta p_{t+1} + b_{p2} x_t + \varepsilon_{pt}, \qquad (7.1)$$

where ε_{pt} is a stochastic error term. Roberts (1995) shows that several New Keynesian models with rational expectations have (7.1) as a common representation—including the models of staggered contracts developed by Taylor (1979*b*, 1980)[2] and Calvo (1983), and the quadratic price adjustment cost model of Rotemberg (1982). GG gives a formulation of the NPCM in line with Calvo's

---

[2] The overlapping wage contract model of sticky prices is also attributed to Phelps (1978).

work: they assume that a firm takes account of the expected future path of nominal marginal costs when setting its price, given the likelihood that its price may remain fixed for multiple periods. This leads to a version of the inflation equation (7.1), where the forcing variable x_t is the representative firm's real marginal costs (measured as deviations from its steady-state value). They argue that the wage share (the labour income share) ws_t is a plausible indicator for the average real marginal costs, which they use in the empirical analysis. The alternative, hybrid version of the NPCM that uses both $\mathsf{E}_t\Delta p_{t+1}$ and lagged inflation as explanatory variables is also discussed later.

7.3   NPCM as a system

Equation (7.1) is incomplete as a model for inflation, since the status of x_t is left unspecified. On the one hand, the use of the term forcing variable, suggests exogeneity, whereas the custom of instrumenting the variable in estimation is germane to endogeneity. In order to make progress, we therefore consider the following completing system of stochastic linear difference equations[3]

$$\Delta p_t = b_{p1}\Delta p_{t+1} + b_{p2}x_t + \varepsilon_{pt} - b_{p1}\eta_{t+1}, \tag{7.2}$$
$$x_t = b_{x1}\Delta p_{t-1} + b_{x2}x_{t-1} + \varepsilon_{xt}, \qquad 0 \le |b_{x2}| < 1. \tag{7.3}$$

The first equation is adapted from (7.1), utilising that $\mathsf{E}_t\Delta p_{t+1} = \Delta p_{t+1} - \eta_{t+1}$, where η_{t+1} is the expectation error. Equation (7.3) captures that there may be feedback from inflation on the forcing variable x_t (output-gap, the rate of unemployment or the wage share) in which case $b_{x1} \neq 0$.

In order to discuss the dynamic properties of this system, re-arrange (7.2) to yield

$$\Delta p_{t+1} = \frac{1}{b_{p1}}\Delta p_t - \frac{b_{p2}}{b_{p1}}x_t - \frac{1}{b_{p1}}\varepsilon_{pt} + \eta_{t+1} \tag{7.4}$$

and substitute x_t with the right-hand side of equation (7.3). The characteristic polynomial for the system (7.3) and (7.4) is

$$p(\lambda) = \lambda^2 - \left[\frac{1}{b_{p1}} + b_{x2}\right]\lambda + \frac{1}{b_{p1}}[b_{p2}b_{x1} + b_{x2}]. \tag{7.5}$$

If neither of the two roots is on the unit circle, unique asymptotically stationary solutions exist. They may be either causal solutions (functions of past values of the disturbances and of initial conditions) or future dependent solutions (functions of future values of the disturbances and of terminal conditions), see Brockwell and Davies (1991: ch. 3) and Gourieroux and Monfort (1997: ch. 12).

The future dependent solution is a hallmark of the NPC. Consider for example the case of $b_{x1} = 0$, so that x_t is a strongly exogenous forcing variable in the NPCM. This restriction gives the two roots $\lambda_1 = b_{p1}^{-1}$ and $\lambda_2 = b_{x2}$.

---

[3] Constant terms are omitted for ease of exposition.

Given the restriction on b_{x2} in (7.3), the second root is always less than one, meaning that x_t is a causal process that can be determined from the backward solution. However, since $\lambda_1 = b_{p1}^{-1}$ there are three possibilities for Δp_t: (1) No stationary solution: $b_{p1} = 1$; (2) A causal solution: $b_{p1} > 1$; (3) A future dependent solution: $b_{p1} < 1$. If $b_{x1} \neq 0$, a stationary solution may exist even in the case of $b_{p1} = 1$. This is due to the multiplicative term $b_{p2}b_{x1}$ in (7.5). The economic interpretation of the term is the possibility of stabilising inter-action between price-setting and product (or labour) markets—as in the case of a conventional Phillips curve.

As a numerical example, consider the set of coefficient values: $b_{p1} = 1$, $b_{p2} = 0.05$, $b_{x2} = 0.7$, and $b_{x1} = 0.2$, corresponding to x_t (interpreted as the output-gap) influencing Δp_t positively, and the lagged rate of inflation having a positive coefficient in the equation for x_t. The roots of (7.5) are in this case $\{0.96, 0.74\}$, so there is a causal solution. However, if $b_{x1} < 0$, there is a future dependent solution since then the largest root is greater than one.

Finding that the existence and nature of a stationary solution is a system property is of course trivial. Nevertheless, many empirical studies only model the Phillips curve, leaving the x_t part of the system implicit. This is unfortunate, since the same studies often invoke a solution of the well-known form[4]

$$\Delta p_t = \left(\frac{b_{p2}}{1 - b_{p1}b_{x2}} \right) x_t + \varepsilon_{pt}. \tag{7.6}$$

Clearly, (7.6) hinges on $b_{p1}b_{x2} < 1$ which involves the coefficient b_{x2} of the x_t process.

If we consider the rate of inflation to be a jump variable, there may be a saddle-path equilibrium as suggested by the phase diagram in Figure 7.1. The drawing is based on $b_{p2} < 0$, so we now interpret x_t as the rate of unem-ployment. The line representing combinations of Δp_t and x_t consistent with $\Delta^2 p_t = 0$ is downward sloping. The set of pairs $\{\Delta p_t, x_t\}$ consistent with $\Delta x_t = 0$ are represented by the thick vertical line (this is due to $b_{x1} = 0$ as above). Point a is a stationary situation, but it is not asymptotically stable. Suppose that there is a rise in x represented by a rightward shift in the vertical curve, which is drawn with a thinner line. The arrows show a potential unstable trajectory towards the north-east away from the initial equilibrium. However, if we consider Δp_t to be a jump variable and x_t as state variable, the rate of inflation may jump to a point such as b and thereafter move gradually along the saddle path connecting b and the new stationary state c.

The jump behaviour implied by models with forward expected inflation is at odds with observed behaviour of inflation. This has led several authors to sug-gest a 'hybrid' model, by heuristically assuming the existence of both forward-and backward-looking agents; see, for example, Fuhrer and Moore (1995). Also Chadha *et al.* (1992) suggest a form of wage-setting behaviour that would

---

[4] That is, subject to the transversality condition $\lim_{n \to \infty} (b_{p1})^{n+1} \Delta p_{t+n+1} = 0$.

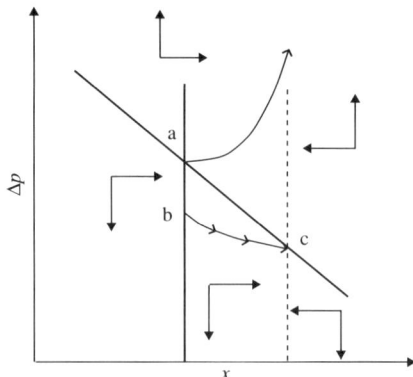

Figure 7.1. Phase diagram for the system for the case of $b_{p1} < 1$, $b_{p2} < 0$, and $b_{x1} = 0$

lead to some inflation stickiness and to inflation being a weighted average of both past inflation and expected future inflation. Fuhrer (1997) examines such a model empirically and finds that future prices are empirically unimportant in explaining price and inflation behaviour compared to past prices.

In the same spirit as these authors, and with particular reference to the empirical assessment in Fuhrer (1997), GG also derive a hybrid Phillips curve that allows a subset of firms to have a backward-looking rule to set prices. The hybrid model contains the wage share as the driving variable and thus nests their version of the NPCM as a special case. This amounts to the specification

$$\Delta p_t = b^f_{p1} \mathsf{E}_t \Delta p_{t+1} + b^b_{p1} \Delta p_{t-1} + b_{p2} x_t + \varepsilon_{pt}. \tag{7.7}$$

Galí and Gertler (1999) estimate (7.7) for the United States in several variants—using different inflation measures, different normalisation rules for the GMM estimation, including additional lags of inflations in the equation and splitting the sample. Their results are robust—marginal costs have a significant impact on short-run inflation dynamics and forward-looking behaviour is always found to be important.

In the same manner as above, equation (8.13) can be written as

$$\Delta p_{t+1} = \frac{1}{b^f_{p1}} \Delta p_t - \frac{b^b_{p1}}{b^f_{p1}} \Delta p_{t-1} - \frac{b_{p2}}{b^f_{p1}} x_t - \frac{1}{b^f_{p1}} \varepsilon_{pt} + \eta_{t+1} \tag{7.8}$$

and combined with (7.3). The characteristic polynomial of the hybrid system is

$$p(\lambda) = \lambda^3 - \left[\frac{1}{b^f_{p1}} + b_{x2} \right] \lambda^2 + \frac{1}{b^f_{p1}} \left[b^b_{p1} + b_{p2} b_{x1} + b_{x2} \right] \lambda - \frac{b^b_{p1}}{b^f_{p1}} b_{x2}. \tag{7.9}$$

Using the typical results for the expectation and backward-looking parameters, $b^f_{p1} = 0.25$, $b^b_{p1} = 0.75$, together with the assumption of an exogenous x_t

process with autoregressive parameter 0.7, we obtain the roots $\{3.0, 1.0, 0.7\}$.[5] Thus, there is no asymptotically stable stationary solution for the rate of inflation in this case.

This seems to be a common result for the hybrid model as several authors choose to impose the restriction

$$b_{p1}^f + b_{p1}^b = 1, \tag{7.10}$$

which forces a unit root upon the system. To see this, note first that a 1–1 reparameterisation of (7.8) gives

$$\Delta^2 p_{t+1} = \left[\frac{1}{b_{p1}^f} - \frac{b_{p1}^b}{b_{p1}^f} - 1 \right] \Delta p_t + \frac{b_{p1}^b}{b_{p1}^f} \Delta^2 p_t - \frac{b_{p2}}{b_{p1}^f} x_t - \frac{1}{b_{p1}^f} \varepsilon_{pt} + \eta_{t+1},$$

so that if (7.10) holds, (7.8) reduces to

$$\Delta^2 p_{t+1} = \frac{(1 - b_{p1}^f)}{b_{p1}^f} \Delta^2 p_t - \frac{b_{p2}}{b_{p1}^f} x_t - \frac{1}{b_{p1}^f} \varepsilon_{pt} + \eta_{t+1}. \tag{7.11}$$

Hence, the homogeneity restriction (7.10) turns the hybrid model into a model of the *change* in inflation. Equation (7.11) is an example of a model that is cast in the difference of the original variable, a so-called differenced autoregressive model (dVAR), only modified by the driving variable x_t. Consequently, it represents a generalisation of the random walk model of inflation that was implied by setting $b_{p1}^f = 1$ in the original NPCM. The result in (7.11) will prove important in understanding the behaviour of the NPCM in terms of goodness of fit, see later.

If the process x_t is strongly exogenous, the NPCM in (7.11) can be considered on its own. In that case (7.11) has no stationary solution for the rate of inflation. A necessary requirement is that there are equilibrating mechanisms elsewhere in the system, specifically in the process governing x_t (e.g. the wage share). This requirement parallels the case of dynamic homogeneity in the backward-looking Phillips curve (i.e. a vertical long-run Phillips curve). In the present context, the message is that statements about the stationarity of the rate of inflation, and the nature of the solution (backward or forward) requires an analysis of the system.

The empirical results of GG and GGL differ from other studies in two respects. First, b_{p1}^f is estimated in the region $(0.65, 0.85)$ whereas b_{p1}^b is one third of b_{p1}^f or less. Second, GG and GGL succeed in estimating the hybrid model without imposing (7.10). GGL (their table 2) report the estimates $\{0.69, 0.27\}$ and $\{0.88, 0.025\}$ for two different estimation techniques. The corresponding roots are $\{1.09, 0.70, 0.37\}$ and $\{1.11, 0.70, 0.03\}$, illustrating that as long as the sum of the weights is less than one the future dependent solution prevails.

---

[5] The full set of coefficient values are: $b_{x1} = 0$, $b_{p1}^f = 0.25$, $b_{p1}^b = 0.75$, $b_{x2} = 0.7$.

7.4   Sensitivity analysis

In the following, we will focus on the results in GGL for the Euro area. Our replication of their estimates is given in (7.12), using the same set of instruments: five lags of inflation, and two lags of the wage share, detrended output, and wage inflation.

$$\Delta p_t = \underset{(0.073)}{0.681 \Delta p_{t+1}} + \underset{(0.072)}{0.281 \Delta p_{t-1}} + \underset{(0.027)}{0.019 ws_t} + \underset{(0.069)}{0.063} \tag{7.12}$$

$$\text{GMM, } T = 107 \ (1971(3) \text{ to } 1998(1))$$
$$\chi_J^2(8) = 8.01[0.43],$$

where $\chi_J^2(\cdot)$ is Hansen's (1982) J-test of overidentifying restrictions. The role of the wage share (as a proxy for real marginal costs) is a definable trait of the NPCM, yet the empirical relevance of ws_t is not apparent in (7.12): it is statistically insignificant. Note also that the sum of the coefficients of the two inflation terms is 0.96. Taken together, the insignificance of ws_t and the near unit-root, imply that (7.12) is almost indistinguishable from a pure time-series model, a dVAR.[6] On the other hand, the formal significance of the forward term, and the insignificance of the J-statistic corroborate the NPCM. The merits of the J-statistic are discussed in Section 7.5: in the rest of this section we conduct a sensitivity analysis with regards to GMM estimation methodology.

The results in (7.12) were obtained by a GMM procedure which computes the weighting matrix once. When instead we iterate over both coefficients and weighting matrix, with fixed bandwidth,[7] we obtain

$$\Delta p_t = \underset{(0.052)}{0.731 \Delta p_{t+1}} + \underset{(0.069)}{0.340 \Delta p_{t-1}} - \underset{(0.029)}{0.042 ws_t} - \underset{(0.070)}{0.102} \tag{7.13}$$

$$\text{GMM, } T = 107 \ (1971(3) \text{ to } 1998(1))$$
$$\chi_J^2(8) = 7.34[0.50].$$

As before, there is clear indication of a unit root (the sum of the two inflation coefficients is now slightly above one). The wage share coefficient is wrongly signed, but it is still insignificantly different from zero, though.

Next, we investigate the robustness with regard to the choice of instruments. We use an alternative output-gap measure ($emugap_t$), which is a simple transformation of the one defined in Fagan *et al.* (2001) as real output relative to potential output, measured by a constant-return-to-scale Cobb–Douglas production function with neutral technical progress. We also omit the two lags

---

[6] See Bårdsen *et al.* (2002b) for a more detailed discussion.
[7] We used the default GMM implementation in Eviews 4.

of wage growth. Apart from yet another sign-change in the ws coefficient, the results respond little to these changes in the set of instruments:

$$\Delta p_t = \underset{(0.06)}{0.60\Delta p_{t+1}} + \underset{(0.06)}{0.35\Delta p_{t-1}} + \underset{(0.03)}{0.03 ws_t} + \underset{(0.06)}{0.08} \qquad (7.14)$$

$$\text{GMM}, \ T = 107 \ (1972(4) \text{ to } 1997(4))$$
$$\chi_J^2(6) = 6.74[0.35].$$

Finally, we investigate the robustness with respect to estimation method. Since the NPCM is a linear model, the only real advantage of choosing GMM as opposed to 2SLS as estimation method is the potential necessity to correct for autocorrelated residuals. Autocorrelation is in line with the rational expectations hypothesis, implied by replacing $E_t\Delta p_{t+1}$ with Δp_{t+1} in estimation—see Blake (1991) and Appendix A.2—but it may also be a symptom of mis-specification, as discussed in Nymoen (2002). As shown below, the estimates are robust with respect to estimation method, even though the standard errors are doubled, since the model suffers from severe autocorrelation:

$$\Delta p_t = \underset{(0.14)}{0.66\Delta p_{t+1}} + \underset{(0.12)}{0.28\Delta p_{t-1}} + \underset{(0.09)}{0.07 ws_t} + \underset{(0.12)}{0.10} \qquad (7.15)$$

$$\text{2SLS}, \ T = 104 \ (1972(2) \text{ to } 1998(1))$$

$\hat{\sigma}_{IV} = 0.28$	RSS $= 7.66$
$F_{AR(1-1)}(1, 99) = 166.93[0.00]$	$F_{AR(2-2)}(1, 99) = 4.73[0.03]$
$F_{ARCH(1-4)}(4, 92) = 2.47[0.05]$	$\chi^2_{normality}(2) = 1.59[0.45]$
$F_{HET x_i x_j}(9, 90) = 2.34[0.02]$	$\chi^2_{ival}(6) = 11.88[0.06]$
$F_{irel}(9, 94) = 70.76[0.00].$	

The p-value of the Sargan specification test, χ^2_{ival}, is 0.06, and indicates that (7.15) could be mis-specified, since some of the instruments could be potential regressors. The F_{irel} is the F-statistic from the first stage regression of Δp_{t+1} against the instrument set and indicates no 'weak instruments' problem, although it is only strictly valid in the case of one endogenous regressor—see Stock *et al.* (2002).[8]

We conclude from the range of estimates that the significance of the wage share is fragile and that its formal statistical significance depends on the exact implementation of the estimation method used. The coefficient of the forward variable on the other hand is pervasive and will be a focal point of the following analysis. Residual autocorrelation is another robust feature, as also noted by GGL. But more work is needed before we can judge whether autocorrelation really corroborates the theory, which is GGL's view, or whether it is a sign of econometric mis-specification.

---

[8] The rule of thumb is a value bigger than 10 in the case of one endogenous regressor.

7.5   Testing the specification

The main tools of evaluation of models like the NPCM have been the GMM test of validity of overidentifying restrictions (i.e. the χ_J^2-test earlier) and measures and graphs of goodness-of-fit.[9] Neither of these tests is easy to interpret. First, the χ_J^2 may have low power. Second, the estimation results reported by GG and GGL yield values of $b_{p1}^f + b_{p1}^b$ close to 1 while the coefficient of the wage share is numerically small. This means that the apparently good fit is in fact no better (or worse) than a model in the double differences (e.g. a random walk); see Bårdsen *et al.* (2002*b*). There is thus a need for other evaluation methods, and in the rest of this chapter we test the NPCM specification against alternative models of the inflation process.

7.5.1   An encompassing representation

The main alternatives to the NPCM as models of inflation are the Standard Phillips Curve Model (PCM) and the Incomplete Competition Model (ICM). They will therefore be important in suggesting ways of evaluating the NPCM from an encompassing perspective. To illustrate the main differences between alternative specifications, consider the following stylised framework—see also Bårdsen *et al.* (2002*a*). Let w be wages and p consumer prices; with a as productivity, the wage share ws is given as real unit labour costs: $ws = ulc - p = w - a - p$; u is the unemployment rate, and gap the output gap, all measured in logs. We abstract from other forcing variables, like open economy aspects. A model of the wage–price process general enough for the present purpose then takes the form

$$\Delta w = \alpha \Delta p^e - \beta ws - \gamma u,$$
$$\Delta p = \delta \Delta p^e + \zeta \Delta w + \eta ws + \vartheta gap,$$

where Δp^e is expected inflation, and the dynamics are to be specified separately for each model. Although the structure is very simple, the different models drop out as non-nested special cases:

1. The NPCM is given as

$$\Delta p_t = \delta_1^f \Delta p_{t+1}^e + \delta_1^b \Delta p_{t-1} + \eta_1 ws_t,$$

where the expectations term Δp_{t+1}^e is assumed to obey rational expectations.

---

[9] For example, in the *Abstract* of GGL the authors state that 'the NPC fits Euro data very well, possibly better than United States data'. Also Galí (2003), responding to critical assessments of the NPCM, states that 'it appears to fit the data much better than had been concluded by the earlier literature'.

2. The PCM is—Aukrust (1977), Calmfors (1977), Nymoen (1990), Blanchard and Katz (1997):

$$\Delta w_t = \alpha_2 \Delta p_t - \gamma_2 u_t$$
$$\Delta p_t = \zeta_2 \Delta w_t + \vartheta_2 gap_t.$$

3. The ICM on equilibrium correction form—Sargan (1964), Layard *et al.* (1991), Bårdsen *et al.* (1998), and Kolsrud and Nymoen (1998):

$$\Delta w_t = \alpha_3 \Delta p_t - \beta_3 (ws - \gamma_2 u)_{t-1}$$
$$\Delta p_t = \zeta_3 \Delta w_t - \delta_1^b [p - \eta_3 (ws + p)]_{t-1} + \vartheta_3 gap_{t-1}.$$

Of course, there exist a host of other, more elaborate, models—a notable omission being non-linear PCMs. However, the purpose here is to highlight that discrimination between the models is possible through testable restrictions. The difference between the two Phillips curve models is that the NPCM has forward-looking expectations and has real unit labour costs, rather than the output gap of the PCM. In the present framework, the ICM differs mainly from the NPCM in the treatment of expectations and from the PCM in the latter's exclusion of equilibrium correction mechanisms that are derived from conflict models of inflation; see Rowthorn (1977), Sargan (1980), Kolsrud and Nymoen (1998), Bårdsen and Nymoen (2003) and Chapter 6. To see this, note that the NPCM can, trivially, be reparameterised as a forward-looking equilibrium-correction model (EqCM) with long-run coefficient restricted to unity:

$$\Delta p_t = \delta_1^f \Delta p_{t+1}^e + \eta_1 \Delta ws_t + \delta_1^b \Delta p_{t-1} - \eta_1 [p - 1(ws + p)]_{t-1}.$$

The models listed in 1–3 are identified, in principle, but it is an open question whether data and methodology are able to discriminate between them on a given data set. We therefore test the various identifying restrictions. This will involve testing against

- richer dynamics
- system representations
- encompassing restrictions.

We next demonstrate these three approaches in practice.

7.5.2 Testing against richer dynamics

In the case of the NPCM, the specification of the econometric model used for testing a substantive hypothesis—forward and lagged endogenous variable—incorporates the alternative hypothesis associated with a mis-specification test (i.e. of residual autocorrelation). Seeing residual correlation as corroborating the theory that agents are acting in accordance with NPCM is invoking a very strong *ceteris paribus* clause. Realistically, the underlying cause of the residual correlation may of course be quite different, for example, omitted variables, wrong functional form or, in this case, a certain form of over-differencing. In fact, likely directions for respecification are suggested by pre-existing results from several decades of empirical modelling of inflation dynamics. For example, variables representing capacity utilisation (output-gap and/or unemployment) have a natural role in inflation models: we use the alternative output-gap measure ($emugap_t$). Additional lags in the rate of inflation are also obvious candidates. As a direct test of this respecification, we move the lagged output-gap ($emugap_{t-1}$) and the fourth lag of inflation (Δp_{t-4}) from the list of instruments used for estimation of (7.14), and include them as explanatory variables in the equation. The results (using 2SLS) are:

$$\Delta p_t = \underset{(0.28)}{0.07\Delta p_{t+1}} + \underset{(0.09)}{0.14 w s_t} + \underset{(0.14)}{0.44\Delta p_{t-1}}$$

$$+ \underset{(0.09)}{0.18\ \Delta p_{t-4}} + \underset{(0.05)}{0.12\ emugap_{t-1}} + \underset{(0.30)}{0.53} \qquad (7.16)$$

$$\text{2SLS, } T = 104 \ (1972(2) \text{ to } 1998(1))$$

$\hat{\sigma}_{IV} = 0.28$ RSS $= 7.52$

$\mathsf{F}_{\mathsf{AR}(1-1)}(1, 97) = 2.33[0.13]$ $\mathsf{F}_{\mathsf{AR}(2-2)}(1, 97) = 2.80[0.10]$

$\mathsf{F}_{\mathsf{ARCH}(1-4)}(4, 90) = 0.80[0.53]$ $\chi^2_{\mathsf{normality}}(2) = 1.75[0.42]$

$\mathsf{F}_{\mathsf{HET}x_i x_j}(20, 77) = 1.26[0.23]$ $\chi^2_{\mathsf{ival}}(4) = 4.52[0.34]$.

When compared to (7.14) and (7.15), four results stand out:

1. The estimated coefficient of the forward term Δp_{t+1} is reduced by a factor of 10, and becomes insignificant.
2. The diagnostic tests indicate no residual autocorrelation or heteroskedasticity.
3. The p-value of the Sargan specification test, χ^2_{ival}, is 0.34, and is evidence that (7.16) effectively represents the predictive power that the set of instruments has about Δp_t.[10]
4. If the residual autocorrelations of the NPCMs above are induced by the forward solution and 'errors in variables', there should be a similar autocorrelation process in the residuals of (7.16). Since there is no detectable residual autocorrelation, that interpretation is refuted, supporting instead that the hybrid NPCM is mis-specified.

Finally, after deleting Δp_{t+1} from the equation, the model's interpretation is clear, namely as a conventional dynamic price-setting equation. Indeed, using the framework of Section 7.5.1, the model is seen to correspond to the ICM price equation, with $\delta_1^f = 0$ (and extended with Δp_{t-4} and $emugap_{t-1}$ as explanatory

---

[10] The full set of instruments is: ws_{t-1}, ws_{t-2}, Δp_{t-2}, Δp_{t-3}, Δp_{t-5}, and $emugap_{t-2}$.

variables). We are therefore effectively back to a conventional dynamic markup equation.

In sum, we find that significance testing of the forward term does not support the NPCM for the Euro data. This conclusion is based on the premise that the equation with the forward coefficient is tested *within* a statistically adequate model, which entails thorough mis-specification testing of the theoretically postulated NPCM, and possible respecification before the test of the forward coefficient is performed. Our results are in accord with Rudd and Whelan (2004), who show that the tests of forward-looking behaviour which Galí and Gertler (1999) and Galí *et al.* (2001) rely on, have very low power against alternative, but non-nested, backward-looking specifications, and demonstrate that results previously interpreted as evidence for the New Keynesian model are also consistent with a backward-looking Phillips curve. Rudd and Whelan develop alternative, more powerful tests, which exhibit a very limited role for forward-looking expectations. A complementary interpretation follows from a point made by Mavroeidis (2002), namely that the hybrid NPCM suffers from underidentification, and that in empirical applications identification is achieved by confining important explanatory variables to the set of instruments, with mis-specification as a result.

7.5.3 Evaluation of the system

The nature of the solution for the rate of inflation is a system property, as noted in Section 7.3. Hence, unless one is willing to accept at face value that an operational definition of the forcing variable is strongly exogenous, the 'structural' NPCM should be evaluated within a system that also includes the forcing variable as a modelled variable.

For that purpose, Table 7.1 shows an estimated system for Euro-area inflation, with a separate equation (the second in the table) for treating the wage share (the forcing variable) as an endogenous variable. Note that the hybrid NPCM equation (first in the table) is similar to (7.14), and thus captures the gist of the results in GGL. This is hardly surprising, since only the estimation method (full information maximum likelihood—FIML in Table 7.1) separates the two NPCMs.

An important feature of the estimated equation for the wage share ws_t is the two lags of the rate of inflation, which both are highly significant. The likelihood-ratio test of joint significance gives $\chi^2(2) = 24.31[0.00]$, meaning that there is clear formal evidence against the strong exogeneity of the wage share. One further implication of this result is that a closed form solution for the rate of inflation cannot be derived from the structural NPCM alone.

The roots of the system in Table 7.1 are all less than one (not shown in the table) in modulus and therefore corroborate a forward solution. However, according to the results in the table, the implied driving variable is $emugap_t$, rather than ws_t which is endogenous, and the weights of the present value

Table 7.1
FIML results for the NPCM system for the
Euro area 1972(2)–1998(1)

$$\Delta p_t = \underset{(0.154)}{0.7696} \Delta p_{t+1} + \underset{(0.131)}{0.2048} \Delta p_{t-1} + \underset{(0.0930)}{0.0323} ws_t$$
$$+ \underset{(0.1284)}{0.0444}$$

$$ws_t = \underset{(0.0296)}{0.8584} ws_{t-1} + \underset{(0.0220)}{0.0443} \Delta p_{t-2} + \underset{(0.0223)}{0.0918} \Delta p_{t-5}$$
$$+ \underset{(0.0067)}{0.0272} emugap_{t-2} - \underset{(0.0447)}{0.2137}$$

$$\Delta p_{t+1} = \underset{(0.0988)}{0.5100} ws_{t-1} + \underset{(0.0907)}{0.4153} \Delta p_{t-1} + \underset{(0.0305)}{0.1814} emugap_{t-1}$$
$$+ \underset{(0.1555)}{0.9843}$$

Note: The sample is 1972(2) to 1998(1), $T = 104$.

$$\hat{\sigma}_{\Delta p_t} = 0.290186$$
$$\hat{\sigma}_{ws} = 0.074904$$
$$\hat{\sigma}_{\Delta p^e_{t+1}} = 0.325495$$
$$F^v_{AR(1-5)}(45, 247) = 37.100[0.0000]^{**}$$
$$F^v_{HETx^2}(108, 442) = 0.94319[0.6375]$$
$$F^v_{HETx_i x_j}(324, 247) = 1.1347[0.1473]$$
$$\chi^{2,v}_{normality}(6) = 9.4249[0.1511]$$

calculation of *emugap$_t$* have to be obtained from the full system. The stat-istics at the bottom of the table show that the system of equations has clear deficiencies as a statistical model, cf. the massive residual autocorrelation detected by $F^v_{AR(1-5)}$. Further investigation indicates that this problem is in part due to the wage share residuals and is not easily remedied on the present information set. However, from Section 7.5.2 we already know that another source of vector autocorrelation is the NPCM itself, and moreover that this mis-specification by and large disappears if we instead adopt equation (7.16) as our inflation equation.

It lies close at hand therefore to suggest another system where we utilise the second equation in Table 7.1, and the conventional price equation that is obtained by omitting the insignificant forward term from equation (7.16). Table 7.2 shows the results of this potentially useful model. No mis-specification is detected, and the coefficients appear to be well determined. In terms of economic interpretation the models resemble an albeit 'watered down' version

Table 7.2
FIML results for a conventional Phillips curve for the
Euro area 1972(2)–1998(1)

$$\Delta p_t = \underset{(0.1202)}{0.2866}\,ws_t + \underset{(0.0868)}{0.4476}\Delta p_{t-1} + \underset{(0.091)}{0.1958}\Delta p_{t-4}$$

$$+ \underset{(0.0259)}{0.1383}\,emugap_{t-1} + \underset{(0.1823)}{0.6158}$$

$$ws_t = \underset{(0.0298)}{0.8629}\,ws_{t-1} + \underset{(0.0222)}{0.0485}\Delta p_{t-2} + \underset{(0.0225)}{0.0838}\Delta p_{t-5}$$

$$+ \underset{(0.0068)}{0.0267}\,emugap_{t-2} - \underset{(0.0450)}{0.2077}$$

Note: The sample is 1972(2) to 1998(1), $T = 104$.

$$\hat{\sigma}_{\Delta p_t} = 0.284687$$

$$\hat{\sigma}_{ws} = 0.075274$$

$$\mathsf{F}^v_{\mathrm{AR}(1-5)}(20, 176) = 1.4669[0.0983]$$

$$\mathsf{F}^v_{\mathrm{HET}x^2}(54, 233) = 0.88563[0.6970]$$

$$\mathsf{F}^v_{\mathrm{HET}x_i x_j}(162, 126) = 1.1123[0.2664]$$

$$\chi^{2,v}_{\mathrm{normality}}(4) = 2.9188[0.5715]$$

$$\chi^2_{\mathrm{overidentification}}(10) = 10.709[0.3807]$$

of the modern conflict model of inflation and one interesting route for further work lies in that direction. That would entail an extension of the information set to include open economy aspects and indicators of institutional developments and of historical events. The inclusion of such features in the information set will also help in stabilising the system.[11]

7.5.4   Testing the encompassing implications

So far the NPCM has mainly been used to describe the inflationary process in studies concerning the United States economy or for aggregated Euro data. Heuristically, we can augment the basic model with import price growth and other open economy features, and test the significance of the forward inflation rate within such an extended NPCM. Recently, Batini *et al.* (2000) have derived an open economy NPCM from first principles, and estimated the model on United Kingdom economy data. Once we consider the NPCM for individual European economies, there are new possibilities for testing—since pre-existing results should, in principle, be explained by the new model (the NPCM). Specifically, and as discussed in earlier chapters, in the United Kingdom there exist models of inflation that build on a different framework than the

---

[11] The largest root in Table 7.2 is 0.98.

NPCM, namely wage bargaining and cointegration; see, for example, Nickell and Andrews (1983), Hoel and Nymoen (1988), Nymoen (1989a), and Blanchard and Katz (1999). Since the underlying theoretical assumptions are quite different, the existing empirical models define an information set that is wider than the set of instruments that are typically employed in the estimation of NPCMs. In particular, the existing studies claim to have found cointegrating relationships between levels of wages, prices, and productivity. These relationships constitute evidence that can be used to test the implications of the NPCM.

Specifically, the following procedure is followed[12]:

1. Assume that there exists a set of variables $\mathbf{z} = [\mathbf{z}_1 \ \mathbf{z}_2]$, where the sub-set $\mathbf{z}_1$ is sufficient for identification of the maintained NPCM model. The variables in $\mathbf{z}_2$ are defined by the empirical findings of existing studies.
2. Using $\mathbf{z}_1$ as instruments, estimate the augmented model

$$\Delta p_t = b_{p1}^f \mathsf{E}_t \Delta p_{t+1} + b_{p1}^b \Delta p_{t-1} + b_{p2} x_t + \cdots + \mathbf{z}_{2,t} \mathbf{b}_{p4}$$

   under the assumption of rational expectations about forward prices.
3. Under the hypothesis that the NPCM is the correct model, $\mathbf{b}_{p4} = \mathbf{0}$ is implied. Thus, non-rejection of the null hypothesis of $\mathbf{b}_{p4} = \mathbf{0}$, corroborates the feed-forward Phillips curve. In the case of the other outcome: non-rejection of $b_{p1}^f = 0$, while $\mathbf{b}_{p4} = \mathbf{0}$ is rejected statistically, the encompassing implication of the NPCM is refuted.

The procedure is clearly related to significance testing of the forward term, but there are also notable differences. As mentioned above, the motivation of the test is that of testing the implication of the rational expectations hypothesis; see Hendry and Neale (1988), Favero and Hendry (1992), and Ericsson and Irons (1995). Thus, we utilise that under the assumption that the NPCM is the correct model, consistent estimation of b_{p1}^f can be based on $\mathbf{z}_1$, and supplementing the set of instruments by $\mathbf{z}_2$ should not significantly change the estimated b_{p1}^f.

In terms of practical implementation, we take advantage of the existing results on wage and price modelling using cointegration analysis which readily imply z_2-variables in the form of linear combinations of levels variables. In other words they represent 'unused' identifying instruments that go beyond information sets used in the Phillips curve estimation. Importantly, if agents are rational, the extension of the information set should not take away the significance of Δp_{t+1} in the NPCM, and $\mathbf{b}_{p4} = \mathbf{0}$.

As mentioned earlier, Batini *et al.* (2000) derive an open economy NPCM consistent with optimising behaviour, thus extending the intellectual rationale of the original NPCM. They allow for employment adjustment costs, hence both future and current employment growth is included (Δn_{t+1} and Δn_t), and

---

[12] David F. Hendry suggested this test procedure to us. Bjørn E. Naug pointed out to us that a similar procedure is suggested in Hendry and Neale (1988).

propose to let the equilibrium markup on prices depend on the degree of foreign competition, *com*. In their estimated equations, they also include a term for the relative price of imports, denoted *rpi* and oil prices *oil*. The wage share variable used is the adjusted share preferred by Batini *et al.* (2000). Equation (7.17) is our attempt to replicate their results, with GMM estimation using their data.[13]

$$
\begin{aligned}
\Delta p_t = -\ &\underset{(0.20)}{0.56} + \underset{(0.09)}{0.33\Delta p_{t+1}} + \underset{(0.04)}{0.32\Delta p_{t-1}} + \underset{(0.06)}{0.07\,gap_t} \\
&+ \underset{(0.01)}{0.02\,com_t} + \underset{(0.05)}{0.13 ws_t} - \underset{(0.01)}{0.004\ rpi_t} - \underset{(0.003)}{0.02\ \Delta oil_t} \\
&- \underset{(0.42)}{0.79\Delta n_{t+1}} + \underset{(0.39)}{1.03\Delta n_t}
\end{aligned}
\tag{7.17}
$$

$$
\text{GMM},\ T = 107\ (1972(3)\text{ to }1999(1)),\quad \hat{\sigma} = 0.0099
$$
$$
\chi_J^2(31) = 24.92[0.77],\quad \mathsf{F}_{\text{irel}}(40,66) = 8.29[0.00].
$$

The terms in the second line represent small open economy features that we noted above. The estimated coefficients are in accordance with the results that Batini *et al.* (2000) report. However, the F_{irel}, which still is the F-statistic from the first stage ordinary least squares (OLS) regression of Δp_{t+1} against the instrument set, indicates that their model might have a potential problem of weak instruments.

In Section 5.6 we saw how Bårdsen *et al.* (1998) estimate a simultaneous cointegrating wage–price model for the United Kingdom (see also Bårdsen and Fisher 1999). Their two equilibrium-correction terms are deviations from a long-run wage-curve and an open economy price markup (see Panel 5 of Table 5.3):

$$
ecmw_t = (w - p - a + \tau 1 + 0.065u)_t, \tag{7.18}
$$
$$
ecmp_t = (p - 0.6\tau 3 - 0.89(w + \tau 1 - a) - 0.11pi)_t, \tag{7.19}
$$

where a denotes average labour productivity, $\tau 1$ is the payroll tax rate, u is the unemployment rate and pi is the price index of imports. The first instrument, $ecmw_t$, is an extended wage share variable which we expect to be a better instrument than ws_t, since it includes the unemployment rate as implied by, for example, bargaining models of wage-setting (see the encompassing representation of Section 7.5.1). The second instrument, $ecmp_t$, is an open economy version of the long-run price markup of the stylised ICM in Section 7.5.1.[14]

---

[13] Although we use the same set of instruments as Batini *et al.* (2000), we are unable to replicate their table 7b, column (b). Inflation is the first difference of log of the gross value added deflator. The *gap* variable is formed using the Hodrick–Prescott (HP) trend; see Batini *et al.* (2000) (footnote 15) for more details.

[14] Inflation Δp_t in equation (7.17) is for the gross value added price deflator, while the price variable in the study by Bårdsen *et al.* (1998) is the retail price index pc_t. However, if the long-run properties giving rise to the *ecms* are correct, the choice of price index should not matter. We therefore construct the two *ecms* in terms of the GDP deflator, p_t, used by Batini *et al.* (2000).

Equation (7.20) shows the results, for the available sample 1976(2)–1996(1), of adding $ecmw_{t-1}$ and $ecmp_{t-1}$ to the NPCM model (7.17):

$$\Delta p_t = -\underset{(0.44)}{1.51} + \underset{(0.13)}{0.03}\Delta p_{t+1} + \underset{(0.08)}{0.24}\Delta p_{t-1} - \underset{(0.11)}{0.02}\,gap_t + \underset{(0.019)}{0.008}\,com_t$$
$$+\ \underset{(0.07)}{0.13}\,ws_t - \underset{(0.03)}{0.01}\,rpi_t - \underset{(0.004)}{0.003}\Delta oil_t + \underset{(0.27)}{0.11}\Delta n_{t+1}$$
$$+\ \underset{(0.19)}{0.87}\Delta n_t - \underset{(0.10)}{0.35}\,ecmw_{t-1} - \underset{(0.12)}{0.61}\,ecmp_{t-1} \qquad (7.20)$$

$$\text{GMM,}\ \ T = 80\ (1976(2)\ \text{to}\ 1996(1)),\quad \hat{\sigma} = 0.0083$$
$$\chi^2_{\mathrm{J}}(31) = 14.39[0.99],\quad \mathsf{F}_{\mathrm{irel}}(42, 37) = 4.28[0.000].$$

The forward term Δp_{t+1} is no longer significant, whereas the ecm-terms, which ought to be of no importance if the NPCM is the correct model, are both strongly significant.[15]

In the same vein, note that our test of GGL's Phillips curve for the Euro area in Section 7.5.2 can be interpreted as a test of the implications of rational expectations. There $\mathbf{z}_2$ was simply made up of Δp_{t-4} and $emugap_{t-1}$ which modelling experience tells us are predictors of future inflation. Thus, from rational expectations their coefficients should be insignificant when Δp_{t+1} is included in the model (and there are good, overidentifying instruments). Above, we observed the converse, namely Δp_{t-4} and $emugap_{t-1}$ are statistically and numerically significant, while the estimated coefficient of Δp_{t+1} was close to zero.

7.5.5   The NPCM in Norway

Consider the NPCM (with forward term only) estimated on quarterly Norwegian data[16]:

$$\Delta p_t = \underset{(0.11)}{1.06}\ \Delta p_{t+1} + \underset{(0.02)}{0.01}\ ws_t + \underset{(0.02)}{0.04}\ \Delta pi_t + \text{dummies} \qquad (7.21)$$

$$\chi^2_{\mathrm{J}}(10) = 11.93[0.29].$$

The closed economy specification has been augmented heuristically with import price growth (Δpi_t) and dummies for seasonal effects as well as special events in the economy described in Bårdsen *et al.* (2002*b*). Estimation is by GMM for the period 1972(4)–2001(1). The instruments used (i.e. the variables in $\mathbf{z}_1$) are lagged wage growth (Δw_{t-1}, Δw_{t-2}), lagged inflation (Δp_{t-1}, Δp_{t-2}), lags of level and change in unemployment (u_{t-1}, Δu_{t-1}, Δu_{t-2}), and changes in

---

[15] The conclusion is unaltered when the two instruments are defined in terms of pc_t, as in the original specification of Bårdsen *et al.* (1998).

[16] Inflation is measured by the official consumer price index (CPI).

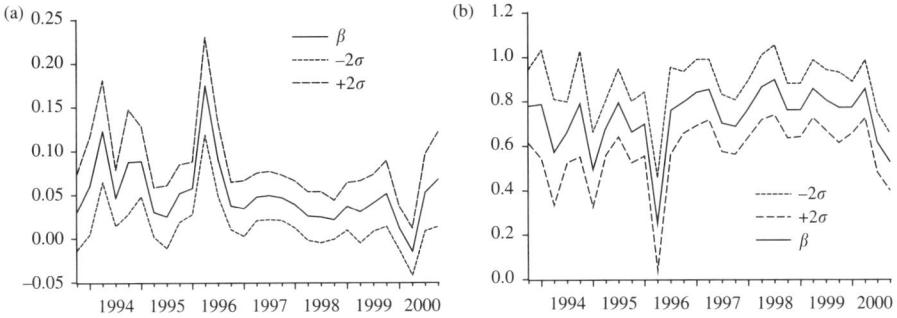

Figure 7.2. Rolling coefficients ± 2 standard errors of the NPCM, estimated on Norwegian data ending in 1993(4)–2000(4). Graph (a) shows the coefficient of ws_t and graph (b) shows the coefficient of Δp_{t+1}.

energy prices (Δpe_t, Δpe_{t-1}), the short term interest rate (ΔRL_t, ΔRL_{t-1}), and the length of the working day (Δh_t).

The coefficient estimates are similar to GG. Strictly speaking, the coefficient of $\mathsf{E}[\Delta p_{t+1} \mid \mathcal{I}_t]$ suggests that a backward solution is appropriate. But more importantly the estimated NPCM once more appears to be a modified random walk model. We also checked the stability of the key parameters of the model by rolling regressions with a fixed window of 85 observations. Figure 7.2 shows that the sample dependency is quite pronounced in the case of Norway.

Next, we define an equilibrium correction term from the results in Bårdsen et al. (2003) and use that variable as the additional instrument, $z_{2,t}$:

$$ecmp_t = p_t - 0.6(w_t - a_t + \tau 1_t) - 0.4 pi_t + 0.5\tau 3_t.$$

The results, using GMM, are

$$\Delta p_t = \underset{(0.125)}{-0.02}\,\Delta p_{t+1} + \underset{(0.025)}{0.04}\,ws_t - \underset{(0.017)}{0.06}\,\Delta pi_t - \underset{(0.020)}{0.10}\,ecmp_{t-1}$$
$$+ \text{dummies}$$

$$\chi^2_{\mathrm{J}}(10) = 12.78[0.24],$$

showing that the implication of the NPCM is refuted by the finding of (1) a highly significant (price) equilibrium correction term defined by an existing study, and (2) the change in the estimated coefficient of Δp_{t+1}, from 1.06 and statistical significance, to -0.02 and no statistical significance.

7.6 Conclusions

Earlier researchers of the NPCM have concluded that the NPCM represents valuable insight into the driving forces of inflation dynamics. Our evaluation gives completely different results. In particular we show that by including variables from the list of instruments as explanatory variables, a statistically adequate model for the Euro area is obtained. In this respecified model, the forward term vanishes, and the Euro area 'inflation equation' can be reinterpreted as a conventional price markup equation. Encompassing implies that a model should be able to explain the results of alternative specifications. In many countries, empirical inflation dynamics is a well researched area, so studies exist that any new model should be evaluated against. Applying the encompassing principle to the NPCM models of United Kingdom inflation as well as Norwegian inflation, leaves no room for the NPCM. The conclusion is that economists should not accept the NPCM too readily.

On the constructive side, our analysis shows that the NPCM can be seen as an equilibrium-correction model augmented by a forward term. This means that although our conclusion refutes the NPCM hypothesis as presently implemented, this does not preclude that forward expectations terms could be found to play a role in explaining inflation dynamics within statistically well-specified models, using the procedures for testing forward terms.

8

Money and inflation

The role of money in the inflation process is an old issue in macro-economics, yet money plays no essential part in the models appearing up to and including Chapter 7 of the book. In this chapter, we explore the relevance of monetary aggregates as explanatory variables for inflation. First, we derive money demand functions for the Euro area and for Norway, and investigate whether these functions can be interpreted as inverted inflation equations. Second, we make a survey of inflation models that have been used in the recent past to analyse Euro area data. Moreover, we evaluate the models' statistical properties and make forecast comparisons. Finally, we make a similar evaluation and comparison of Norwegian inflation models. The P*-model, which emphasises disequilibria of monetary aggregates as the main driving force behind inflation, plays an important part in both cases. For the case of Norway, we also test an inflation equation—derived as the reduced form of the dynamic Incomplete Competition Model—for neglected monetary effects.

8.1 Introduction

The monetarist view of inflation—that inflation is always and everywhere a monetary phenomenon (Friedman 1963, p. 17)—runs contrary to the inflation models we have considered in the preceding chapters. Despite the notable differences that exist between them, they all reflect the view that inflation is best understood as reflecting imbalances in product and labour markets. This view is inconsistent with a simple quantity theory of inflation, but not with having excess demand for money as a source of inflation pressure.

In Section 8.2, we review briefly some results from the theory of money demand and show that this theory forms the basis for empirically stable money demand functions for the Euro area (Section 8.3) as well as for Norway (Section 8.4). Using criteria formulated by Hendry and Ericsson (1991), we evaluate the claim that these stable money demand functions in reality are inverted inflation equations.

In Section 8.5, we survey models of inflation which have been recently used in the literature to explain Euro-area inflation. A reduced-form inflation equation, derived from the wage–price block of the macroeconometric model area wide model (AWM) of the European Central Bank, forms a baseline for comparing competing models of inflation. The P*-model of inflation suggested in Hallman *et al.* (1991) is one serious contender. The P*-model specifies a direct effect from the lagged price gap, defined as the lagged price level minus the long-run equilibrium price level which is implied by a long-run quantity equation. Trecroci and Vega (2002) and Gerlach and Svensson (2003) find support for this model formulation on Euro-area data, as do Tödter and Reimers (1994) on German data.[1] The above models are estimated and evaluated against each other within a common framework in Section 8.6, along with a hybrid New Keynesian Phillips curve model and the inflation equation derived from a version of the Incomplete Competition Model (ICM).

In Section 8.7, we present a reduced-form representation of the ICM inflation model for Norway, which (much in the same way as the AWM inflation model for the Euro-area data) forms a benchmark against which we evaluate several variants of the P*-model, a hybrid Phillips curve and the inverted money demand inflation equation of Section 8.4. The focus remains on monetary aggregates: in Section 8.7.4 we test the robustness of the ICM inflation model for neglected monetary effects based on a sequence of omitted variable tests. Section 8.7.7 concludes and compares the findings on the two data sets.

8.2   Models of money demand

8.2.1   The velocity of circulation

Models of the velocity of circulation are derived from the 'equation of exchange' identity often associated with the quantity theory of money (Fisher 1911) which on logarithmic form can be written:

$$m_t + v_t = p_t + y_t, \tag{8.1}$$

where m_t is money supply, v_t is money velocity, y_t is a scaling variable (e.g. real output), and p_t is the price level. We define the inverse velocity of money as $m_t - y_t - p_t = -v_t$ (small letters denote variables in logarithms). A simple

---

[1] In other studies, such direct effects from money aggregates (or measures derived from them) are rejected, cf. for example, de Grauwe and Polan (2001) who argue that the seemingly strong link between inflation and the growth rate of money is almost wholly due to the presence of high (or hyper-) inflation countries in the sample. Similarly, Estrella and Mishkin (1997) reject the idea that broad money is useful as an information variable and provide a good signal of the stance of monetary policy, based on their analysis of United States and German data.

theory of money demand is obtained by adding the assumption that the velocity is constant, implying that the corresponding long-run money demand relationship is a linear function of the scaling variable y_t, and the price level p_t. The stochastic specification can be written as:

$$m_t - y_t - p_t = \gamma_0 + \varepsilon_t \qquad (8.2)$$

assuming that $E[\varepsilon_t|\mathcal{I}_{t-1}] = 0$ on some appropriate information set $\mathcal{I}_{t-1}$. The price homogeneity restriction in (8.2) implies that real money, $(m_t - p_t)$, will be determined by the scaling variable, y_t, which has a unit elasticity. The constancy of γ_0 is, however, pervasively rejected in the empirical literature, cf. for example, Rasche (1987) who discusses the trending behaviour of velocity v_t. Bordo and Jonung (1990) and Siklos (1993) analyse the properties of the velocity in a '100 years perspective' and they explain the changes in velocity over this period by institutional changes, comparing evidence from several countries. Klovland (1983) has analysed the demand for money in Norway during the period from 1867 to 1980, and he argues along similar lines that institutional and structural factors such as the expansion of the banking sector and the increased degree of financial sophistication seems to be linked with the variations in velocity across this period.

Bomhoff (1991) has proposed a model where the inverse velocity is time dependent, that is, $-v_t = \gamma_t$, and he applies the Kalman filter to model the velocity changes as a function of a shift parameter, a deterministic trend, and some relevant interest rate variable R_t, with the additional assumption that there are stochastic shocks in the shift and trend parameters. This allows for a very flexible time-series representation of velocity, which can be shown to incorporate the class of equilibrium-correction models which we will discuss later. A maintained hypothesis in the velocity models is that the long-run income elasticity is one. This hypothesis has been challenged from a theoretical perspective, for example, in 'inventory models' (Baumol 1952; Tobin 1956) and in 'buffer stock models' (Miller and Orr 1966; Akerlof 1979). The empirical evidence is such that this issue remains an open empirical question. A commonly used generalisation of the velocity model yields a money demand function of the following type:

$$m_t = f_m(p_t, y_t, R_t, \Delta p_t), \qquad (8.3)$$

where the model is augmented with the overall inflation rate Δp_t, which measures the return to holding goods, and the yields on financial assets, represented by a vector of interest rates, R_t.

The choice of explanatory variables in equations like (8.3) varies a great deal between different theoretical and empirical studies. A typical mainstream relationship, which is often found in empirical studies of long-run real money balances, is the following semi-logarithmic specification:

$$m_t - p_t = \gamma_y y_t + \gamma_R R_t + \gamma_{\Delta p} \Delta p_t + \text{constant}. \qquad (8.4)$$

8.2.2   Dynamic models

The equilibrium-correction model provides a flexible dynamic specification for the money demand function. This entails explicit and separate modelling of the short-run dynamic specification and the long-run cointegrating relationship for m_t, which allows us to distinguish between shocks which will only cause temporary effects on money holdings and shocks with persistent long-run effects. Furthermore, the economic variables which exert the strongest short-run effects in money holdings, say, in the first quarters following the shock, need not be the same as the variables which drive money holdings in the long run. This is consistent with the models of Miller and Orr (1966) and Akerlof (1979), who study optimal inventories when changes in the cash balances are stochastic, leading to (s, S) target/threshold models. In these models, the short-run elasticity with respect to income and interest rates can be negligible as long as targets and thresholds remain constant, while the long-run elasticities follow from the long-run cointegrating relationship.

A simple equilibrium correction specification for m_t using the vector z_t as explanatory variables is

$$\Delta m_t = \sum_{i=1}^{q-1} \delta_i \Delta m_{t-i} + \sum_{i=0}^{q-1} \gamma_i' \Delta z_{t-i} + \alpha_m (m_{t-1} - \beta' z_{t-1}) + \varepsilon_t, \qquad (8.5)$$

$$\varepsilon_t \sim \text{i.i.d.}(0, \sigma^2).$$

The parameter α_m captures a feedback effect on the change in money holdings, Δm_t, from the lagged deviation from the long-run target money holdings, $(m - m^*)_{t-1}$. The target m_t^* is defined as a linear function of the forcing variables z_t, that is, as $m_t^* = \beta' z_t$. Compared to a partial adjustment model, the equilibrium-correction model allows for richer dynamics in terms of more flexible dynamic responses in money balances to shocks in the forcing variables.

Equation (8.5) can be obtained from an unrestricted Autoregressive Distributed Lag model in the levels of the variables by imposing the appropriate set of equilibrium-correction restrictions. The duality between equilibrium correction and cointegration (Engle and Granger 1987) makes the equilibrium-correction specification (8.5) an attractive choice for the modelling of non-stationary time-series, for example, variables which are I(1). If the forcing variables z_t are weakly exogenous with respect to the parameters in the money demand equation, there will be no loss of information in modelling the change in money holdings Δm_t in the context of a conditional single-equation model like (8.5).

8.2.3   Inverted money demand equations

In reviewing the lineages of the Phillips curve in Chapter 4, we saw that the relationship between wage growth and the level of economic activity (or unemployment) has a prominent position in the new classical macroeconomics literature; see, for example, Lucas and Rapping (1969, 1970) and Lucas (1972). Two issues were in focus. First, according to this literature, the causality of Phillips' original model is reversed: if a correlation between inflation and unemployment exists at all, the causality runs from inflation to the level of activity and unemployment. Since price and wage growth are then determined from outside the Phillips curve, the rate of unemployment would typically be explained by the rate of wage growth (and/or inflation). Second, given this inversion of the Phillips curve, the determination of the price level in Lucas and Rapping's model is based on a quantity theory relationship, where they condition on an exogenous or autonomously determined money stock.

Later we investigate the relationship between money and inflation from this monetarist perspective. Obviously, a causal relationship between money and inflation can be analysed from several angles. The most direct approach would be to model inflation as a function of some monetary aggregates. However, we shall first look to estimated versions of the money demand functions we introduced earlier, in order to see if they can be interpreted as inverted equations for price growth. This amounts to inverting the money demand relationship to obtain a relationship for price growth in the same way as the Phillips curve was inverted to explain unemployment earlier.

In their study of money demand in the United Kingdom and the United States, Hendry and Ericsson (1991) estimate a money demand relationship for the United Kingdom under the assumption that it represents a conditional model for money growth with output, prices, and interest rates as the main explanatory factors. The model is well specified with stable parameters. Inversion of this model to an inflation equation yields a non-constant representation, with several signs of model mis-specification. Noting that the price level p_t is included among the explanatory variables in z_t, Hendry and Ericsson (1991) estimate an inverted money demand relationship of the type

$$\Delta p_t = \hat{\beta}_0 \Delta m_t + \hat{\beta}_1 \Delta m_{t-1} + \hat{\xi}_0' \Delta z_t + \hat{\xi}_1' \Delta z_{t-1} + \hat{\kappa}_m (m_{t-1} - \beta' z_{t-1}) + \hat{\varepsilon}_t.$$

(8.6)

In the following section we repeat this exercise: first, on data for the Euro area and second, on data for Norway.

8.3 Monetary analysis of Euro-area data

8.3.1 Money demand in the Euro area 1980–97

In this section, we establish that money demand in the Euro area can be modelled with a simple equilibrium correction model. We base the empirical results on the work by Coenen and Vega (2001) who estimate the aggregate demand for broad money in the Euro area. In Table 8.1 we report a model which is a close approximation to their preferred specification for the quarterly growth

Table 8.1

Empirical model for $\Delta(m - p)_t$ in the Euro area based on
Coenen and Vega (2001)

$$\Delta\widehat{(m - p)}_t = \underset{(0.067)}{-0.74} + \underset{(0.040)}{0.08\Delta\Delta y_t} + \underset{(0.074)}{0.19}\; \frac{\Delta\mathrm{RS}_t + \Delta\mathrm{RS}_{t-1}}{2}$$

$$\underset{(0.08)}{-\,0.36\Delta\mathrm{RL}_{t-1}} \underset{(0.050)}{-\,0.53}\; \frac{\Delta\mathrm{pan}_t + \Delta\mathrm{pan}_{t-1}}{2} \underset{(0.002)}{-\,0.01\,\mathrm{dum}86_t}$$

$$\underset{(0.012)}{-\,0.14}\,[(m - p) - 1.140y + 1.462\Delta\mathrm{pan} + 0.820(\mathrm{RL} - \mathrm{RS})]_{t-2}$$

$$\hat{\sigma} = 0.23\%$$

Diagnostic tests

$$
\begin{aligned}
\mathsf{F}_{\mathsf{AR}}(5, 55) &= 0.97[0.44] \\
\mathsf{F}_{\mathsf{ARCH}}(4, 52) &= 0.29[0.89] \\
\chi^2_{\mathsf{normality}}(2) &= 0.82[0.66] \\
\mathsf{F}_{\mathsf{HET}_{\mathsf{x}^2}}(12, 47) &= 0.65[0.79] \\
\mathsf{F}_{\mathsf{HET}_{\mathsf{x}_i \mathsf{x}_j}}(24, 35) &= 0.59[0.91] \\
\mathsf{F}_{\mathsf{RESET}}(1, 59) &= 0.16[0.69]
\end{aligned}
$$

Note: The sample is 1980(4)–1997(2), quarterly data.

rate in aggregated real broad ($M3$) money holdings, $\Delta(m - p)_t$, over the original sample period 1980(4)–1997(2). We condition on the estimated long-run real money demand relationship (8.7) in Coenen and Vega (2001):

$$(m - p)_t = 1.14y_t - 1.462\Delta\mathrm{pan} - 0.820(\mathrm{RL} - \mathrm{RS})_t, \qquad (8.7)$$

where $(m - p)_t$ denotes (log of) real $M3$ money holdings, y_t is (log of) real GDP, RS_t is the short interest rate, RL_t is the long interest rate, and $\Delta\mathrm{pan}_t$ denotes the annualised quarterly change in the GDP deflator.[2]

The money demand relationship for the Euro area appears to be fairly well specified with stable parameters as indicated by the plot of recursive residuals and Chow tests in Figure 8.1. The question is: can this model be turned into a model of inflation by inversion?

8.3.2   Inversion may lead to forecast failure

Assuming that the monetary authorities can control the stock of money balances in the economy, it would be appealing if one could obtain a model of inflation from the established money demand relationship above. We follow Hendry and Ericsson (1991) and invert the empirical money demand relationship in Table 8.1 to a model for quarterly inflation Δp_t. Since the model

---

[2] The Euro-area data are seasonally adjusted.

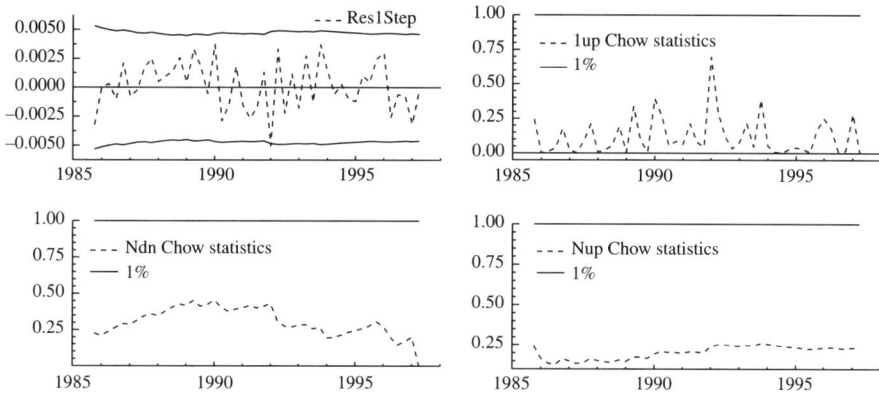

Figure 8.1. Estimation of money demand in the Euro area,
1985(4)–1997(2)—recursive residuals and Chow tests

in Table 8.1 explains quarterly changes in real money holdings, we can simply move Δm_t to the right-hand side of the equation and re-estimate the relationship over the selected period 1980(1)–1992(4), saving 20 observations for post-sample forecasts.

Recalling the empirical findings of Hendry and Ericsson (1991) and the fact that we started out with a money demand relationship with stable parameters over this period, one might expect to see a badly specified inflation relationship with massive evidence of model mis-specification including clear evidence of parameter non-constancy—at least enough to indicate that there is little to learn about the inflation process from this relationship.

The results in Table 8.2 are surprising: it turns out that the inverted relationship is fairly stable over the selected sample period as well, and it is well specified according to the tests reported in the table. Figure 8.2 shows that the inflation model has stable parameters and, except in one quarter (1987(2)), recursive Chow tests indicate that the model is reasonably constant. So, the non-invertibility of the money demand relationship reported in Hendry and Ericsson (1991) does not seem to apply for the Euro area in this period. The model has significantly positive effects on inflation from real money growth and from changes in output growth, $\Delta\Delta y_t$. Also, lagged changes in long-term interest rates have a positive effect on inflation, while changes in short interest rates have a negative impact.

The picture changes completely when we test the model outside the selected sample: Figure 8.3 shows one-step forecasts from this model over the period from 1993(1) to 1998(4). The model seems to provide a textbook illustration of forecast failure.[3] The forecast failure is caused by parameter instability which

_____

[3] See Clements and Hendry (1998) and Chapter 11.

Table 8.2
Inverted model for Δp_t in the Euro area based on Coenen and Vega (2001)

$$\widehat{\Delta p_t} = \underset{(0.059)}{0.96} + \underset{(0.074)}{0.46\Delta m_t} + \underset{(0.031)}{0.003\Delta\Delta y_t} - \underset{(0.060)}{0.17} \frac{\Delta RS_t + \Delta RS_{t-1}}{2}$$

$$+ \underset{(0.073)}{0.30\Delta RL_{t-1}} + \underset{(0.039)}{0.46} \frac{\Delta pan_t + \Delta pan_{t-1}}{2} + \underset{(0.002)}{0.004\Delta dum86_t}$$

$$+ \underset{(0.011)}{0.17}[(m-p) - 1.140y + 1.462\Delta pan + 0.820(RL - RS)]_{t-2}$$

$$\hat{\sigma} = 0.16\%$$

Diagnostic tests

$$\begin{aligned}
F_{AR(1-4)}(4,37) &= 1.02[0.41] \\
F_{ARCH(1-4)}(4,33) &= 0.39[0.81] \\
\chi^2_{normality}(2) &= 0.53[0.77] \\
F_{HET\chi^2}(14,26) &= 1.12[0.38] \\
F_{RESET}(1,40) &= 5.96[0.02]^*
\end{aligned}$$

Note: The sample is 1980(4)–1992(4), quarterly data.

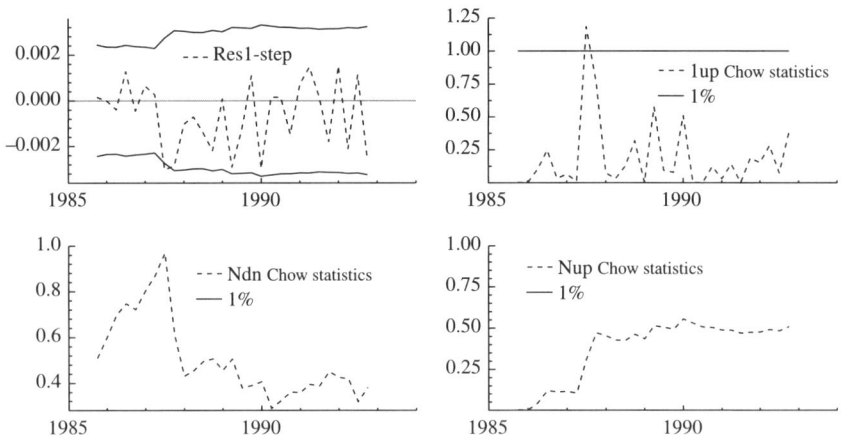

Figure 8.2. Inverted money demand equation for the Euro area
1985(4)–1992(4)—recursive residuals and Chow tests

takes the form of a structural break as the Euro-area inflation rate starts to fall in the early 1990s. This is demonstrated by the plots of recursive residuals and Chow tests in Figure 8.4 which are obtained when we re-estimate the model over the entire sample until 1997(2). The sample evidence for the entire period thus shows that while we find a constant empirical relationship for money conditional on prices, the inverse relationship is all but stable and we have established non-invertibility. Hence, as pointed out in Hoover (1991), these results indicate that

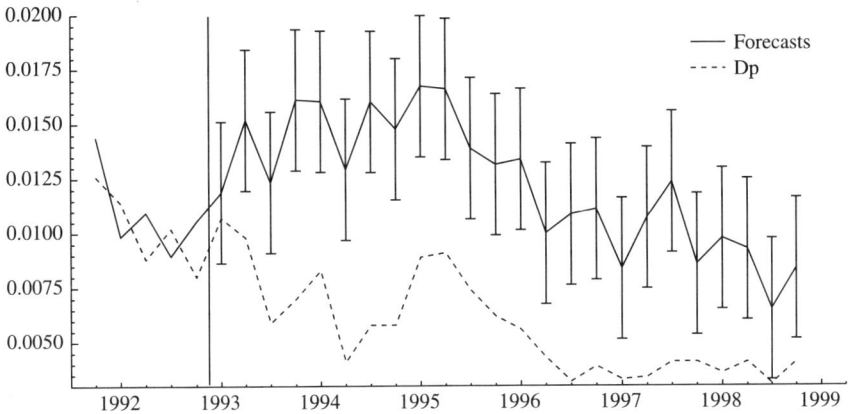

Figure 8.3. *Post-sample* forecast failure when the inverted money demand equation for the Euro area is used to forecast inflation 1993(1) to 1998(4)

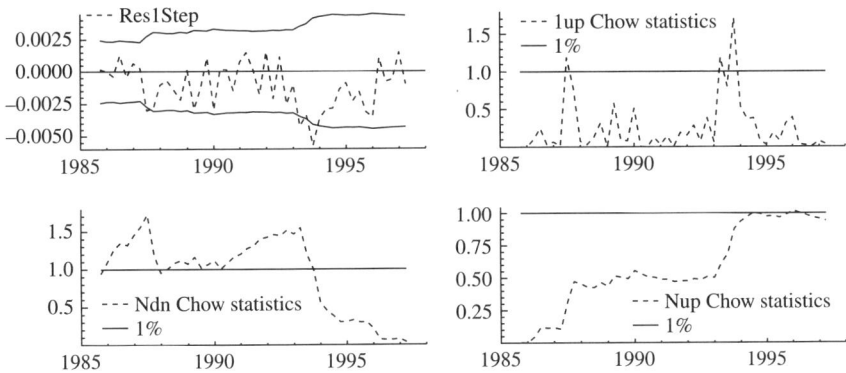

Figure 8.4. Instabilities in the inverted money demand equation for the Euro area after 1993—recursive residuals and Chow tests

causality runs from prices to money rather than from money to prices also in the case of the Euro area.

8.4 Monetary analysis of Norwegian data

8.4.1 Money demand in Norway—revised and extended data

The demand for broad money in Norway has previously been analysed by Eitrheim (1998) using seasonally unadjusted data from 1969(1) to 1993(4). In that study a cointegrating relationship for money was derived jointly with cointegrating relationships for wages and consumer prices, and the analysis showed that in the long run, real money balances adjust dynamically to absorb shocks in the real GDP level and the relative price of financial assets (the yield spread) and the relative price of goods (the own real interest rate). In the short run, money balances were also affected by shocks in the exchange rate and private wealth. Evidence for prices being weakly exogenous was also found with respect to the parameters in the money demand relationship, which by implication support the interpretation that it is money holdings that adjust endogenously to changes in the forcing variables in the long run.

In the empirical models in this section we condition on the long-run cointegrating relationship for money balances found in Eitrheim (1998). Assuming homogeneity of degree one in the price level, this relationship can be formulated as:

$$m_t - p_t = \beta_y y_t + \beta_{rbt}(\text{RB}_t - \text{RT}_t) + \beta_{rtd4p}(\text{RT}_t - \Delta_4 p_t),$$

where y_t is (log of) real output (GDP), p_t is the consumer price index, hence $\Delta_4 p_t$ is the annual rate of headline inflation, RB_t is the yield on assets outside money (government bonds with six years maturity), and RT_t is the own interest rate on money (the time deposits rate). The yield spread $(\text{RB}_t - \text{RT}_t)$ represents the nominal opportunity cost of holding money relative to other financial assets, while the 'own real interest rate' $(\text{RT}_t - \Delta_4 p_t)$ can be interpreted as a measure of the return on money relative to consumer goods.

This long-run equation is grafted into a simplified equilibrium correction model for quarterly money growth with only one lag, which means that (8.5) can be written

$$\Delta m_t = \delta_1 \Delta m_{t-1} + \gamma_0' \Delta z_t + \gamma_1' \Delta z_{t-1} + \alpha_m(m_{t-1} - \beta' z_{t-1}) + \varepsilon_t,$$
$$\varepsilon_t \sim \text{i.i.d.}(0, \sigma^2). \tag{8.8}$$

Note that since $\Delta_4 m_t = \Delta m_t + \Delta_3 m_{t-1}$ we arrive at a relationship for annual money growth $\Delta_4 m_t$ by adding $\Delta_3 m_{t-1}$ to both sides of (8.8). If the coefficient on $\Delta_3 m_{t-1}$ is close to one, the annual representation is a simple isomorphic transformation of a similar quarterly model.

Re-estimating a money demand model for Norway   Compared to Eitrheim (1998), we report results for seven years of new observations. Also, since then, Norwegian National Accounts data for the entire sample period have been substantially revised in order to comply with new international standards, and there has been a major revision in the Monetary Statistics data for broad money holdings.[4] One of the changes in the new definition of broad money is

---

[4] Concepts and definitions used by Norges Bank to compile Monetary Statistics are now in line with the guidelines in the Monetary and Financial Statistics Manual (MFSM) of the International Monetary Fund (IMF).

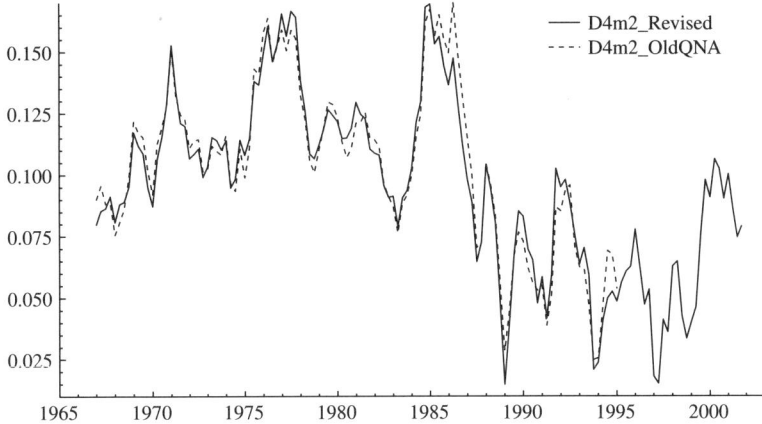

Figure 8.5. Money demand (1969(1)–2001(1))—revised (solid line) and old (dotted line) observations of the percentage growth in $M2$ over four quarters

that unused overdraft facilities and building loans are now excluded. Figure 8.5 shows the revised data along with the data which were analysed in Eitrheim (1998). Despite the exclusion of unused overdraft facilities and building loans, it does not seem that the pattern of annual growth rates in the monetary aggregate has been significantly altered.

Table 8.3 shows the results from re-estimating the model specification in Eitrheim (1998). Despite the revisions of the data for the money and output variables, the old relationship seems to hold up reasonably well on the extended data set. In the short run, money growth is influenced by shocks to the exchange rate (e_t is (log of) the nominal exchange rate of Norwegian Kroner) and by changes in nominal household wealth (wh_t). We have also included a dummy variable for the release of tax-exempted savings deposits, $M2D914 = 1$ in 1991(4) else 0, as well as a variable, $S4_t * \Delta_4 RT_t$, which is intended to pick up the effect from changes in accrued interest earnings, which are capitalised at the end of each year.

Some of the coefficients lose their previous significance, but the re-estimated model passes all mis-specification tests reported in Table 8.3. The estimated $\hat{\sigma}$ is 1.13% compared with 0.93% in Eitrheim (1998), so the data fit has deteriorated. From recursive plots (not reported here) of the parameter estimates of the short-run effect from shocks in exchange rates ($\Delta\Delta e_t$), it is possible to trace instabilities which may be linked to changes occurring in the exchange rate system in Norway after 1997. After leaving a fixed exchange rate system in 1992 in favour of a managed float, the Norwegian Krone has seen several episodes with more or less free float following speculative attacks, notably in 1997 and 1998. It is not surprising if the currency substitution effect on money holdings did change on those occasions.

Table 8.3

Re-estimating the money demand model for Norway in Eitrheim (1998) on revised and extended data (seven years of new observations)

$$\widehat{\Delta_4 m_t} = \underset{(0.0446)}{-0.0449}(\Delta\Delta e_{t-1} + \Delta\Delta e_{t-3}) + \underset{(0.0393)}{0.1383}\Delta wh_{t-2} + \underset{(0.1296)}{1.0825}\Delta_3 m_{t-1}$$

$$+ \underset{(0.1701)}{0.0257}(\Delta m_{t-1} + \Delta m_{t-3}) - \underset{(0.0769)}{0.3107}(\Delta m_{t-2} - \Delta m_{t-4})$$

$$- \underset{(0.0197)}{0.1026}(m_{t-1} - p_{t-1} - 0.8y_{t-1} + 2.25(\text{RB} - \text{RM})_{t-1}$$

$$- (\text{RM} - \Delta_4 p)_{t-1}) + \underset{(0.0120)}{0.0278}M2D914_t + \underset{(0.1623)}{0.1505}S4 * \Delta_4\text{RT}_t$$

$$+ \underset{(0.0037)}{0.0186}(S1_t + S3_t) - \underset{(0.0718)}{0.3756}$$

$$\hat{\sigma} = 1.13\%$$

Diagnostic tests

$$F_{\text{AR}(1-5)}(5, 114) \quad = 1.0610[0.3858]$$
$$F_{\text{ARCH}(1-4)}(4, 111) = 1.7918[0.1355]$$
$$\chi^2_{\text{normality}}(2) \qquad = 0.5735[0.7507]$$
$$F_{\text{HET}x^2}(16, 102) \quad = 1.4379[0.1391]$$
$$F_{\text{RESET}}(1, 118) \qquad = 0.6260[0.4304]$$

Note: The sample is 1969(1)–2001(1), quarterly data.
Long run: $m_t = p_t + 0.8y_t - 2.25(\text{RB} - \text{RM})_t + (\text{RM} - \Delta_4 p)_t + ecmmd_t$.

An improved model for the period 1969(1)–2001(1) It turns out to be possible to achieve a slight improvement on the re-estimated model in Table 8.3. Table 8.4 shows a model where the short-run dynamics are simplified. We have introduced a step dummy Sdum97Q1 to pick up a permanent shift as of 1997(1) in the coefficient for exchange rate shocks. Finally, based on experiments with single equation models with unrestricted variables in levels, we have simplified the long-run relationship for money omitting the real interest rate $(\text{RM} - \Delta_4 p)_t$, but keeping the yield spread $(\text{RB} - \text{RM})_t$. Insignificant parameters on the revised and extended dataset are omitted from the model. The estimated $\hat{\sigma}$ for the improved model is 1.09%. We have also compared parameter constancy forecast tests for the re-estimated and improved money demand models over the period 1995(1) to 2001(1).[5] The tests for forecast stability in the re-estimated model are, $\chi^2_{\text{forecast}}(25) = 63.211[0.0000]^{**}$ and $F_{\text{chow}}(25, 94) = 1.9332[0.0123]^*$ while the corresponding values in the improved

[5] We started the forecast comparison in 1995(1) since the old national accounts data end in 1994(4). The sample period in Eitrheim (1998) ends in 1994(4), but the qualitative results would be the same if we start the forecasting exercise in 1994(1).

Table 8.4
Improved model for annual money growth, $\Delta_4 m$, for Norway

$$\widehat{\Delta_4 m_t} = -0.0800(\Delta\Delta e_{t-1} + \Delta\Delta e_{t-3})$$
$$\phantom{\widehat{\Delta_4 m_t} = }(0.0489)$$

$$+ \; 0.1493 \text{Sdum97Q1}(\Delta\Delta e_{t-1} + \Delta\Delta e_{t-3}) + \; 0.1145\Delta wh_{t-2}$$
$$(0.0886) \phantom{\text{Sdum97Q1}(\Delta\Delta e_{t-1} + \Delta\Delta e_{t-3}) + \;}(0.0367)$$

$$+ \; 1.1134\Delta_3 m_{t-1} - 0.3235(\Delta m_{t-2} - \Delta m_{t-4})$$
$$(0.0394) \phantom{\Delta_3 m_{t-1} - }(0.0464)$$

$$-0.1084(m_{t-1} - p_{t-1} - 0.9y_{t-1} + 2.5(\text{RB}_{t-1} - \text{RM}_{t-1}))$$
$$(0.0186)$$

$$+ \; 0.0300 M2D914_t + \; 0.0175(S1_t + S3_t) - 0.5272$$
$$(0.0111) (0.0021) (0.0898)$$

$$\hat{\sigma} = 1.09\%$$

Diagnostic tests

$$\begin{aligned}
F_{AR(1-5)}(5, 115) &= 0.7026[0.6226] \\
F_{ARCH(1-4)}(4, 112) &= 0.5574[0.6940] \\
\chi^2_{\text{normality}}(2) &= 2.4736[0.2903] \\
F_{HET\chi^2}(14, 105) &= 1.6997[0.0664] \\
F_{RESET}(1, 119) &= 0.2022[0.6538]
\end{aligned}$$

Note: The sample is 1969(1)–2001(1), quarterly data.
Long run: $m_t = p_t + 0.9y_t - 2.5(\text{RB} - \text{RM})_t + ecmmd_t$.

model are $\chi^2_{\text{forecast}}(25) = 36.293[0.0673]$ and $F_{\text{chow}}(25, 95) = 1.3452[0.1547]$.[6]
Hence, the parameter forecast stability has been improved in the revised money demand model in Table 8.4. In Sections 8.7.3 and 8.7.4, we use the equilibrium correction term, $ecmmd_t$, of Table 8.4 to test for neglected monetary effects in models explaining inflation in Norway.

8.4.2 Monetary effects in the inflation equation?

We find no effect of inflation in the money demand equations for Norway. Hence it does not make sense to interpret the money demand functions as inverted inflation equations. We have, however, experimented with a model where we consider money in real terms $(m_t - p_t)$, the real interest rate on money and the yield spread as potential explanatory variables for inflation. These are the variables that enter the cointegrating relationship of the money demand equation in Eitrheim (1998), cf. Table 8.3.

This gives us a model which has several aspects in common with the inverted money demand relationship for the Euro area in Section 8.3.2. In addition to

---

[6] Test statistics marked * and ** indicate significance at the 5% and 1% level.

Table 8.5

The MdInv model of inflation, including variables (in levels) from
the money demand relationship

$$\widehat{\Delta_4 p_t} = \underset{(0.0350)}{1.1021\Delta_3 p_{t-1}} + \underset{(0.0696)}{0.2211\Delta p_{t-2}} + \underset{(0.0109)}{0.0436\Delta pe_t}$$

$$+ \underset{(0.0115)}{0.0587\Delta y_{t-2}} + \underset{(0.0170)}{0.0272\Delta wh_{t-3}} \underset{(0.0103)}{-0.0208(m_{t-1} - p_{t-1})}$$

$$+ \underset{(0.0097)}{0.0155\,y_{t-1}} - \underset{(0.0202)}{0.0099(\mathrm{RT}_{t-1} - \Delta_4 p_{t-1})} - \underset{(0.0422)}{0.0262(\mathrm{RB}_{t-1} - \mathrm{RM}_{t-1})}$$

$$- \underset{(0.0010)}{0.0120\,P\mathrm{dum}_t} - \underset{(0.0577)}{0.0586}$$

$$\hat{\sigma} = 0.45\%$$

Diagnostics tests

$$
\begin{aligned}
F_{\mathrm{AR}(1-5)}(5, 113) &= 1.6482[0.1530]\\
F_{\mathrm{ARCH}(1-4)}(4, 110) &= 1.2934[0.2771]\\
\chi^2_{\mathrm{normality}}(2) &= 7.3731[0.0251]^*\\
F_{\mathrm{HET}_{x^2}}(20, 97) &= 2.6762[0.0007]^{**}\\
F_{\mathrm{HET}_{x_i x_j}}(65, 52) &= 1.5171[0.0606]\\
F_{\mathrm{RESET}}(1, 117) &= 7.6875[0.0065]^{**}
\end{aligned}
$$

Note: The sample is 1969(1)–2001(1), quarterly data.

the monetary variables we have included short-run effects of changes in energy
prices Δpe_t, changes in output, and a composite dummy variable $P\mathrm{dum}_t$ which
inter alia captures the effect of income policies in the late 1970s and 1980s. Even
though the model fits the data reasonably well with an estimated standard error
of $\hat{\sigma} = 0.45\%$, the model does nonetheless fail in several of the mis-specification
tests, reported in Table 8.5. The model captures the persistence in inflation
through the included lags in price growth, and the effects from Δpe_t and $P\mathrm{dum}_t$
are reasonable, but the effects from the included monetary variables are more
difficult to interpret. First, the results in Table 8.5 indicate that the effect
on inflation of the real interest rate is insignificant. Assuming that the real
interest rate is stationary, it is the other monetary variables from the long-run
cointegrating relationship (i.e. the money demand equation in Eitrheim (1998))
that represent an excess money effect on inflation. A priori one would expect
a positive rather than a negative effect from these variables on inflation.

Second, as will be clear when we present the forecasting properties of this
relationship along with other inflation models for the Norwegian economy, that
this model—which we have dubbed MdInv—suffers from severe parameter non-
constancies in the early 1980s and around 1994. This is a likely explanation why
the model—when estimated on data up to 1990(4)—badly mispredicts inflation

over the period 1991(1)–2000(4) (see Section 8.7.6). Thus, we conclude as we did for the Euro-area data: if we try to construct an inverted money demand relationship for Norway, forming an inflation equation based on the information set used in the money demand models in the preceding section, we find evidence of severe parameter non-constancy and resulting forecast failure.

8.5 Inflation models for the Euro area

In Section 8.3 we found that an inverted money demand function did not provide a sound basis for explaining inflation in the Euro area. Still, there may be a case for models of inflation that conceive of inflation primarily as a monetary phenomenon. In this section, we compare and evaluate four inflation models which have been used to analyse data for the Euro area. These include the P*-model, which relates the steady-state of the price level to the quantity theory of money, a hybrid New Keynesian Phillips curve model (NPCM) of inflation (see Chapter 7) and two reduced form inflation equations: one derived from the dynamic version of the Incomplete Competition Model (ICM) we developed in Chapters 5 and 6, and the other from the wage–price block of the AWM of the European Central Bank.

Many researchers addressing inflation in the Euro area have opted for approaches like the P*-model or the NPCM, which either amounts to modelling inflation as a single equation or as part of very small systems. By contrast, the price block of the AWM, as described in Fagan *et al.* (2001), is defined within a full-blown macroeconometric model for the Euro area, even though the equations for wage growth and inflation are estimated by single equation methods. Moreover, the AWM is providing the most commonly used data set for the Euro area, and hence it is an obvious benchmark and point of reference for the comparison.[7]

In the following we shall give an outline of the wage–price block of the AWM (Section 8.5.1), brief reminders of the ICM (Section 8.5.2) and the NPCM (Section 8.5.3) which are described elsewhere in this book and, finally, a more detailed presentation of the P*-model (Section 8.5.4).

---

[7] The aggregated data underlying AWM are constructed by using a set of fixed purchasing power parity (PPP) exchange rates between the national currencies, calculated for the year 1995, to convert all series to a common currency (i.e. Euro). An alternative aggregation method has been suggested by Beyer *et al.* (2001) (see also Beyer *et al.* (2000)). They argue that aggregation across individual countries is problematic because of past exchange rate changes. Hence, a more appropriate method, which aggregates exactly when exchange rates are fixed, consists in aggregating weighted within-country *growth rates* to obtain euro-zone growth rates and cumulating this euro-zone growth rate to obtain aggregated levels. The aggregate of the implicit deflator price index coincides with the implicit deflator obtained from the aggregated nominal and real data.

8.5.1   The wage–price block of the Area Wide Model

The unique feature of the AWM is that it treats the Euro area as a single economy. Since the Euro was introduced only on 1 January 1999 and the information
set underlying the estimation of the model—as documented in Fagan *et al.*
(2001)—is a constructed data set covering the period 1970(1)–1998(4), the
counterfactual nature of this modelling exercise is evident.

The AWM is used for forecasting purposes and the model has been specified
to ensure that a set of structural economic relationships holds in the long run.
It is constrained to be consistent with the neoclassical steady-state in which the
long-run output is determined via a production function by exogenous techno-
logical progress and the available factors of production, where the growth rate
of labour force is exogenous. Money is neutral in the long run and the model's
long-run properties is further pinned down by an exogenous NAIRU.

Our focus is on the modelling of inflation, which is modelled jointly with
wage growth in the AWM. Whereas the long-run equilibria are largely determined by a priori considerations through the output production function and
the exogenous growth rates in factor productivity, the labour force and the
NAIRU, the short run is modelled empirically as (single equation) equilibrium-
correction models. The empirical models are re-estimated in Jansen (2004) on
an extended data set (1970(1)–2000(4)) and the results do not deviate much
from those in Fagan *et al.* (2001); see appendix B in Jansen (2004).

Wages are modelled as a Phillips curve in levels, with wage growth depending on the change in productivity, current, and lagged inflation—in terms of
the consumption deflator p_t—and the deviation of the unemployment u_t from
its NAIRU level $\bar{u}_t$, that is, $(u_t - \bar{u}_t)$ defines the equilibrium-correction term,
$ecmw_t^{\mathrm{AWM}}$. Inflation and productivity growth enter with unit coefficients, so the
equation is expressed with the change in the wage share Δws_t, which equals
the change in real unit labour cost, $\Delta ulc_t - \Delta p_t$, as left-hand side variable.
ulc_t is nominal unit labour cost and, as before, natural logarithms of variables
are denoted by lower-case symbols.

The output price or GDP at factor costs, q_t, is a function of trend unit
labour costs, $\overline{ulc}_t$, both in the long run (levels) and the short run (changes).
The equilibrium-correction term equals $(q_t - (\overline{ulc}_t - (1-\beta)))$, where $(1-\beta)$ is the
elasticity of labour in the output production function, thus linking the long-run
real equilibrium to the theoretical steady-state. The markup is also influenced
by an output gap and import price inflation (Δpi_t) has short-run effects on
Δq_t. Finally, consumer price inflation (i.e. the consumption deflator) Δp_t is
determined by the GDP deflator at market prices, and import prices, both in
the short run and in the long run (with estimated weights equal to 0.94 and 0.06,
respectively). There is also a small effect of world market raw materials prices
in this equation. Noting that the GDP deflator at market prices by definition
equals GDP at factor prices corrected for the rate of indirect taxation $(q_t + t_t)$,
we find by substituting for q_t that the equilibrium correction term for Δp_t can

be written as

$$ecmp_t^{\text{AWM}} = p_t + 0.59 \cdot 0.94 - 0.94\overline{ulc}_t - 0.06pi_t - 0.94t_t. \qquad (8.9)$$

8.5.2 The Incomplete Competition Model

The dynamic version of the ICM is presented in Chapters 5 and 6 and an example of empirical estimation is discussed in greater detail within the framework of a small econometric model for Norway in Chapter 9 (Section 9.2). We shall therefore be brief in the outline of the ICM for the Euro area; details are given in Jansen (2004).

The econometric approach follows a stepwise procedure, where the outcome can be seen as a product of interpretation and formal testing: we first consider an information set of wages, prices, and an appropriate selection of conditioning variables like the output gap, unemployment, productivity, import prices, etc. It turns out that the data rejects the long-run restrictions from theory in this case. Only when we model the long-run steady-state equations with prices and unit labour costs as the endogenous variables do we find empirical support for the theory restrictions. The final outcome is steady-state equations of the following restricted form:

$$ulc_t = p_t - \varpi u_t, \qquad (8.10)$$
$$p_t = (1 - \phi)ulc_t + \phi pi_t + t3_t, \qquad (8.11)$$

where $t3_t$ is indirect taxes. We note that only two parameters, ϖ and ϕ, enter unrestrictedly in (8.10) and (8.11).

8.5.3 The New Keynesian Phillips Curve Model

Recall the definition in Chapter 7: the NPCM states that inflation is explained by expected inflation one period ahead $\mathsf{E}(\Delta p_{t+1} \mid \mathcal{I}_t)$, and excess demand or marginal costs x_t (e.g. the output gap, the unemployment rate, or the wage share in logs):

$$\Delta p_t = b_{p_1}\mathsf{E}(\Delta p_{t+1} \mid \mathcal{I}_t) + b_{p2}x_t. \qquad (8.12)$$

The 'hybrid' NPCM, which heuristically assumes the existence of both forward- and backward-looking agents and obtains if a subset of firms has a backward-looking rule to set prices, nests (8.12) as a special case. This amounts to the specification

$$\Delta p_t = b_{p1}^f\mathsf{E}(\Delta p_{t+1} \mid \mathcal{I}_t) + b_{p1}^b\Delta p_{t-1} + b_{p2}x_t. \qquad (8.13)$$

Our analysis in Chapter 7 leads to a rejection of the NPCM as an empirical model of inflation for the Euro area and we conclude that the profession should not accept the NPCM too readily. Still, the model maintains a dominant position in modern monetary economics and it is widely used in analyses of Euro-area data.

With reference to the original contributions by Galí and Gertler (1999) and Galí *et al.* (2001), Smets and Wouters (2003) estimate a New Keynesian Phillips curve as part of a stochastic dynamic general equilibrium model for the Euro area. The inflation equation is estimated as part of a simultaneous system with nine endogenous variables in a Bayesian framework using Markov-chain Monte Carlo methods, and the authors find parameter estimates which are in line with Galí *et al.* (2001) for a hybrid version of the New Keynesian Phillips curve (with weights 0.72 and 0.28 on forward and lagged inflation, respectively).

Also, Coenen and Wieland (2002) investigate whether the observed inflation dynamics in the Euro area (as well as in the United States and Japan) are consistent with microfoundations in the form of staggered nominal contracts and rational expectations. On Euro-area data, they find that the fixed period staggered contract model of Taylor outperforms the New Keynesian Phillips curve specification based on Calvo-style random duration contracts and they claim support for the hypothesis of rational expectations.[8]

8.5.4   The P*-model of inflation

In the P*-model (Hallman *et al.* 1991) the long-run equilibrium price level is defined as the price level that would result with the current money stock, m_t, provided that output was at its potential (equilibrium level), y_t^*, and that velocity, $v_t = p_t + y_t - m_t$, was at its equilibrium level v_t^*:

$$p_t^* \equiv m_t + v_t^* - y_t^*. \tag{8.14}$$

The postulated inflation model is given by

$$\Delta p_t = \mathsf{E}(\Delta p_t \mid \mathcal{I}_{t-1}) + \alpha_p(p_{t-1} - p_{t-1}^*) + \beta_z z_t + \varepsilon_t, \tag{8.15}$$

where the main explanatory factors behind inflation are inflation expectations, $\mathsf{E}(\Delta p_t \mid \mathcal{I}_{t-1})$, the price gap, $(p_{t-1} - p_{t-1}^*)$, and other variables denoted z_t. Note that if we replace the price gap in (8.15) with the output gap, we obtain the NPCM (8.12) discussed in the previous section with the expectations term backdated one period.

In order to calculate the price gap one needs to approximate the two equilibria for output, y_t^*, and velocity, v_t^*, respectively. The price gap, $(p_t - p_t^*)$, is obtained by subtracting p_t from both sides of (8.14) and applying the identity $v_t \equiv p_t + y_t - m_t$. It follows that the price gap is decomposed into the velocity

---

[8] Coenen and Wieland adopt a system approach, namely an indirect inference method due to Smith (1993), which amounts to fitting a constrained VAR in inflation, the output gap and real wages, using the Kalman filter to estimate the structural parameters such that the correlation structure matches those of an unconstrained VAR in inflation and the output gap.

gap, $(v_t - v_t^*)$, minus the output gap, $(y_t - y_t^*)$:

$$(p_t - p_t^*) = (v_t - v_t^*) - (y_t - y_t^*). \qquad (8.16)$$

The P*-model can alternatively be expressed in terms of the real money gap, $rm_t - rm_t^*$, where $rm_t^* = m_t - p_t^*$. The inverse relationship holds trivially between the real money gap and price gap, that is, $(rm_t - rm_t^*) = -(p_t - p^*)$, and thus the P*-model predicts that there is a direct effect on inflation from the lagged real money gap $(rm - rm^*)_{t-1}$. Moreover, in the P*-model, fluctuations in the price level around its equilibrium, p_t^*, are primarily driven by fluctuations in velocity and output.

Another defining characteristic of recent studies adopting the P*-model is that inflation is assumed to be influenced by $\Delta_4 pgap_t$, which is the change in the difference between the actual inflation $\Delta_4 p_t$ and a reference or target path $\Delta_4 \tilde{p}_t$, and also by an analogous variable for money growth, $\Delta_4 mgap_t$. The reference path for money growth $\Delta_4 \tilde{m}_t$ is calculated in a similar way as suggested in Gerlach and Svensson (2003), referred to below. If we know the inflation target (or reference path for inflation in the case when no explicit target exists), we can calculate the corresponding reference path for money growth as follows (see Bofinger 2000):

$$\Delta_4 \tilde{m}_t = \Delta_4 \tilde{p}_t + \Delta_4 y_t^* - \Delta_4 v_t^*. \qquad (8.17)$$

In our empirical estimates of the P*-model below we have simply let the reference value for inflation, $\Delta_4 \tilde{p}_t$, vary with the actual level of smoothed inflation and $\Delta_4 pgap_t$ is defined accordingly. The heuristic interpretation is that the monetary authorities changed the reference path according to the actual behaviour, adapting to the many shocks to inflation in this period and we calculate $\Delta_4 \tilde{p}_t$ with a Hodrick–Prescott (HP) filter[9] with a large value of the parameter which penalises non-smoothness, that is, we set $\lambda = 6400$ to avoid volatility in $\Delta_4 \tilde{p}_t$. Likewise, we apply the HP-filter to derive measures for the equilibrium paths for output, y_t^*, and velocity, v_t^*, and in doing so, we use $\lambda = 1600$ to smooth output series y_t^* and $\lambda = 400$ to smooth velocity v_t^*. $\Delta_4 \tilde{m}_t$ follows from (8.17), as does $\Delta_4 mgap_t$.

Gerlach and Svensson (2003) estimate a variant of the P*-model (8.15), and they find empirical support for the P*-model on aggregated data for the Euro area. In this study Gerlach and Svensson introduce and estimate a measure for the inflation target in the Euro area as a gradual adjustment to the (implicit) inflation target of the Bundesbank, and they interpret the gradual adjustment as a way of capturing a monetary policy convergence process in the Euro area throughout their estimation period (1980(1)–2001(2)).

Gerlach and Svensson (2003) find a significant effect of the energy component of consumer price index on inflation measured by the total consumer price index, and when they include the output gap in (8.15), in addition to the

---

[9] See Hodrick and Prescott (1997).

real money gap, both gaps come out equally significant, indicating that each is an important determinant of future price changes. By contrast, they find that the Eurosystem's money-growth indicator, defined as the gap between current $M3$ growth and its reference value, has little predictive power beyond that of the output gap and the real money gap.

Trecroci and Vega (2002) re-estimate the AWM equation for the GDP deflator at factor prices for the period 1980(4)–1997(4), and they find that (an earlier version of) the Gerlach and Svensson P* equation (without the output gap) outperforms the AWM price equation (for q_t) in out of sample forecasts for the period 1992(1)–1997(4) at horizons ranging from 1 to 8 periods ahead.[10] Likewise, Nicoletti Altimari (2001) finds support for the idea that monetary aggregates contain substantial information about future price developments in the Euro area and that the forecasting performance of models with money-based indicators improves as the forecast horizon is broadened.

8.6   Empirical evidence from Euro-area data

In this section, we present estimated reduced form versions of the AWM and ICM inflation equations in order to evaluate the models and to compare forecasts based on these equations with forecasts from the inflation models referred to in Section 8.5, that is, the P*-model and the NPCM. The models are estimated on a common sample covering 1972(4)–2000(3), and they are presented in turn below, whereas data sources and variable definitions are found in Jansen (2004).

8.6.1   The reduced form AWM inflation equation

We establish the reduced form inflation equation from the AWM by combining the wage and price equations of the AWM (see appendix B of Jansen 2004). The reduced form of the equation is modelled from general to specific: we start out with a fairly general information set which includes the variables of the wage and price block of the AWM: three lags of inflation, Δp_t, as well as of changes in trend unit labour costs, $\Delta \overline{ulc}_t$, and two lags of the changes in: the wage share, Δws_t, the world commodity price index, Δp_t^{raw}; the GDP deflator at factor prices, Δq_t, unemployment, Δu_t, productivity, Δa_t, import prices, Δpi_t, and indirect taxes, $\Delta t3_t$. The output gap is included with lagged level (gap_{t-1}) and change $(\Delta \text{gap}_{t-1})$. The dummies from the wage and price block of AWM, $\Delta I82.1$, $\Delta I82.1$, $I92.4, I77.4 \, I78.1$, $I81.1$, and $\Delta I84.2$,[11] are included and a set of centred seasonal dummies (to mop up remaining seasonality in

---

[10] Trecroci and Vega estimate the P*-model within a small VAR, which previously has been analysed in Coenen and Vega (2001).

[11] The first three are significant in all estimated equations reported below, the last two which originate in the AWM wage equation are always insignificant.

the data, if any). Finally, we include into the reduced form information set two equilibrium-correction terms from the structural price and wage equations, $ecmp_t^{AWM}$ and $ecmw_t^{AWM}$, defined in Section 8.5.1.

The parsimonious reduced form AWM inflation equation becomes:

$$
\begin{aligned}
\widehat{\Delta p_t} = {}& \underset{(0.017)}{0.077} + \underset{(0.06)}{0.19\Delta p_{t-3}} + \underset{(0.05)}{0.08\overline{\Delta ulc}_{t-1}} + \underset{(0.08)}{0.34\Delta q_{t-1}} \\
& - \underset{(0.04)}{0.07\Delta a_{t-2}} + \underset{(0.01)}{0.07\Delta pi_{t-1}} + \underset{(0.28)}{0.82\Delta t3_{t-1}} \\
& - \underset{(0.011)}{0.051 ecmp_{t-1}^{AWM}} - \underset{(0.0015)}{0.01\ ecmw_{t-1}^{AWM}} + \text{dummies}
\end{aligned}
$$

$$\sigma = 0.00188 \tag{8.18}$$

$$1972(4)\text{--}2000(3)$$

$$F_{AR(1-5)}(5, 94) = 0.41[0.84] \quad F_{ARCH(1-4)}(4, 91) = 0.43[0.78]$$
$$\chi^2_{\text{normality}}(2) = 1.01[0.60] \quad F_{HET\chi^2}(23, 75) = 1.35[0.17]$$
$$F_{RESET}(1, 98) = 0.06[0.80]$$

All restrictions imposed on the general model leading to (8.18), are accepted by the data, both sequentially and when tested together. We note that the effects of the explanatory variables are much in line with the structural equations reported in appendix B in Jansen (2004) and that both equilibrium-correction terms are highly significant. If we deduct the respective means of the equilibrium-correction terms on the right-hand side, the constant term reduces to 0.5%, which is significantly different from zero with a *t*-value of 5.36. The fit is poorer than for the structural inflation equation, which is mainly due to the exclusion of contemporary variables in the reduced form. If we include contemporary values of Δpi_t, Δa_t, and Δp_t^{raw}, the standard error of the equation improves by 30% and a value close to the estimated σ of the inflation equation in appendix B in Jansen (2004) obtains. Figure 8.6 contains recursive estimates of the model's coefficients. We note that there is a slight instability in the adjustment speed for the two equilibrium terms in the period 1994–96.

8.6.2 The reduced form ICM inflation equation

We derive a reduced form inflation equation for the ICM much in the same vein as for the AWM. The information set for this model is given by all variables included in the estimation of the price–unit labour cost system in Jansen (2004). The information set differs from that of the AWM on the following points: lags of changes in unit labour costs, Δulc_t, are used instead of lags of changes in trend unit labour costs; the changes in the wage share, Δws_t, the world commodity price index, Δp_t^{raw}, and the GDP deflator at factor prices, Δq_t, are not included; and the equilibrium-correction terms are those of the ICM,

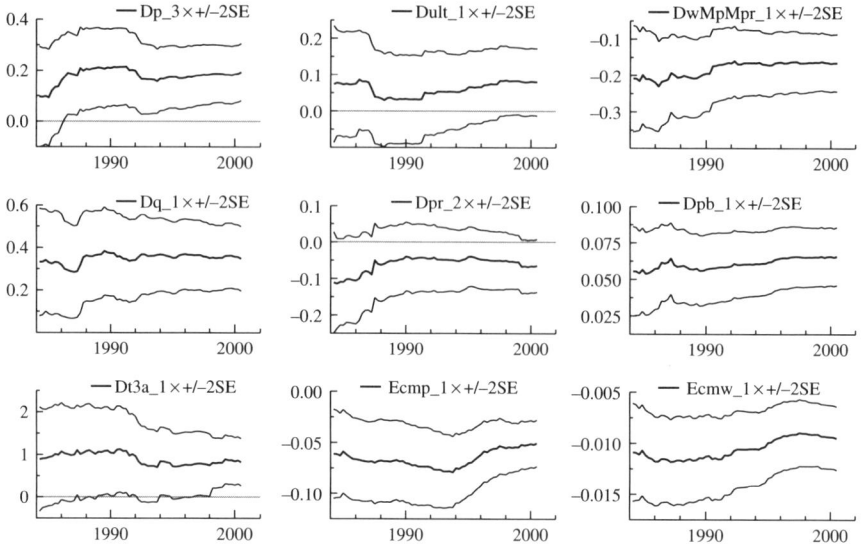

Figure 8.6. Recursive estimates for the coefficients of the (reduced form) AWM inflation equation

$ecmp_t^{\text{ICM}}$ and $ecmulc_t^{\text{ICM}}$, which are derived from the estimated steady-state equations (cf. (8.10) and (8.11)).

$$ulc = p - \underset{(0.02)}{0.11}u$$

$$p = 0.91ulc + \underset{(0.03)}{0.09}pi + t3.$$

After imposing valid restrictions on the general model, the final reduced form ICM inflation equation becomes:

$$\widehat{\Delta p_t} = \underset{(0.006)}{0.014} + \underset{(0.10)}{0.41}\Delta p_{t-1} + \underset{(0.08)}{0.16}\Delta p_{t-2} + \underset{(0.01)}{0.03}\Delta pi_{t-1}$$

$$+ \underset{(0.02)}{0.06}\,\text{gap}_{t-1} + \underset{(0.04)}{0.14}\Delta\text{gap}_{t-1}$$

$$- \underset{(0.016)}{0.078}ecmp_{t-1}^{\text{ICM}} - \underset{(0.007)}{0.031}ecmulc_{t-1}^{\text{ICM}} + \text{dummies}$$

$$\sigma = 0.00205 \tag{8.19}$$

$$1972(4)\text{--}2000(3)$$

$F_{\text{AR}(1-5)}(5,96) = 0.62[0.68]$ $F_{\text{ARCH}(1-4)}(4,93) = 0.18[0.95]$
$\chi^2_{\text{normality}}(2) = 0.16[0.92]$ $F_{\text{HET}x^2}(20,80) = 0.64[0.87]$
$F_{\text{RESET}}(1,100) = 2.98[0.09]$

We observe that the reduced form inflation equation of the ICM is variance encompassed by the corresponding AWM equation. Again, all restriction

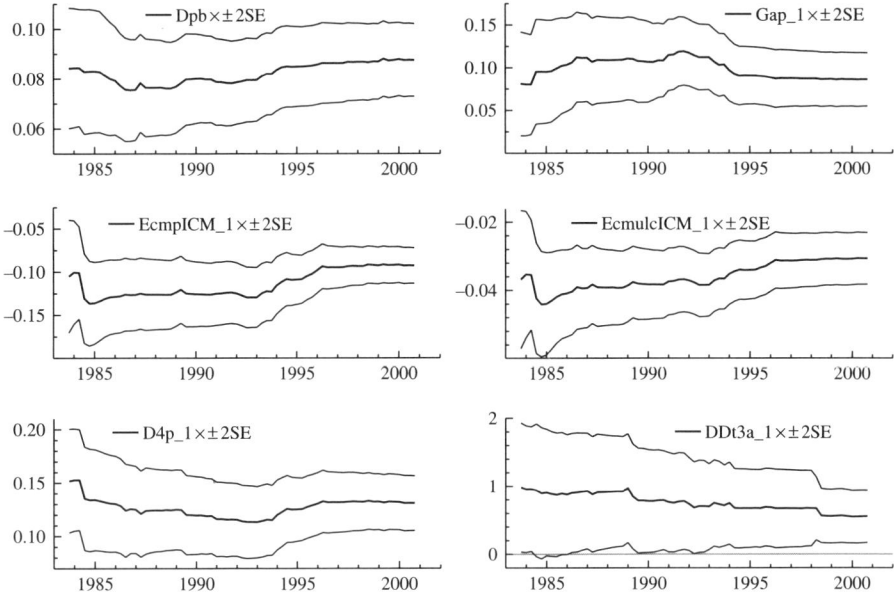

Figure 8.7. Recursive coefficient estimates of the reduced form ICM

imposed on the general model to obtain (8.19) are accepted by the data, both sequentially and when tested together. The reduced form inflation equation picks up the combined effects from the price and the unit labour cost structural equations, the latter is seen through the significant effects of Δp_{t-1}, gap_{t-1}, and the equilibrium-correction term $ecmulc_{t-1}^{\text{ICM}}$ in (8.19). Both equilibrium-correction terms are highly significant. If we deduct the respective means of the equilibrium-correction terms on the right-hand side, the constant term reduces to 0.6%, which is significantly different from zero with a t-value of 4.68. Figure 8.7 contains recursive estimates of the coefficients in (8.19). We note that the speed of adjustment towards the steady-state for the two equilibrium-correction terms is more stable than in the case of AWM.

8.6.3   The P*-model

The estimation of the P*-model in Section 8.5.4 requires additional data relative to the AWM data set. We have used a data series for broad money ($M3$) obtained from Gerlach and Svensson (2003) and Coenen and Vega (2001), which is shown in Figure 8.8.[12] It also requires transforms of the original data: Figures 8.9 and 8.10 show the price gap $(p - p^*)_t$ and the real money gap

---

[12] The series is extended with data from an internal ECB data series for $M3$ (M.U2.M3B0.ST.SA) which matches the data of Gerlach and Svensson (2003) with two exceptions, as is seen from Figure 8.8.

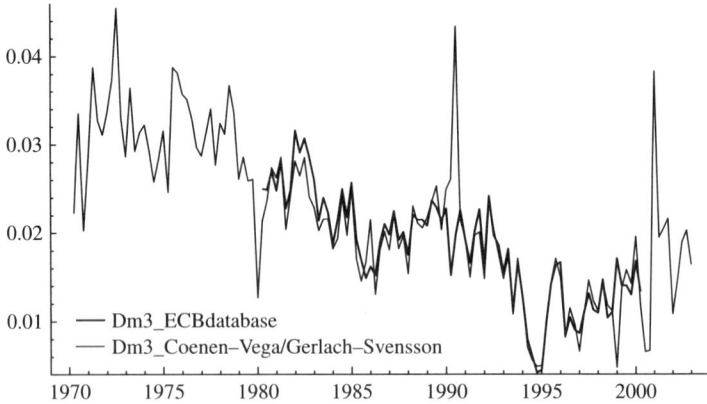

Figure 8.8. The $M3$ data series plotted against the shorter $M3$ series obtained from Gerlach and Svensson (2003), which in turn is based on data from Coenen and Vega (2001). Quarterly growth rate

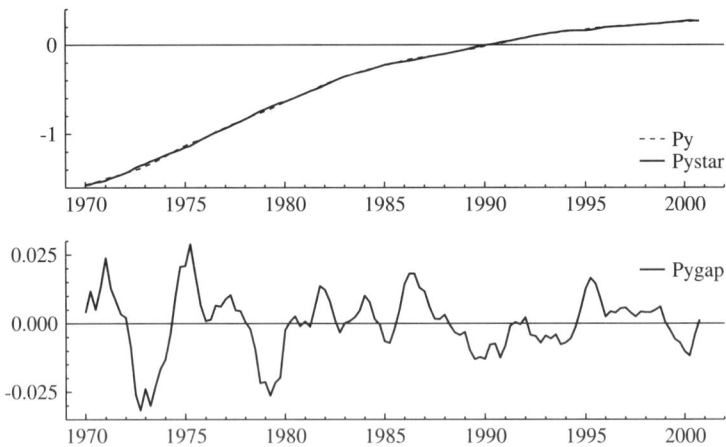

Figure 8.9. The upper graphs show the GDP deflator and the equilibrium price level (p*), whereas the lower graph is their difference, that is, the price gap, used in the P*-model

$(rm - rm^*)_t$ along with the corresponding level series. As noted in Section 8.5.4 we have applied HP-filters to derive measures for y_t^* and v_t^*.[13] Then p_t^* can be calculated from (8.14), as well as the price and real money gaps.

The reference path for inflation is trend inflation from a smoothed HP filter, as described in Section 8.5.4. In Figure 8.11 we have plotted trend inflation together with an alternative which is the same series with the reference path

[13] We use $\lambda = 1600$ to smooth the output series y_t^* and $\lambda = 400$ to smooth the velocity v_t^*.

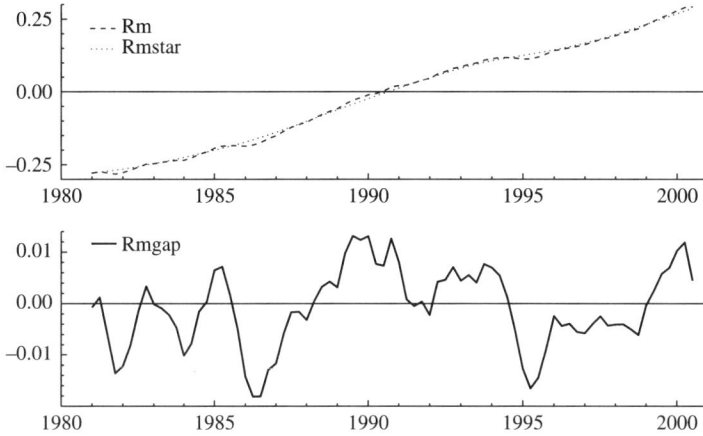

Figure 8.10. The upper graphs show real money and the equilibrium real money, whereas the lower graph is their difference, that is, the real money gap, used in the P*-model

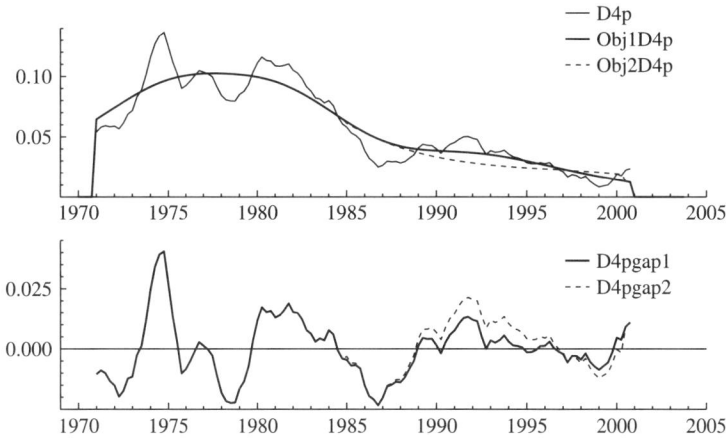

Figure 8.11. The upper figure plots annual inflation against two alternative measures of the reference path for inflation. The solid line shows the HP trend of inflation and the dotted line shows the case where the Gerlach–Svensson target variable is substituted for the HP trend for the subsample 1985(1)–2000(2). The lower graphs show the corresponding D4pgap variables in the same cases

for the price (target) variable of Gerlach and Svensson (2003) substituted in for the period 1985(1)–2000(2). It is seen that the alternative reference path series share a common pattern with the series we have used. Figure 8.12 shows the corresponding graphs for the reference path of money growth.

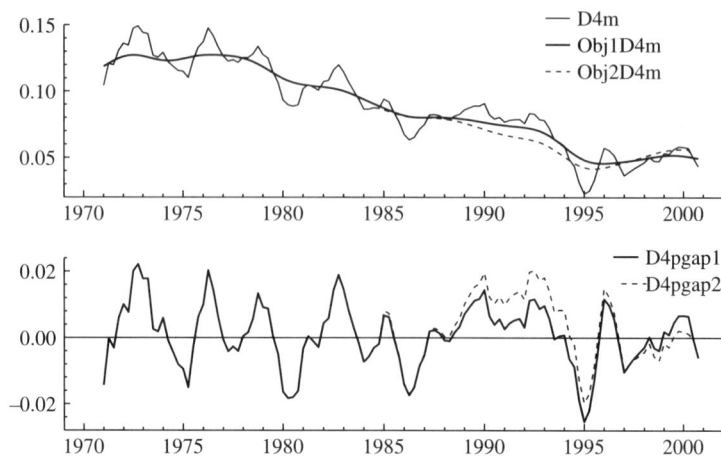

Figure 8.12. The upper figure shows actual annual money growth plotted against the alternative measures of the reference path for money growth. The solid line is the reference path derived from the HP trend of inflation and the dotted line is the alternative, which is derived from inflation reference path with the Gerlach–Svensson target variable substituted for the HP trend for the subsample 1985(1)–2000(2). The lower graphs show the corresponding D4mgap variables in the same cases

The P*-model is estimated in two versions: one version is related to the standard formulation of the P*-model as discussed in Section 8.5.4, in which inflation is explained by the real money gap $(rm - rm^*)$ and the differences between actual price and money growth from their reference (target) paths, $\Delta_4 pgap_t$ and $\Delta_4 mgap_t$.[14]

In order to retain comparability across the inflation models, we differ from previous studies by using the private consumption deflator rather than, for example, the GDP deflator of Trecroci and Vega (2002) or a consumer prices index like the one constructed by Gerlach and Svensson (2003). We also include four lags of inflation, two lags of output growth, Δy_t, and an interest rate spread gap $sgap_t$ (defined as the deviations of the actual spread from a HP trend spread). The other version, P* enhanced, is modelled general to specific, where the general specification is based on the information set of AWM with $(rm - rm^*)_t$, $\Delta_4 pgap_t$, $\Delta_4 mgap_t$, and $sgap_t$ substituted for the equilibrium-correction terms $ecmp_t^{AWM}$ and $ecmw_t^{AWM}$.

[14] We have considered two alternative reference paths for inflation: it is either trend inflation from a smoothed HP filter, or as the same series with the reference path for the price (target) variable of Gerlach and Svensson (2003) substituted in for the period 1985(1)–2000(2). It is seen that the alternative reference path series share a common pattern. Here we report results based on the first alternative.

After we have imposed valid restrictions, the first version based on the narrower information set becomes:

$$\widehat{\Delta p_t} = -\underset{(0.0005)}{0.0015} + \underset{(0.08)}{0.60\Delta p_{t-1}} + \underset{(0.09)}{0.24\Delta p_{t-2}} + \underset{(0.07)}{0.19\Delta p_{t-4}}$$

$$+ \underset{(0.04)}{0.18\Delta y_{t-1}} - \underset{(0.02)}{0.05\Delta_4 pgap_{t-1}} - \underset{(0.03)}{0.04\Delta_4 mgap_{t-1}}$$

$$+ \underset{(0.03)}{0.09\,(rm - rm^*)_{t-1}} - \underset{(0.0003)}{0.0006\,sgap_{t-1}} + dummies$$

$$\sigma = 0.00211 \tag{8.20}$$

$$1972(4)\text{--}2000(3)$$

$F_{AR(1-5)}(5, 95) = 0.52[0.76] \quad F_{ARCH(1-4)}(4, 92) = 0.68[0.61]$
$\chi^2_{normality}(2) = 0.42[0.81] \quad F_{HETx^2}(21, 78) = 0.81[0.70]$
$F_{RESET}(1, 99) = 7.27[0.008^{**}]$

We find that money growth deviation from target $\Delta_4 mgap_{t-1}$ is insignificant which is in line with results reported in Gerlach and Svensson (2003). The other explanatory variables specific to the P*-model comes out significant and with expected signs. The model shows signs of mis-specification through the significant RESET-test.

The enhanced P*-model—based on the broader information set—is given by:

$$\widehat{\Delta p_t} = -\underset{(0.0005)}{0.0004} + \underset{(0.07)}{0.27\Delta p_{t-3}} + \underset{(0.04)}{0.15\widehat{\Delta ulc}_{t-1}} + \underset{(0.06)}{0.49\Delta q_{t-1}}$$

$$+ \underset{(0.04)}{0.10\Delta a_{t-1}} - \underset{(0.04)}{0.12\Delta a_{t-2}} + \underset{(0.27)}{1.08\Delta t3_{t-1}} - \underset{(0.02)}{0.03\Delta_4 pgap_{t-1}}$$

$$- \underset{(0.025)}{0.04\,\Delta_4 mgap_{t-1}} + \underset{(0.02)}{0.11\,(rm - rm^*)_{t-1}} + dummies$$

$$\sigma = 0.00190 \tag{8.21}$$

$$1972(4)\text{--}2000(3)$$

$F_{AR(1-5)}(5, 93) = 0.65[0.66] \quad F_{ARCH(1-4)}(4, 90) = 0.74[0.56]$
$\chi^2_{normality}(2) = 3.83[0.15] \quad F_{HETx^2}(25, 72) = 0.76[0.77]$
$F_{RESET}(1, 97) = 0.01[0.93]$

The model reduction is supported by the data, and the enhanced P* is well specified according to the standard diagnostics reported. We find the P*-model based on the broader information set variance encompasses the P*-model derived from the narrower set of variables, with a reduction of the estimated σ of equation (8.21) of 10% compared with the estimated σ of equation (8.20).

A striking feature of the enhanced P*-model is that the short-run explanatory variables in the first two lines are nearly identical to its counterpart in the AWM reduced form inflation equation (Δa_{t-1} substituting for Δpi_{t-1}) with coefficients of the same order of magnitude. The real money gap $(rm - rm^*)_{t-1}$ is highly significant, whereas $sgap_t$ drops out. Also, the P*-specific explanatory

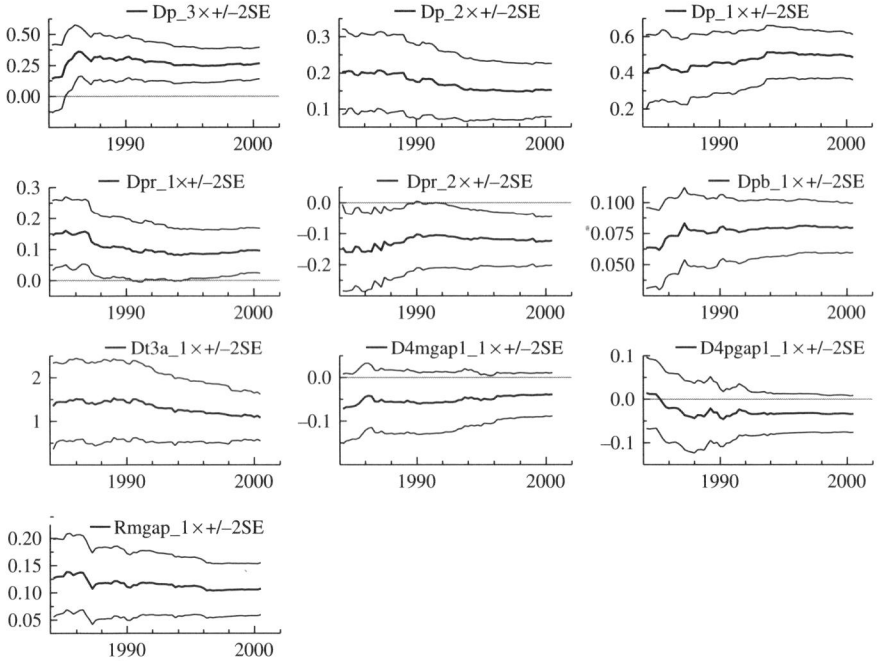

Figure 8.13. Recursive coefficient estimates of the P*-model based
on the broad information set

variables, $\Delta_4 pgap_{t-1}$ and $\Delta_4 mgap_{t-1}$—the deviations from target—are insignificant at the 5% level, but are retained to represent the P* mechanisms.

Figure 8.13 shows that the coefficient estimates of the enhanced P*-model are recursively stable.

8.6.4 The New Keynesian Phillips curve

We estimate a hybrid NPCM as described in Section 8.5.3 (cf. Chapter 7 for further details). Using the instruments of Galí *et al.* (2001)[15]—five lags of inflation, Δp_t, and two lags in the wage share, ws_t, and output gap (gap)—we are able to replicate the results for the hybrid model in Chapter 7, which in turn are representative for the empirical results reported in Galí *et al.* (2001). We have chosen to estimate a small simultaneous model where the inflation lead Δp_{t+1} and the wage share ws_t are specified as functions of the instruments

---

[15] Rudd and Whelan (2004) show that including Δp_{t-1} among the instruments leads to an upward bias in the coefficient of the forward variable; see also Roberts (2001). We have, however, maintained the use of the Galí *et al.* (2001) instruments simply to get as close as possible to the estimation procedure adopted by the 'proprietors' of the NPCM in the same way as we have tried to do in the cases of AWM price block and the P*-model earlier.

and full information maximum likelihood estimation[16] then yields the following
inflation equation:

$$\widehat{\Delta p_t} = -0.0008 + 0.72\Delta p_{t+1} + 0.31\Delta p_{t-1} + 0.002 + \text{dummies}$$
$$\quad\quad\quad (0.006)\quad (0.07)\quad\quad\quad (0.07)\quad\quad\quad (0.008)$$

$$\sigma = 0.00232 \quad\quad\quad\quad\quad\quad\quad\quad\quad\quad\quad\quad\quad\quad (8.22)$$

$$1972(4)\text{--}2000(3)$$

Single equation diagnostics
$$F_{AR(1-5)}(5, 96) = 4.55[0.001^{**}]\ \ F_{ARCH(1-4)}(4, 97) = 0.87[0.48]$$
$$\chi^2_{normality}(2) = 5.16[0.08]\quad\quad F_{HETx^2}(18, 86) = 1.56[0.09]$$
Systems diagnostics
$$F^v_{AR(1-5)}(45, 262) = 9.45[0.000^{**}]$$
$$\chi^{2,v}_{normality}(6) = 8.64[0.19]$$
$$F^2_{HETx^2}(108, 471) = 1.38[0.01^*]$$

In (8.22) we have augmented the NPCM equation with the significant dummies from the other models. Increasing the information set by adding more
instruments does not change the estimates for the NPCM equation. The dummies reduce the estimated σ for the NPCM by 10%, but this is still 10–20%
higher than the other three model classes. The highly significant $F^v_{AR(1-5)}$-test
in (8.22) is not only due to first-order autocorrelation (which is consistent with
the New Keynesian Phillips curve theory[17]), but reflects also higher order autocorrelation. Figure 8.14 underscores that the coefficients of the forward and the
backward terms of the NPCM are recursively stable, as is also the wage share
coefficient at a zero value.

8.6.5 Evaluation of the inflation models' properties

In this section, we summarise the statistical properties of the different inflation models, in order to make more formal comparisons. In Table 8.6 we have
collected the p-values for the mis-specification tests for residual autocorrelation, autoregressive conditional heteroskedasticity, non-normality, and wrong
functional form. With the exception of the normality tests which are $\chi^2(2)$,
we report F-versions of all tests, as in the previous sections. We also report k,
the number of estimated coefficients, and $\sigma_{\Delta p}\%$, the estimated standard error.
    One way of condensing this information is to perform encompassing tests.[18]
In Table 8.7 we consider AWM as the incumbent model, the one we want to

---

[16] Our estimation method thus differs from those in Chapter 7, where we estimate the
hybrid model using generalised method of moments (GMM) as well as by two-stage least
squares. Note that we in Chapter 7, like Galí *et al.* (2001), use the GDP deflator while in
this section the inflation variable is the consumption deflator.

[17] First-order autocorrelation may also have other causes, as pointed out Chapter 7.

[18] For an introduction to the encompassing principle, see Mizon and Richard (1986) and
Hendry and Richard (1989).

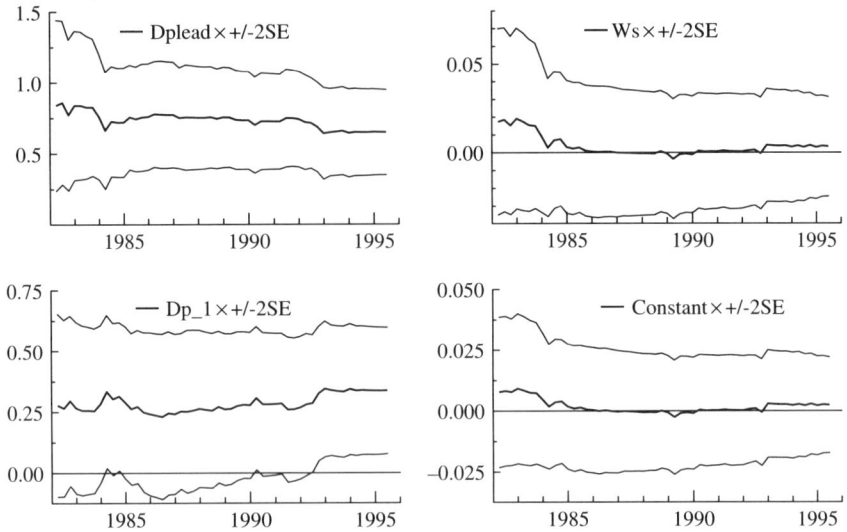

Figure 8.14. Recursive coefficient estimates of the hybrid New Keynesian Phillips curve model (estimated by instrumental variables)

Table 8.6
Mis-specification tests

Δp model	k	$\sigma_{\widehat{\Delta p}}\%$	p-values				
			$F_{AR(1-5)}$	$F_{ARCH(1-5)}$	$\chi^2_{normality}$	$F_{HET\times^2}$	F_{RESET}
AWM	13	0.19	0.84	0.78	0.60	0.17	0.80
ICM	11	0.21	0.68	0.95	0.92	0.87	0.09
P*	12	0.21	0.76	0.61	0.81	0.70	0.008**
P*_enh	14	0.19	0.66	0.56	0.15	0.77	0.93
NPCM	7	0.23	0.00**	0.48	0.08	0.01*	

compare with its competitors, while ICM has this role in Table 8.8. In these tables, we show the p-values for alternative encompassing tests. In the case of the first table, the statistic $F_{Enc,1}$ tests the AWM against each of the three alternatives[19] using joint F-tests for parsimonious encompassing of each of the two models in question against their minimal nesting model. The adjacent test, $F_{Enc,2}$, is based on pairs of model residuals from the AWM (M_1) and from each of the alternative inflation models M_j. In each case, we regress $\hat{\varepsilon}_{1,t}$ against the difference between the residuals of model j and model 1 respectively, $\hat{\varepsilon}_{jt} - \hat{\varepsilon}_{1t}$. Under the null hypothesis that model M_1, the AWM, encompasses model M_j, the coefficient of this difference has zero expectation. The hypothesis that model

---

[19] For technical reasons the NPCM was not included in these tests.

Table 8.7

Encompassing tests with AWM as incumbent model

Δp model	k	$\sigma_{\widehat{\Delta p}}\%$	$F_{Enc\,GUM}(j,83)$		p-values for two types of encompassing tests			
			j	p-value	$F_{Enc,1}$		$F_{Enc,2}$	
					M_1 vs. M_j	M_j vs. M_1	M_1 vs. M_j	M_j vs. M_1
AWM	13	0.19	16	0.08				
ICM	11	0.21	18	0.00**	0.75	0.006**	0.24	0.00**
P*	12	0.21	17	0.00**	0.06	0.00**	0.03*	0.00**
P*_enh	14	0.19	15	0.04*	0.11	0.04*	0.009**	0.005**
NPCM	7	0.23	22	0.00**				

Table 8.8

Encompassing tests with ICM as incumbent model

Δp model	k	$\sigma_{\widehat{\Delta p}}\%$	$F_{Enc\,GUM}(j,83)$		p-values for two types of encompassing tests			
			j	p-value	$F_{Enc,1}$		$F_{Enc,2}$	
					M_1 vs. M_j	M_j vs. M_1	M_1 vs. M_j	M_j vs. M_1
ICM	11	0.21	18	0.00**				
AWM	13	0.19	16	0.08	0.006**	0.75	0.00**	0.24
P*	12	0.21	17	0.00**	0.002**	0.000**	0.017*	0.001**
P*_enh	14	0.19	15	0.04*	0.003**	0.26	0.000**	0.013*
NPCM	7	0.23	22	0.00**				

M_j encompasses M_1 is tested by running the regression of the residuals from model M_j, $\hat{\varepsilon}_{j,t}$, on the same difference (with changed sign). The simple F-test of the hypothesis that the difference has no (linear) effect is reported in the table. Following Mizon and Richard (1986) and Hendry and Richard (1989), a congruent encompassing model can account for the results obtained by rival models, and hence encompassing tests form a richer basis for model comparison than ordinary goodness-of-fit measures.

Tables 8.7 and 8.8 show results from the two encompassing tests explained above, and in addition we report a test for parsimonious encompassing. We have embraced all five models in forming their minimal nesting model, and report p-values of F_{EncGum} tests in the fourth column of the two tables.[20] We see that only the AWM parsimoniously encompasses the general unrestricted

---

[20] It should be noted that the encompassing tests F_{EncGum}, reported in Tables 8.7 and 8.8, are based on two-stage least squares estimation of the NPCM. This gives estimates of the inflation equation that are close to, but not identical to, those in equation (8.22), since full information maximum likelihood (FIML) takes account of the covariance structure of the system. In order to form the minimal nesting model it was necessary to estimate the NPCM on a single equation form to make it comparable with the other (single equation) models.

model (GUM[21]). For all the other models we reject the corresponding set of restrictions relative to the GUM (at the 5% level). In some cases, neither of the pair of models encompasses the other. When both tests lead to rejection this is *prima facie* evidence that both models are mis-specified; see Ericsson (1992).

8.6.6 Comparing the forecasting properties of the models

Figure 8.15 shows graphs of 20 quarters of one-step ahead forecasts with $+/-$ two forecast errors to indicate the forecast uncertainty for the five models we have estimated. It is difficult to tell from the diagrams by means of 'eyeball' econometrics whether there are any differences between them. So there is a need for formal tests: Table 8.9 provides a summary of the forecasting properties of the different inflation models as it reports root mean squared forecast errors (RMSFEs) along with their decomposition into forecast error bias and standard errors. The models are re-estimated on a sample up to the start of the forecasting horizon, and then used to forecast quarterly inflation until 2000(3). Two horizons are considered: a 36-period horizon starting in 1991(4), and a 20-period horizon starting in 1995(4). The first three lines of Table 8.9 show the RMSFE of inflation from the AWM, and its decomposition into mean forecasting *bias* and standard deviation *sdev*. The other rows of the table shows the same three components of the RMSFE-decomposition for each of the other inflation models, measured relative to the results for the AWM, such that, for example, a number greater than one indicates that the model has a larger RMSFE than the AWM. For one-step forecasts 20 quarters ahead, we find that all competing models beat the AWM on the RMSFE—and *bias*—criteria, whereas AWM is superior according to *sdev*.

Tables 8.10 and 8.11 show the results from forecast encompassing tests, regressing the forecast errors of model 1, $\hat{\varepsilon}_{1t}$, against the difference between the forecast errors of model j and model 1 respectively, $\hat{\varepsilon}_{jt} - \hat{\varepsilon}_{1t}$.[22] Under the null that there is no explanatory power in model j beyond what is already reflected in model 1, the expected regression coefficient is zero. In the tables we report p-values when we run the forecast encompassing test in both directions. The AWM is used as benchmark (model 1) in Table 8.10 and the table contains evidence that AWM forecast encompasses three out of four competitors over 20 quarters (and the fourth—the P*-model enhanced—comes close to being encompassed at the 5% level), while the reverse is not true. Over 36 quarters there is clear evidence that the AWM forecast encompasses the NPCM, but is

---

[21] Strictly speaking, the generic GUM is the union of all information sets we have used to create the general models in Sections 8.6.1–8.6.4. In the minimal nesting (parsimonious) GUM, we have left out all variables that are not appearing in any of the five final equations and it is more precise to call this a pGUM.

[22] Again, the forecast encompassing tests are based on two-stage least squares estimates of the NPCM.

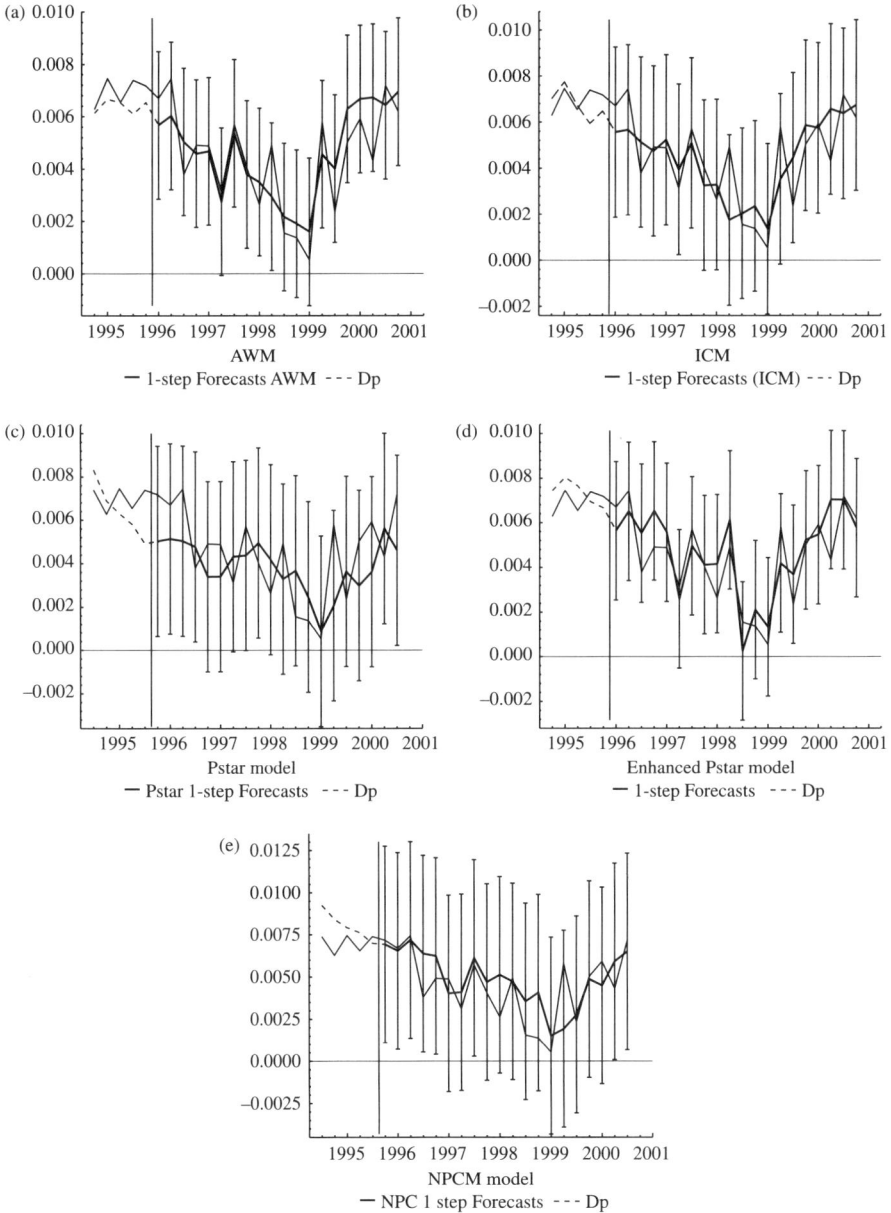

Figure 8.15. Forecasts of quarterly inflation in the Euro area with five
different models: over the period 1995(4)–2000(3). The models are:
(a) the AWM; (b) the ICM; (c) the P*-model; (d) the enhanced P*-model;
and (e) the NPCM. The bars show 2× forecast errors

Table 8.9
Forecasting the quarterly rate of inflation. RMSFE and its
decomposition: bias, standard deviations, and RMSFE of
different inflation models, relative to the AWM

$\Delta_4 p$ model		Forecasting $\widehat{\Delta p}$	
		1991(4)–2000(3)	1995(4)–2000(3)
AWM	RMSFE	0.0022	0.0021
	Bias	0.0011	0.0016
	sdev	0.0019	0.0014
ICM	Rel. RMSFE	1.08	0.82
	Rel. bias	1.28	0.42
	Rel. sdev	1.01	1.14
P*	Rel. RMSFE	0.92	0.88
	Rel. bias	0.55	0.38
	Rel. sdev	1.02	1.26
P*_enh	Rel. RMSFE	0.76	0.73
	Rel. bias	0.09	0.13
	Rel. sdev	0.88	1.10
NPCM	Rel. RMSFE	1.11	0.73
	Rel. bias	0.20	0.06
	Rel. sdev	1.29	1.12

Table 8.10
Forecast encompassing tests over 36 and 20 periods, ending in 2000(3)

Model	k	$\sigma_{\widehat{\Delta p}}\%$	$F_{\text{Enc GUM}}(j, 63)$		Forecast encompassing tests: p-values			
			j	p-value	1991(4)–2000(3)		1995(4)–2000(3)	
					M_1 vs. M_j	M_j vs. M_1	M_1 vs. M_j	M_j vs. M_1
AWM	13	0.19	16	0.08				
ICM	11	0.21	18	0.00**	0.08	0.06	0.96	0.03*
P*	12	0.12	17	0.00**	0.04*	0.02*	0.38	0.003**
P*_enh	14	0.19	15	0.04*	0.002**	0.42	0.88	0.067
NPCM	7	0.23	22	0.00**	0.21	0.00**	0.35	0.03*

The AWM is used as benchmark.

itself overwhelmingly forecast encompassed by the enhanced P*-model (based on the same broad information set).

In Table 8.11 the ICM is used as benchmark (model 1). The ICM is not forecast encompassing any competitor over 20 quarters, but is, as noted above, itself forecast encompassed by the AWM. Over 36 quarters ICM forecast encompasses

Table 8.11

Forecast encompassing tests over 36 and 20 periods, ending in 2000(3)

Model	k	$\sigma_{\widehat{\Delta p}}\%$	$F_{\text{Enc GUM}}(j, 63)$		Forecast encompassing tests: p-values			
			j	p-value	1991(4)–2000(3)		1995(4)–2000(3)	
					M_1 vs. M_j	M_j vs. M_1	M_1 vs. M_j	M_j vs. M_1
ICM	11	0.21	18	0.00**				
AWM	13	0.19	16	0.08	0.06	0.08	0.03*	0.96
P*	12	0.12	17	0.00**	0.11	0.06	0.87	0.06
P*_enh	14	0.19	15	0.04*	0.001**	0.18	0.09	0.22
NPCM	7	0.23	22	0.00**	0.64	0.00**	0.10	0.17

The ICM is used as benchmark.

the NPCM, and—like the AWM—it is forecast encompassed by the enhanced version of the P*-model.

An important caveat applies to the results in this section. In interpreting the favourable results for the P*-model it should be borne in mind that the forecasts made for the P*-specifications are greatly helped by the two-sided filters used to define the equilibrium values for, say rm^*, as described in Section 8.5.4.[23]

8.6.7   Summary of findings—Euro-area data

The model comparisons in this section do not allow us to draw decisive conclusions. Some caveats no doubt apply: the presumptions of a clearly defined monetary policy for the economy under study, which are underlying the P*-model as it is laid out in Gerlach and Svensson (2003), is not favoured by adopting an observation period which starts nearly 30 years before the intro-duction of the Euro.[24] Likewise, the ICM—with its focus on the labour market influx on inflation—is probably a better model description of the national economies than for the Euro area.

That said—from the model evaluation and the forecast comparisons—some comparative advantages seem to emerge in favour of the (reduced form) AWM inflation equation: it is the only model that encompasses a GUM and it forecast encompasses the competitors when tested on 20 quarters of one-step ahead forecasts. The P*-model—based on the extended (AWM) information set—forecast encompasses the other models based on 36 quarters of one-step forecasts. In that context the NPCM appears to be a particularly poor model.

---

[23] A more realistic approach would have been to let the estimates of the equilibrium values be derived from some backward-looking filter. Such a procedure would better capture the relevant information available to the forecaster when forecasts are made.

[24] This point is, however, not relevant to the P*-model in its original tapping (see Hallman *et al.* 1991), where weight is put on the quantity equation and the stability of the money demand function. Fagan and Henry (1998) suggest that money demand may be more stable at the aggregated Euro-area level than at the national levels.

The results of the forecast competition are in accordance with the model evaluation in the preceding sections. The ICM is likely to suffer in forecasting due to recursive instability in the long-run coefficients (table 2 in Jansen 2004) as well as in the short-run coefficients (Figure 8.7). Generally, we find that the models that are derived from the wider information sets (AWM and P* enhanced) do better in forecasting than those based on a narrower information set, mainly prescribed by theory, like the P*-model proper and the NPCM.

8.7   Empirical evidence for Norway

In this and the following sections, we compare an inflation equation which is a reduced form of the dynamic ICM for Norway with variants of the P*-model and the hybrid NPCM, as defined in Section 8.5. We also include in this comparison the inverted money demand function MdInv of Section 8.4.2 and we show that models based on the P*-formulation are more successful than the alternatives in capturing effects on inflation from monetary aggregates. The models are estimated on a common sample covering 1969(1)–2001(1), and they are presented in turn below, whereas data sources and variable definitions are found in Eitrheim (2003).

8.7.1   The Incomplete Competition Model

The dynamic version of the ICM is the main model in this book to explain the formation of wages and prices in the Norwegian economy. In Chapter 9, we follow a stepwise procedure to estimate this model, as described for the Euro area in Section 8.5. In the case of Norway the long-run restrictions from theory are supported by the data when we consider an information set of wages (w_t) and consumer prices (p_t) and the conditioning variables, output growth (Δy_t), unemployment (u_t), productivity (a_t), import prices (pi_t), payroll taxes ($t1_t$), and indirect taxes ($t3_t$).

Empirical results for the simultaneous wage–price sub-system are presented in Section 9.2.2.[25] This wage–price model is an updated version of the core model reported in Bårdsen *et al.* (2003), and FIML-results for the dynamic wage and price equations are reported in (9.5)–(9.6). The estimated long-run equations are:

$$ecmw = w - p - a + 0.11u$$
$$ecmp = p - 0.73(w + t1 - a) + 0.27pi + 0.5t3.$$

These equations are embedded in the model as equilibrium-correction terms. For the purpose of comparing the ICM with the alternative inflation models in

---

[25] The final inflation equation also includes short-run effects of changes in the length of the working day (Δh_t) and seasonal dummies.

Table 8.12
Annual CPI inflation in Norway, $\Delta_4 p_t$. The reduced form ICM model

$$
\begin{aligned}
\widehat{\Delta_4 p_t} = \;& \underset{(0.0173)}{0.0419 \Delta_3 w_{t-1}} - \underset{(0.0260)}{0.0667 \Delta w_{t-2}} + \underset{(0.0304)}{1.0296 \Delta_3 p_{t-1}} + \underset{(0.0570)}{0.1308 \Delta p_{t-2}} \\
& + \underset{(0.0136)}{0.0662 \Delta_2 y_{t-1}} + \underset{(0.0098)}{0.0235 \Delta p i_t} - \underset{(0.0064)}{0.0595 \, EqCP_{t-1}} \\
& - \underset{(0.0096)}{0.0185 \, EqCW_{t-1}} + \underset{(0.0088)}{0.0416 \Delta p e_t} - \underset{(0.0154)}{0.0355 \Delta_4 a_t} - \underset{(0.0540)}{0.0930 \Delta h_t} \\
& - \underset{(0.0007)}{0.0106 \, P\mathrm{dum}_t} - \underset{(0.0011)}{0.0025 \, W\mathrm{dum}_t} - \underset{(0.0016)}{0.0066 \, CS1_t} - \underset{(0.0073)}{0.0146}
\end{aligned}
$$

$\hat{\sigma} = 0.35\%$

Diagnostic tests

$$
\begin{aligned}
F_{\text{AR(1-5)}}(5, 109) &= 0.6800 [0.6395] \\
F_{\text{ARCH(1-4)}}(4, 106) &= 0.2676 [0.8982] \\
\chi^2_{\text{normality}}(2) &= 4.7510 [0.0930] \\
F_{\text{HET}\chi^2}(27, 86) &= 1.3303 [0.1620] \\
F_{\text{RESET}}(1, 113) &= 0.0165 [0.8979]
\end{aligned}
$$

Note: The sample is 1969(1)–2001(1), quarterly data.

this chapter we derive a *reduced form* representation of the simultaneous wage–price sub-system. This reduced form version is modelled general to specific with PcGets—see Hendry and Krolzig (2001)—starting out with a general model with 34 variables from which the reduced form of the wage–price sub-system in equations (9.5)–(9.6) is one among many potential model simplifications. In order to further challenge the ICM reported in Section 9.2.2, we also included a wide set of variables from the previous sections: (lags in) household wealth, variables which capture exchange rate changes, and a measure of excess money derived from the long-run money demand relationship.

None of the 'outside' variables were found to be significant in the simplified relationship suggested by the Gets procedure.[26] On the other hand, according to Table 8.12, all key variables in the reduced form representation of (9.5)–(9.6) turn out to be significant, including both the equilibrium correcting terms above. The reported ICM in Table 8.12 is well specified according to the reported mis-specification tests.

8.7.2 The New Keynesian Phillips curve

In Chapter 7, Section 7.5.5, we considered the pure version of the NPCM for Norway (see Equation (7.21)). The model was estimated by GMM over the period 1972(4)–2001(1). The instruments used were lagged wage growth $(\Delta w_{t-1}, \Delta w_{t-2})$, lagged inflation $(\Delta p_{t-1}, \Delta p_{t-2})$, lags of the level and change

---

[26] In Section 8.7.4 we corroborate this finding using formal tests of neglected monetary effects in the ICM, and in Section 8.7.5 we report similar findings from encompassing tests.

in unemployment (u_{t-1}, Δu_{t-1}, Δu_{t-2}), and changes in energy prices (Δpe_t, Δpe_{t-1}), the short-term interest rate (ΔRL_t, ΔRL_{t-1}) and the change in the length of the working day (Δh_t). In this chapter we focus on the hybrid version of the NPCM. We use the same set of instruments, except that the first lag of inflation Δp_{t-1} now enters as a regressor.

The results, using both GMM and 2SLS as estimation methods, are given in Table 8.13. First, using GMM, expected inflation becomes insignificant when adding lagged inflation. This is in stark contrast to the results of the pure NPCM of equation (7.21) in Chapter 7. Second, the sum of the coefficients is notably lower than the results for the Euro area. Third, when 2SLS is used, the Sargan specification test $\chi^2_{\text{ival}}(6) = 48.14[0.00]$ indicates a mis-specified model.

Table 8.13

Estimation of the hybrid **NPCM** of inflation on Norwegian data

GMM results

$$\widehat{\Delta p_t} = \underset{(0.1437)}{0.2263\,\Delta p_{t+1}} + \underset{(0.1005)}{0.3960\,\Delta p_{t-1}} + \underset{(0.0175)}{0.0794\,ws_t} - \underset{(0.0111)}{0.0512}$$

$$- \underset{(0.0110)}{0.0823\,\Delta pm_t} - \underset{(0.0017)}{0.0137\,P\text{dum}_t} + \underset{(0.0014)}{0.0005\,S1_t} + \underset{(0.0015)}{0.0024\,S2_t}$$

$$+ \underset{(0.0020)}{0.0030\,S3_t}$$

$$\hat{\sigma} = 0.70\%$$

Diagnostic tests

$$\chi^2_{\text{J}}(11) = 11.12[0.43]$$

2SLS results

$$\widehat{\Delta p_t} = \underset{(0.1637)}{0.5392\,\Delta p_{t+1}} + \underset{(0.1192)}{0.1701\,\Delta p_{t-1}} + \underset{(0.0175)}{0.0438\,ws_t} - \underset{(0.0110)}{0.0242}$$

$$+ \underset{(0.0182)}{0.0131\,\Delta pm_t} - \underset{(0.0020)}{0.0162\,P\text{dum}_t} + \underset{(0.0016)}{0.0070\,S1_t}$$

$$+ \underset{(0.0021)}{0.0066\,S2_t} + \underset{(0.0024)}{0.0076\,S3_t}$$

$$\hat{\sigma} = 0.54\%$$

Diagnostic tests

$$\chi^2_{\text{ival}}(6) = 48.14[0.00]$$
$$F_{\text{AR}(1\text{-}5)}(5, 99) = 14.18[0.00]$$
$$F_{\text{ARCH}(1\text{-}4)}(4, 96) = 1.17[0.33]$$
$$\chi^2_{\text{normality}}(2) = 1.40[0.50]$$
$$F_{\text{HETx}^2}(13, 90) = 1.89[0.04]$$

Note: The sample is 1972(4)–2000(4), quarterly data.

8.7.3 Inflation equations derived from the P*-model

The P*-model is presented in Section 8.5.4. The basic variables of the model are calculated in much the same way for Norway as for the Euro area in the previous section. Figure 8.16 shows the price gap $(p - p^*)_t$ and the real money gap $(rm - rm^*)_t$ along with the corresponding level series using Norwegian data. The price gap is obtained from equation (8.16) after first applying the HP filter to calculate equilibria for output (y^*) and velocity (v_t^*), respectively. As for the Euro area we have used $\lambda = 1600$ to smooth the output series y_t^* and $\lambda = 400$ to smooth velocity v_t^*. Then p_t^* can be calculated from (8.14), as well as the price- and real money gaps. It is easily seen from the figure that $(p - p^*)_t = -(rm - rm^*)_t$.

The reference path for money growth $\Delta_4 \widetilde{m}_t$ is calculated in a similar way as in Section 8.5.4. Recall that if we know $\Delta_4 \widetilde{p}_t$, the inflation target (or reference path for inflation in the case when no explicit target exist), we can use equation (8.17), that is, $\Delta_4 \widetilde{m}_t = \Delta_4 \widetilde{p}_t + \Delta_4 y_t^* - \Delta_4 v_t^*$, to calculate the corresponding reference path for money growth. The equilibrium paths for output, y_t^*, and velocity, v_t^*, are defined above (calculated by the HP-filter). We let the reference value for inflation vary with the actual level of smoothed inflation for the larger part of the sample period, from 1969(1) to 1995(4). The heuristic interpretation is that the monetary authorities changed the reference path according to the actual behaviour, adapting to the many shocks to inflation in this period and we calculate the reference value of inflation with a HP-filter with a large value of the parameter which penalises non-smoothness, that is, we set $\lambda = 6400$ to avoid volatility in $\Delta_4 \widetilde{p}_t$. For the period from 1996(1) to 2001(1) (end of sample)

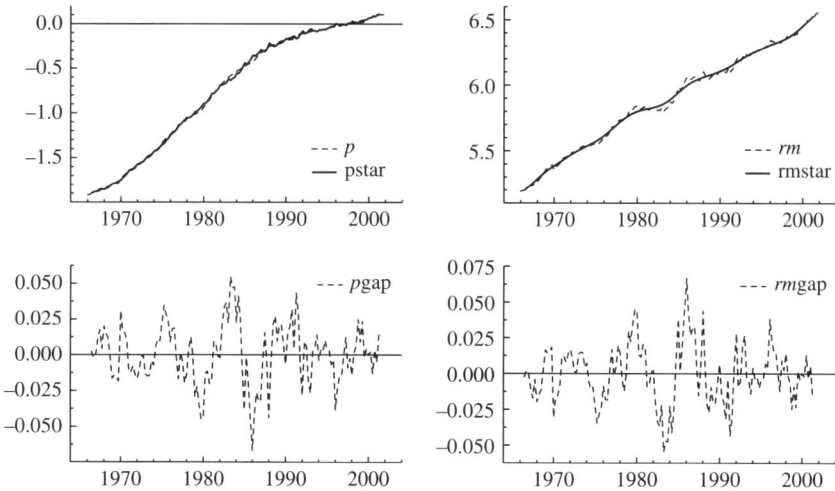

Figure 8.16. Price and real money gaps. Norwegian data.

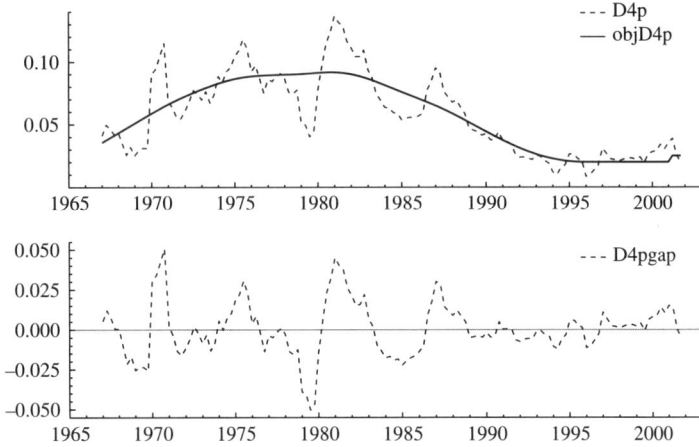

Figure 8.17. Inflation objective and gap. Norwegian data.

we have set the reference value to 2% which is consistent with the actual level of inflation in this period as well as corresponding to the upper limit of inflation in the Euro area in this period. Although Norway formally followed a managed float exchange rate regime in this period, there were substantial deviations from the target exchange rate level in this period, and towards the end of the century the monetary policy regime in Norway was for all practical purposes equivalent to an inflation targeting regime with a target geared towards the Euro-area inflation target (Figure 8.17). Finally we define $\Delta_4 pgap_t$ as the change in the difference between the actual inflation $\Delta_4 p_t$ and the reference path $\Delta_4 \tilde{p}_t$, and $\Delta_4 mgap_t$ is defined accordingly as $\Delta_4 m_t - \Delta_4 \tilde{m}_t$.

The basic version of the P*-model for Norway corresponds to the model we have reported in Section 8.6.3 for the Euro area. In addition to the potential effect from the real money gap $(rm - rm^*)_{t-1}$, we have also included lagged values of the reference money growth gap indicator, $\Delta_4 mgap_{t-1}$ (see Figure 8.18), the deviation from the reference value of inflation (inflation gap for short), $\Delta_4 pgap_{t-1}$, and the yield spread deviation from its trend value, $RBRMgap_{t-1}$. We also follow Gerlach and Svensson (2003) in including variables which account for temporary shocks to inflation from changes in energy prices, Δpe_t, and output growth, Δy_t. Moreover, we augment the equation with the change in household wealth (Δwh_t) and the dummies $Wdum_t$ and $Pdum_t$. As shown in Table 8.14 changes in energy prices and output growth come out as significant explanatory factors, while the empirical support is less convincing for the gap variables. Only the real money gap, $(rm - rm^*)_{t-1}$, and the inflation gap, $\Delta_4 pgap_{t-1}$, are significant at the 10% level. The real money gap has a positive effect, and the inflation gap a negative effect on inflation. When we include the gap variables in this model one at a time, only the real money gap and the inflation gap come out as significant at the 5% level. The reported

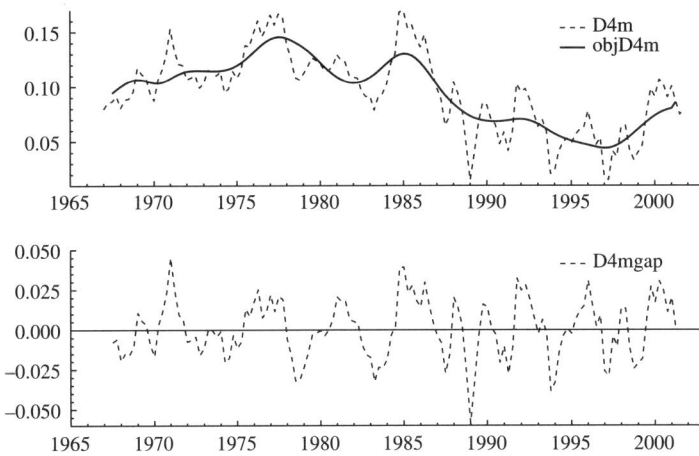

Figure 8.18. Money growth objective and gap. Norwegian data.

Table 8.14
The P*-model for annual CPI inflation, $\Delta_4 p_t$

$$\widehat{\Delta_4 p_t} = + \underset{(0.0191)}{1.2763} \Delta_3 p_{t-1} + \underset{(0.0107)}{0.0436} \Delta pe_t + \underset{(0.0109)}{0.0481} \Delta y_{t-2} + \underset{(0.0173)}{0.0303} \Delta wh_{t-3}$$

$$- \underset{(0.0323)}{0.0824} \Delta_4 pgap_{t-1} + \underset{(0.0292)}{0.0491} rmgap_{t-1} + \underset{(0.0474)}{0.0217} gdpgap_{t-1}$$

$$- \underset{(0.0271)}{0.0024} \Delta_4 mgap_{t-1} - \underset{(0.0548)}{0.0202} RBRMgap_{t-1} - \underset{(0.0010)}{0.0116} Pdum_t + \underset{(0.0010)}{0.0006}$$

$\hat{\sigma} = 0.46\%$

Diagnostic tests

$$F_{AR(1-5)}(5, 113) = 2.1491[0.0647]$$
$$F_{ARCH(1-4)}(4, 110) = 2.3686[0.0570]$$
$$\chi^2_{normality}(2) = 4.9067[0.0860]$$
$$F_{HET_{x^2}}(20, 97) = 3.7178[0.0000]^{**}$$
$$F_{HET_{x_i x_j}}(65, 52) = 1.7848[0.0160]^*$$
$$F_{RESET}(1, 117) = 5.5016[0.0207]^*$$

Note: The sample is 1969(1)–2001(1), quarterly data.

mis-specification tests indicate that the model only barely passes the tests for residual autocorrelation, ARCH and normality tests at a 5% significance level, but fails to meet the tests of zero heteroskedasticity and the RESET test of functional form. The findings that the money growth indicator $\Delta_4 mgap_{t-1}$ is insignificant whereas the real money gap, $(rm - rm^*)_{t-1}$, picks up a significant effect are in line with our results for the Euro area as well as those found in Gerlach and Svensson (2003).

Table 8.15
The enhanced P*-model (P*_enh) for annual CPI inflation, $\Delta_4 p_t$

$$
\begin{aligned}
\widehat{\Delta_4 p_t} = & + \underset{(0.0259)}{1.0653} \Delta_3 p_{t-1} + \underset{(0.0552)}{0.2606} \Delta p_{t-2} + \underset{(0.0087)}{0.0496} \Delta p e_t \\
& + \underset{(0.0133)}{0.0302} \Delta_2 y w_{t-1} - \underset{(0.0137)}{0.0574} \Delta y_{t-1} + \underset{(0.0252)}{0.0650} \Delta m_t \\
& - \underset{(0.0112)}{0.0647} (m_{t-1} - p_{t-1} - 0.9 y_{t-1} + 2.5(\mathrm{RB}_{t-1} - \mathrm{RM}_{t-1})) \\
& + \underset{(0.0178)}{0.1234} rm\mathrm{gap}_{t-1} + \underset{(0.0496)}{0.1373} \mathrm{RBRMgap}_{t-1} + \underset{(0.0012)}{0.0024} CS2 \\
& - \underset{(0.0007)}{0.0121} P\mathrm{dum}_t - \underset{(0.0012)}{0.0033} W\mathrm{dum}_t - \underset{(0.0536)}{0.3113} \\
\hat{\sigma} = & \; 0.35\%
\end{aligned}
$$

Diagnostics

$$
\begin{aligned}
\mathsf{F}_{\mathrm{AR}(1-5)}(5, 111) &= 1.2569[0.2877] \\
\mathsf{F}_{\mathrm{ARCH}(1-4)}(4, 108) &= 0.7746[0.5441] \\
\chi^2{}_{\mathrm{normality}}(2) &= 1.8738[0.3918] \\
\mathsf{F}_{\mathrm{HET}_{x^2}}(23, 92) &= 1.6716[0.0452]^* \\
\mathsf{F}_{\mathrm{HET}_{x_i x_j}}(82, 33) &= 0.6567[0.9353] \\
\mathsf{F}_{\mathrm{RESET}}(1, 115) &= 0.6076[0.4373]
\end{aligned}
$$

Note: The sample is 1969(1)–2001(1), quarterly data.

As for the Euro area, we have tried to improve on the P*-model by including a wider set of variables from the other inflation models. Most importantly, we have lifted the equilibrium-correction term *ecmmd* from the (improved) money demand function in Section 8.4.1 (see Table 8.4) into the P*-model. The model is derived general to specific using the liberal PcGets modelling strategy and it is seen from Table 8.15 that this model which we have dubbed the enhanced P*-model (P*-enhanced for short) improves strongly on the previous P*-model: the model fits the data better, and the estimated standard error is reduced from $\hat{\sigma} = 0.46\%$ in Table 8.14 to $\hat{\sigma} = 0.35\%$. The model is also well designed and with the exception of the $\mathsf{F}_{\mathrm{HET}_{x^2}}$-test it passes all the reported mis-specification tests.

8.7.4 Testing for neglected monetary effects on inflation

The ICM equation for aggregate consumer price inflation in Table 8.12 contains three key sources of inflation impulses to a small open economy: imported inflation including currency depreciation (a pass-through effect), domestic cost pressure (unit labour costs), and excess demand in the product market. Monetary shocks or financial market shocks may of course generate inflation impulses in situations where they affect one or more of the variables associated with these inflation channels. In this section, we will investigate another possibility, namely that shocks in monetary or financial variables have direct effects on inflation which have been neglected in the ICM. Results for Denmark in Juselius (1992) indicate that 'monetary variables' are important explanatory variables in an empirical model for Danish inflation and that they have clearly significant direct effects. In the following, we test the robustness of the ICM inflation equation in Table 8.12 with respect to neglected monetary effects on inflation, simply by subjecting this equation to a sequence of omitted variables tests. In turn we test the significance of current and lagged money growth $(\Delta m_t, \ldots, \Delta m_{t-4})$, real interest rate $(\text{RT} - \Delta_4 p)_{t-1}$, interest rate spread $(\text{RB} - \text{RT})_{t-1}$, the lagged equilibrium correction term from the broad money demand equation $ecmmd_{t-1}$, credit gap $(cr - cr^*)_{t-1}$, and all gap variables from the P*-models above $(ygap_{t-1}, \Delta_4 mgap_{t-1}, \Delta_4 pgap_{t-1}, \text{RBRTgap}_{t-1}, rmgap_{t-1})$.

The results in Table 8.16 show that neither of these variables are significant when they are added to the ICM price equation. The same results hold for these variables irrespective of whether we test their significance simultaneously or include the variables one at a time. The lagged equilibrium correction term for broad money, $ecmmd_{t-1}$, is clearly insignificant when it is added to the price equation. This is an important result, since it provides corroborative evidence that prices are weakly exogenous for the parameters in the long-run money demand relationship. This is a plausible finding from a theoretical point of view, and it is also in line with empirical evidence found in a series of previous studies, including Hoover (1991), Bårdsen (1992),

Table 8.16

Omitted variable tests (OVT) for neglected monetary effects
on inflation in the 'reduced form' ICM price equation

Money growth, Interest rates, excess money and credit	
$\Delta m, \ldots, \Delta m_{t-4}$	$F_{\text{OVT}}(5,109) = 0.2284[0.9494]$
$(\text{RT} - \Delta_4 p)_{t-1}$	$F_{\text{OVT}}(1,113) = 0.0328[0.8565]$
$(\text{RB} - \text{RT})_{t-1}$	$F_{\text{OVT}}(1,113) = 0.3075[0.5803]$
$(m - m^*)_{t-1}$	$F_{\text{OVT}}(1,113) = 0.1302[0.7189]$
$(cr - cr^*)_{t-1}$	$F_{\text{OVT}}(1,113) = 0.5173[0.4735]$
'Gap' variables from the P-star model	
$gdpgap_{t-1}$	$F_{\text{OVT}}(1,113) = 0.4476[0.5049]$
$\Delta_4 mgap_{t-1}$	$F_{\text{OVT}}(1,113) = 1.5663[0.2133]$
$\Delta_4 pgap_{t-1}$	$F_{\text{OVT}}(1,113) = 0.0114[0.9152]$
RBRMgap_{t-1}	$F_{\text{OVT}}(1,113) = 0.1164[0.7336]$
$rmgap_{t-1}$	$F_{\text{OVT}}(1,113) = 2.0426[0.1557]$
Joint all five above	$F_{\text{OVT}}(5,109) = 0.4685[0.7990]$

Hendry and Ericsson (1991), Engle and Hendry (1993), and Hendry and Mizon (1993).

8.7.5 Evaluation of inflation models' properties

The models above are estimated both for annual inflation ($\Delta_4 p_t$) and quarterly inflation (Δp_t) for all the inflation models, except for the NPCM where the forward-looking term on the right-hand makes the quarterly model the obvious choice. As with the Euro-area data, we shall seek to evaluate the different inflation models by comparing some of their statistical properties. In Table 8.17 we report p-values for mis-specification tests for residual autocorrelation, autoregressive conditional heteroskedasticity, non-normality and wrong functional form. With the exception of the normality tests which are $\chi^2(2)$, we have reported F-versions of all tests.

None of the models reported in the upper part of Table 8.17 fails on the $F_{AR(1-5)}$ or $F_{ARCH(1-5)}$ tests, hence there seems to be no serial correlation nor ARCH in the model residuals, but we see that the MdInv and the P*-model fail either on the F_{HETx^2} test and/or the F_{RESET} test for wrong functional form. The results for the NPCM reported at the bottom of Table 8.17 indicate strong serial correlation, but as we have seen in Chapter 7, models with forward-looking expectational terms have moving average residuals under the null hypothesis that they are correctly specified. The fit of the other models vary within the range of $\hat{\sigma} = 0.35\%$ for the ICM and the enhanced P*-model to $\hat{\sigma} = 0.46\%$ for the P*-model.

In Table 8.18 we show p-values for the encompassing tests we employed on the Euro-area data in Section 8.7.5. Recall that the statistics $F_{Enc,1}$ tests the ICM against each of the six alternatives using a joint F-test for parsimonious encompassing of each of the two models in question against their minimal nesting model. The adjacent test, $F_{Enc,2}$ is based on pairs of model residuals from

Table 8.17
Mis-specification tests

$\Delta_4 p$ model	k	$\sigma_{\widehat{\Delta_4 p}}\%$	p-values				
			$F_{AR(1-5)}$	$F_{ARCH(1-5)}$	$\chi^2_{normality}$	F_{HETx^2}	F_{RESET}
ICM	15	0.35	0.64	0.90	0.09	0.16	0.90
MdInv	11	0.45	0.15	0.28	0.03*	0.00**	0.01**
P*	11	0.46	0.07	0.06	0.09	0.00**	0.02*
P*_enh	13	0.35	0.29	0.54	0.39	0.05*	0.44
Δp model	k	$\sigma_{\widehat{\Delta p}}\%$	$F_{AR(1-5)}$	$F_{ARCH(1-4)}$	$\chi^2_{normality}$	F_{HETx^2}	
NPCM	9	0.54	0.00**	0.33	0.50	0.04*	

Table 8.18
Encompassing tests with ICM as incumbent model (M_1)

$\Delta_4 p$ model	k	$\sigma_{\widehat{\Delta_4 p}}\%$	$F_{\text{Enc GUM}}(j, 63)$		p-values for two types of encompassing tests			
			j	p-value	$F_{\text{Enc},1}$		$F_{\text{Enc},2}$	
					M_1 vs. M_j	M_j vs. M_1	M_1 vs. M_j	M_j vs. M_1
ICM	15	0.35	51	0.67				
MdInv	11	0.45	55	0.00^{**}	0.38	0.00^{**}	0.50	0.00^{**}
P*	11	0.46	55	0.00^{**}	0.35	0.00^{**}	0.39	0.00^{**}
P*_enh	13	0.35	51	0.70	0.00^{**}	0.01^{**}	0.00^{**}	0.00^{**}
Δp model	k	$\sigma_{\widehat{\Delta p}}\%$						
NPCM	9	0.54	57	0.00^{**}			0.27	0.00^{**}

the ICM (M_1) and from each of the alternative inflation models M_j. In each case we regress $\hat{\varepsilon}_{1,t}$ against the difference between the forecast errors of model j and model 1 respectively, $\hat{\varepsilon}_{jt} - \hat{\varepsilon}_{1t}$. Under the null hypothesis that model M_1, the ICM, encompasses model M_j, the coefficient of this difference should be expected to be zero and vice versa for the opposite hypothesis that model M_j encompasses M_1.

We see from Table 8.18 that the ICM outperforms most of the alternative models on the basis of the encompassing tests. We have formed a minimal nesting model for all the models, and report p-values of F_{EncGum} tests against the minimal nesting model in the fourth column of the table. We see that the ICM and the enhanced P*-model parsimoniously encompasses the GUM. For the MdInv and the P*-model, that is, the models where we have added a set of variables from the 'monetary' information set, we obtain outright rejection of the corresponding set of restrictions relative to the GUM.[27] Also for the NPCM we clearly reject these restrictions.[28] Looking to the other tests, $F_{\text{Enc},1}$ and $F_{\text{Enc},2}$, we find that for the ICM and the enhanced P*-model, neither model encompasses the other. The tests show that the ICM clearly encompasses the other three models.

[27] It should be noted that the encompassing tests F_{EncGum}, reported in Table 8.18, are based on two-stage least squares estimation of the NPCM. In order to form the minimal nesting model it was necessary to estimate NPCM on a single equation form to make it comparable to the other (single equation) models.

[28] Strictly speaking, the generic GUM is the union of all information sets we have used to create the general models in Sections 8.7.1–8.7.3. In the minimal nesting (parsimonious) GUM we have left out all variables that are not appearing in any of the five final equations and it is more precise to call this a pGUM.

8.7.6   Comparing the forecasting properties of the models

Table 8.19 provides a summary of the forecasting properties of the inflation models. We report results for forecasting exercises where the models are
re-estimated on a sample up to the start of the forecasting horizon, and then
used to forecast quarterly and annual inflation until 2000(4). Three different
horizons are considered: a 40-period horizon with forecasts starting in 1991(1),
a 24-period horizon with forecasts starting in 1995(1) and a 12-period horizon
with forecasts starting in 1999(1).[29] The first three lines of Table 8.19 show the
RMSFE of inflation from the ICM, and its decomposition into mean forecasting
bias and standard deviation sdev. The other rows of the table shows the same
three components of the RMSFE-decomposition for each of the other inflation
models, measured relative to the results for the ICM, such that, for example,
a number greater than one indicates that the model has a larger RMSFE than
the ICM. For the forecast error bias we see that since the ICM has a very low
bias on the 40-period forecast horizon, the relative values for the other models
take on quite large values. Take the MdInv-model as an example: this performs
poorly on this long period, in part because of parameter instabilities shortly
after the start of the forecast period. Again, this can be interpreted as a result
from forecast breakdown, and this is confirmed by the relative bias (Rel. bias) of
the MdInv-model, which is around 15 for quarterly inflation and 187 for annual
inflation on the 40 period forecast horizon. The MdInv-model does much better
relative to the ICM on the two shorter forecasting horizons, which is consistent
with better parameter constancy over these horizons.

Table 8.20 shows the results from forecast encompassing tests, regressing
the forecast errors of model 1, $\hat{\varepsilon}_{1t}$, against the difference between the forecast
errors of model 2 and model 1 respectively, $\hat{\varepsilon}_{2t} - \hat{\varepsilon}_{1t}$. Under the null that
there is no explanatory power in model 2 beyond what is already reflected in
model 1, the expected regression coefficient is zero. In the table, we report p-
values when we run the forecast encompassing test in both directions. The table
shows that whereas the forecast encompassing tests are unable to discriminate
effectively between the competing models on the longest horizon, there is a clear
tendency toward ICM encompassing the competitors on the shorter forecast
horizons. These conclusions are confirmed and reinforced in the Table 8.21
which summarises the performance of the models of quarterly inflation, which
allows us also to include the NPCM in the contest. In this case the ICM is

---

[29] From the previous sections we have seen that many of the models automatically provide
forecasts of annual inflation since $\Delta_4 p_t$ is the left-hand side variable. In all models of this
type we have included $\Delta_3 p_{t-1}$ unrestrictedly as a right-hand side variable. If the coefficient
of $\Delta_3 p_{t-1}$ is close to one, the annual representation is a simple isomorphic transformation
of a similar quarterly model. The NPCM is only estimated with quarterly inflation, Δp_t, as
left-hand side variable. Thus, for the purpose of model comparison we have re-estimated all
models with Δp_t as left-hand side variable.

Table 8.19

Forecasting annual and quarterly rates of inflation. RMSFE and its decomposition. Bias, standard deviations, and RMSFE of different inflation models, relative to the ICM

$\Delta_4 p$ model		Forecasting $\widehat{\Delta p}$			Forecasting $\widehat{\Delta_4 p}$		
		91(1)–00(4)	95(1)–00(4)	98(1)–00(4)	91(1)–00(4)	95(1)–00(4)	98(1)–00(4)
ICM	RMSFE	0.0024	0.0025	0.0026	0.0024	0.0024	0.0025
	Bias	0.0004	0.0015	0.0017	0.0001	0.0014	0.0017
	sdev	0.0024	0.0020	0.0020	0.0024	0.0020	0.0019
MdInv	Rel. RMSFE	1.45	1.16	1.22	1.25	0.98	1.03
	Rel. bias	15.25	1.14	0.60	187.45	1.76	0.22
	Rel. sdev	1.66	2.26	1.93	2.48	1.15	1.28
P*	Rel. RMSFE	5.07	4.97	4.04	1.92	1.25	1.14
	Rel. bias	30.09	7.90	5.90	43.33	0.82	0.06
	Rel. sdev	1.52	1.88	1.98	1.14	1.40	1.52
P*_enh	Rel. RMSFE	1.19	1.28	1.06	1.00	1.13	0.98
	Rel. bias	4.62	1.45	0.55	15.23	1.21	0.32
	Rel. sdev	0.94	1.18	1.30	0.84	1.09	1.28
NPCM	Rel. RMSFE	3.15	2.52	2.73	3.17	2.64	2.84
	Rel. bias	9.84	1.58	2.62	42.80	1.61	2.53
	Rel. sdev	2.14	2.56	2.46	2.11	2.62	2.67

Table 8.20

Forecast encompassing tests based on forecasting annual inflation rates over 40, 24, and 12 periods ending in 2000(4). The ICM model is used as benchmark (M_1).

Model	k	$\sigma_{\widehat{\Delta_4 p}}$%	\multicolumn{2}{c}{$F_{Enc\,GUM}(j, 63)$}	\multicolumn{6}{c}{Forecast encompassing tests: p-values}
					\multicolumn{2}{c}{1991(1)–2000(4)}	\multicolumn{2}{c}{1995(1)–2000(4)}	\multicolumn{2}{c}{1998(1)–2000(4)}
			j	p-value	M_1 vs. M_j	M_j vs. M_1	M_1 vs. M_j	M_j vs. M_1	M_1 vs. M_j	M_j vs. M_1
ICM	15	0.35	51	0.67						
MdInv	11	0.45	55	0.00**	0.02*	0.00**	0.18	0.00**	0.72	0.00*
P*	11	0.46	55	0.00**	0.02**	0.00**	0.09	0.00**	0.91	0.00**
P*_enh	13	0.35	51	0.70	0.00**	0.06**	0.01*	0.00**	0.29	0.04*
NPCM	9				0.40	0.00**	0.45	0.00**	0.66	0.00**

Table 8.21

Forecast encompassing tests based on forecasting quarterly inflation rates over 40, 24, and 12 periods ending in 2000(4). The ICM model is used as benchmark (M_1)

Model	k	$\sigma_{\widehat{\Delta p}}\%$	$F_{\text{Enc GUM}}(j:63)$		Forecast encompassing tests: p-values					
					1991(1)–2000(4)		1995(1)–2000(4)		1998(1)–2000(4)	
			j	p-value	M_1 vs. M_j	M_j vs. M_1	M_1 vs. M_j	M_j vs. M_1	M_1 vs. M_j	M_j vs. M_1
ICM	15	0.33	52	0.64						
MdlInv	11	0.47	56	0.00**	0.02*	0.00**	0.14	0.00**	0.84	0.00*
P*	11	0.77	56	0.00**	0.35	0.00**	0.14	0.00**	0.46	0.00**
P*_enh	13	0.36	52	0.51	0.00**	0.01**	0.01*	0.00**	0.26	0.01*
NPCM	9	0.54	57	0.00**	0.78	0.00**	0.39	0.00**	0.64	0.00**

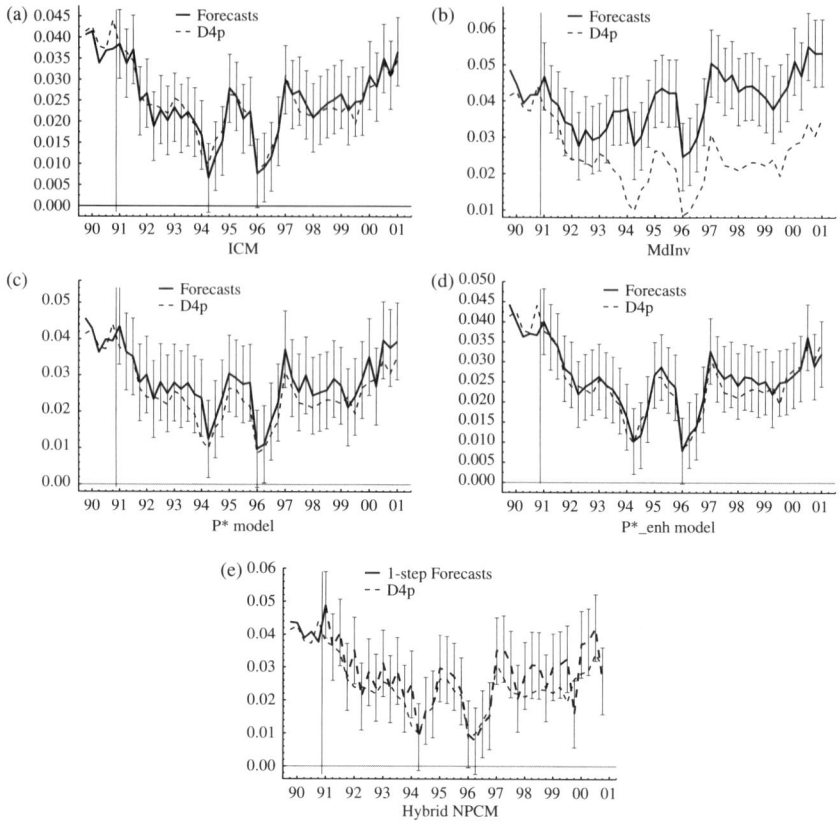

Figure 8.19. Forecasting annual CPI inflation in Norway, $\Delta_4 p_t$, over the period 1991(1)–2000(4) using five different models. The models are: (a) the ICM; (b) the MdInv model; (c) the P*-model; (d) the enhanced P*-model; and (e) the hybrid NPCM. The bars show $2\times$ forecast errors

encompassing the NPCM on all horizons, and it is also encompassing the P*-model on the longest horizon, from 1991(1) to 2000(4).[30]

8.7.7 Summary of the findings—Norway vs. Euro area

The overall conclusion from the comparisons of inflation models for the Norwegian economy is that monetary measures do not play an important part in explaining and/or predicting Norwegian inflation. The preferred specifications of money demand do not include inflation as a significant explanatory

[30] The caveat mentioned in Section 8.6.6 of the P*-model being greatly helped by the use of two-sided HP-filters is also relevant for the case of Norway.

variable and hence the money demand equation cannot be interpreted as an inverted inflation equation. An attempt to model an inflation equation as an inverted money demand function shows clear signs of mis-specification and the MdInv model is demonstrated to be inferior to all other competitors based on in-sample evaluations as well as in forecasting (Figure 8.19). Also the P*-model, which embody several aggregates which monetarist theorists predict would explain inflation, fails to do so. Only when we augment the P*-model with the equilibrium-correction term for broad money, $ecmmd_{t-1}$, does the model (P*-enhanced) appear to perform adequately in explaining and predicting the inflation series.

It is shown elsewhere in this book that the ICM gives a data congruent representation of Norwegian wage–price formation. In this chapter, it transpires that a reduced form representation of the ICM seems to perform better than the rival models it is compared to based on in-sample evaluations as well as forecasting. Moreover, there are no signs of the neglected monetary effects in the reduced form ICM inflation equation. In conclusion, the support for the ICM inflation model is much stronger in the case of the small open economy Norway, than in the case of a large aggregated economy, as is the Euro area. The AWM reduced form inflation equation emerges as the strongest contender amongst the Euro-area inflation equations and the enhanced P*-model is almost equally good.[31]

---

[31] Recall the caveat in Section 8.6.6—that the P*-model is unduly helped by the use of two-sided filters—which further strengthens the case for the AWM.

9

Transmission channels and model properties

In this chapter, we develop an econometric model for forecasting of inflation in Norway, an economy that recently opted for inflation targeting. We illustrate the estimation methodology advocated earlier, by estimating and evaluating a model of prices, wages, output, unemployment, the exchange rate, and interest rates on government bonds and bank loans. The model is built up sequentially. We partition the simultaneous distribution function into a small model of wages and prices, and several marginal models for the rest of the economy. The choice of model framework for the wage and price model follows from the analysis in earlier chapters. We use the model to analyse the transmission mechanism and to address monetary policy issues related to inflation targeting.

9.1 Introduction

On 29 March 2001 Norway adopted inflation targeting. Rather than stabilising the exchange rate by pegging the Norwegian Krone to the Euro (or previously a basket of foreign currencies) the central bank became committed to an inflation target of 2.5%. This was in line with an international trend, as countries like Canada, New Zealand, Sweden, and the United Kingdom had already changed their monetary policy towards an explicit inflation target; cf. Bernanke *et al.* (1999).

   Research on monetary policy has focused on the conditional inflation forecast as the operational target for monetary policy, yet the literature is dominated by either theoretical or calibrated models—examples are Ball (1999), Batini and Haldane (1999), Røisland and Torvik (2004), Walsh (1999), Svensson (2000), Woodford (2000, 2003) and Holden (2003). True to the approach taken in this book we will argue that econometric evaluation of models is useful, not only as an aid in the preparation of inflation forecasts, but also as a way of testing, quantifying, and elucidating the importance of transmission mechanisms in the inflationary process. In this way, inflation targeting moves the quality of econometric methodology and practice into the limelight of the economic policy debate.

Inflation is a many-faceted phenomenon in open economies, and models that include only a few dimensions, for example, the output gap and expectations of the future rate of inflation, are likely to fail in characterising the data adequately, as demonstrated in Chapter 7. Econometric work that views inflation as resulting from disequilibria in many markets fares much better (see Hendry 2001*b* and Juselius 1992). Our starting point is therefore that, at a minimum, foreign and domestic aspects of inflation have to be modelled jointly, and that the inflationary impetus from the labour market—the battle of markups between unions and monopolistic firms—needs to be represented, for example, as in the Incomplete Competition Model (ICM) which also stands out as the preferred model in Chapter 8.

The approach taken in this chapter to construct a small model of inflation is illustrated in Figure 9.1.

The focus is on the simultaneous wage–price model $D_y(\mathbf{y}_t \mid \mathbf{z}_t, \mathbf{Y}_{t-1}, \mathbf{Z}_{t-1})$, where $\mathbf{y}_t = [w_t p_t]'$, the vector $\mathbf{z}_t$ contains all conditioning variables, and $(\mathbf{Y}_{t-1}, \mathbf{Z}_{t-1})$ collects all lagged values of $\mathbf{y}_t$ and $\mathbf{z}_t$. The variables in $\mathbf{z}_t$ are partitioned into $[\mathbf{z}_{1,t} \ \mathbf{z}_{2,t} \ \mathbf{z}_{3,t}]'$, where $\mathbf{z}_{1,t}$ denote feedback variables, $\mathbf{z}_{2,t}$ are non-modelled variables, and $\mathbf{z}_{3,t}$ are monetary policy instruments. Lagged values are partitioned correspondingly, $\mathbf{Z}_{t-1} = (\mathbf{Z}_{1,t-1}, \mathbf{Z}_{2,t-1}, \mathbf{Z}_{3,t-1})$.

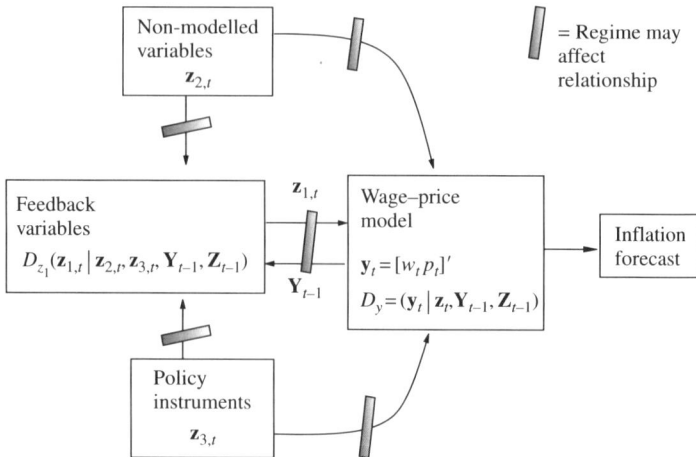

Figure 9.1. Model-based inflation forecasts

The feedback variables $\mathbf{z}_{1,t}$ include unemployment, output, productivity, and import prices.[1] Figure 9.1 indicates that the marginal models, $D_{\mathbf{z}_1}(\mathbf{z}_{1,t} \mid \mathbf{z}_{2,t}, \mathbf{z}_{3,t}, \mathbf{Y}_{t-1}, \mathbf{Z}_{t-1})$, are not only functions of lagged wages and prices, but may also depend on both the non-modelled explanatory variables $\mathbf{z}_{2,t}$ and on the policy variables $\mathbf{z}_{3,t}$. The feedback variables are treated as weakly exogenous variables in the wage–price model. This is a testable property that we address in Section 9.4 after modelling the feedback relationships.

The conditional non-modelled variables $\mathbf{z}_{2,t}$ consist of domestic tax-rates and world prices. The crucial question for the policy instruments $\mathbf{z}_{3,t}$ is whether there exists a single reaction function for the interest rate. Norway was pegging its exchange rate to different currency baskets throughout the sample period, which is 1972(4) 2001(1).[2] For a substantial part of this time period the country saw frequent devaluations, particularly in the 1980s. Finding an empirically constant reaction function from inflation forecasts to interest rates is therefore a non-starter. Hence, we treat the short-run interest rate as a strongly exogenous policy variable, meaning that there is no reaction function in the model linking the inflation forecast to the interest rate.[3] The important monetary feedback variable is the exchange rate, determining import prices for given foreign prices. The exchange rate depends on inflation, the short-run interest rate and foreign variables.[4]

Section 9.2 sets out the core model of inflation as a wage–price system, conditional on output, productivity, unemployment, and the exchange rate. After evaluating steady-state properties, we derive a dynamic model for wage and price growth. We enlarge this core model to include relationships for output, productivity, unemployment, and exchange rates in Section 9.3, and the exogeneity assumptions underlying such a modelling strategy are examined in Section 9.4. Equipped with the core model and the marginal models we next establish a small econometric model. Despite aggregation of aggregate demand, it is seen that the simultaneous model captures essential features of the transmission mechanisms in the inflationary process for the small open economy. It provides a testing bed for the impact of policy changes on the economy. In particular, it highlights the behaviour of exchange rates, which is central to the conduct of monetary policy in small open economies. The exchange rate behaviour is characterized by a data-consistent empirical model

---

[1] We model the mainland economy only, although the oil sector accounts for close to 20 per cent of total GDP. The oil activities, including the huge oil investments, are driven by factors that are exogenous to the mainland economy, which we have chosen to focus on.

[2] In other words, a formal inflation target was introduced at the end of the last quarter included in the sample.

[3] In Chapter 10 we analyse the performance of different monetary reaction functions.

[4] A precursor to the model can be found in Bårdsen et al. (2003). Other comparable econometric studies are Sgherri and Wallis (1999), Jacobson et al. (2001), and Haldane and Salmon (1995)—albeit with different approaches and focus.

with short-run interest rate and inflation effects, and convergence towards purchasing power parity (PPP) in the long run.

Section 9.5 contains a discussion of the main monetary policy channels in the model, that is, both the interest rate and the exchange rate channels. We also evaluate the properties of the model for inflation forecasting, while we study the effects of an exogenous change in the interest rate in Section 9.6. In Section 9.7 we sum up our experiences so far.

9.2   The wage–price model

We first model the long-run equilibrium equations for wages and prices based on the framework of Chapter 5. As we established in Section 5.4 the long-run equations of that model can be derived as a particular identification scheme for the cointegrating equations; see (5.19)–(5.20). Second, we incorporate those long-run equations as equilibrium correcting terms in a dynamic two-equation simultaneous core model for (changes in) wages and prices.

9.2.1   Modelling the steady state

From equations (5.19)–(5.20), the variables that contain the long-run real wage claims equations are collected in the vector $[w_t \; p_t \; a_t \; pi_t \; u_t]'$. The wage variable w_t is average hourly wages in the mainland economy, excluding the oil producing sector and international shipping. The productivity variable a_t is defined accordingly—as mainland economy value added per man hour at factor costs. The price index p_t is the official consumer price index. Import prices pi_t are measured as the deflator of total imports. The unemployment variable u_t is the rate of open unemployment, excluding labour market programmes.

In addition to the variables in the wage-claims part of the system, we include (as non-modelled and without testing) the payroll-tax $t1_t$, indirect taxes $t3_t$, energy prices pe_t, and output y_t—the changes in which represent changes in the output gap, if total capacity follows a trend. Institutional variables are also included. Wage compensation for reductions in the length of the working day is captured by changes in the length of the working day Δh_t—see Nymoen (1989b). The intervention variables $W\mathrm{dum}_t$ and $P\mathrm{dum}_t$ are used to capture the impact of incomes policies and direct price controls. This system, where wages and prices enter with three lags and the other main variables enter with one or two lags, is estimated over 1972(4)–2001(1).

We impose restrictions on the steady-state equations (5.19)–(5.20), by assuming no wedge and normal cost pricing. We also find empirical support that changes in indirect taxes are off-set in long-run inflation with a factor of 50%.

Table 9.1
The estimated steady-state equations

The estimated steady-state equations (9.1)–(9.2)

$$w = p + a - 0.11u$$
$$(0.01)$$

$$p = 0.73(w + t1 - a) + 0.27pi + 0.5t3$$
$$(0.08)$$

Cointegrated system

46 parameters	w_t	p_t	*System*
$\chi^2_{normality}$ (2)	4.21[0.12]	2.48[0.29]	
$F_{HET\chi^2}$ (22, 83)	1.01[0.46]	1.28[0.21]	
$\chi^2_{overidentification}$ (8)			13.21[0.10]
$\chi^2_{normality}$ (4)			5.14[0.27]
$F^v_{HET\chi^2}$ (66, 138)			0.88[0.72]

Note: Reference: see Table 9.2. The numbers in [..] are p-values. The sample is 1972(4)–2001(1), 114 observations.

We end up with a restricted form where only ϑ and ϕ enter unrestrictedly:

$$w = p + a - \vartheta u, \tag{9.1}$$
$$p = (1 - \phi)(w - a + t1) + \phi pi + 0.5t3, \tag{9.2}$$

with estimation results in Table 9.1.

The results are qualitatively the same as the results for Norway in Bårdsen *et al.* (1998) for a sample covering 1966(4)–1993(1) and the near identical results in Bårdsen *et al.* (2003), where the sample covers 1966(4)–1996(4).[5] Figure 9.2 records the stability over the period 1984(1)–2001(1) of the coefficient estimates in Table 9.1 with ±2 standard errors (±2se in the graphs), together with the tests of constant cointegrating vectors over the sample. We note that the eight overidentifying long-run restrictions are accepted by the data at all sample sizes. The estimated wage responsiveness to the rate of unemployment is approximately 0.1, which is close to the finding of Johansen (1995*a*) for manufacturing wages. This estimated elasticity is numerically large enough to represent a channel for economic policy on inflation.

On the basis of Table 9.1 we conclude that the steady-state solution of our system can be represented as

$$w = p + a - 0.1u, \tag{9.3}$$
$$p = 0.7(w + t1 - a) + 0.3pi + 0.5t3. \tag{9.4}$$

---

[5] Compared to the previous findings, the weight on productivity and tax corrected wages is increased and the effect of indirect taxes reduced in the price equation.

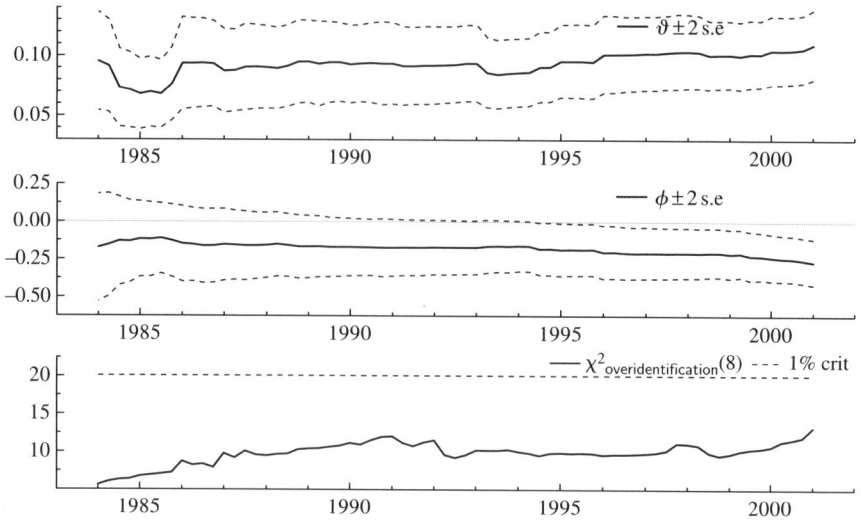

Figure 9.2. Identified cointegration vectors. Recursively estimated parameters (elasticity of unemployment in the wage equation and the elasticity of the import price in the price equation) and the $\chi^2(8)$ test of the overidentifying restrictions of the long-run system in Table 9.1

9.2.2 The dynamic wage–price model

When modelling the short-run relationships we impose the estimated steady state from (9.3) to (9.4) on a subsystem for $\{\Delta w_t, \Delta p_t\}$ conditional on $\{\Delta a_t, \Delta y_t, \Delta u_{t-1}, \Delta pi_t, \Delta t1_t, \Delta t3_t\}$ with all variables entering with two additional lags. In addition to energy prices Δpe_t, we augment the system with $\{\Delta h_t, W\text{dum}_t, P\text{dum}_t\}$ to capture short-run effects. Seasonal$_t$ is a centred seasonal dummy. The diagnostics of the unrestricted $I(0)$ system are reported in the upper part of Table 9.2.

The short-run model is derived general to specific by deleting insignificant terms, establishing a parsimonious statistical representation of the data in $I(0)$-space, following Hendry and Mizon (1993) and is found below:

$$
\begin{aligned}
\widehat{\Delta w_t} = &-0.124 + \underset{(0.109)}{0.809}\Delta p_t - \underset{(0.123)}{0.511}\Delta h_t + \underset{(0.017)}{0.081}\Delta a_t \\
&- \underset{(0.021)}{0.163}\,(w_{t-1} - p_{t-1} - a_{t-1} + 0.1u_{t-2}) + \underset{(0.002)}{0.024}\,\text{Seasonal}_{t-2} \\
&- \underset{(0.003)}{0.020}W\text{dum}_t + \underset{(0.004)}{0.023}\,P\text{dum}_t
\end{aligned}
\tag{9.5}
$$

$$\sigma = 0.0089.$$

Table 9.2
Diagnostics for the unrestricted $I(0)$ wage–price
system and the model

Unrestricted $I(0)$ system	
52 parameters	
$F^v_{AR}(1-5)$ $(20, 154)$	$0.68[0.85]$
$\chi^{2,v}_{normality}$ (4)	$4.39[0.36]$
$F^v_{HET\times^2}$ $(141, 114)$	$0.81[0.88]$
Final model	
19 parameters	
$\chi^2_{overidentification}$ (33)	$33.72[0.43]$
$F^v_{AR(1-5)}$ $(20, 188)$	$1.45[0.10]$
$\chi^{2,v}_{normality}$ (4)	$6.82[0.15]$
$F^v_{HET\times^2}$ $(141, 165)$	$1.23[0.10]$

Note: References: overidentification test
(Anderson and Rubin 1949, 1950; Koopmans *et
al.* 1950; Sargan 1988), AR-test (Godfrey 1978;
Doornik 1996), Normality test (Doornik and
Hansen 1994), and Heteroskedasticity test
(White 1980; Doornik 1996). The numbers in [..]
are *p*-values.
The sample is 1972(4)–2001(1), 114 observations.

$$
\begin{aligned}
\widehat{\Delta p_t} = \ & 0.006 + 0.141\Delta w_t + 0.100\Delta w_{t-1} + 0.165\Delta p_{t-2} - 0.015\Delta a_t \\
& (0.001) \quad (0.026) \qquad (0.021) \qquad\quad (0.048) \qquad\quad (0.006) \\
& + 0.028\Delta y_{t-1} + 0.046\Delta y_{t-2} + 0.026\Delta pi_t + 0.042\Delta pe_t \\
& \quad (0.012) \qquad\quad (0.012) \qquad\quad (0.008) \qquad\quad (0.007) \\
& - 0.055\,(p_{t-3} - 0.7(w_{t-2} + tl_{t-1} - a_{t-1}) - 0.3pi_{t-1} - 0.5t3_{t-1}) \\
& \quad (0.006) \\
& - 0.013\,Pdum_t \\
& \quad (0.001)
\end{aligned}
\tag{9.6}
$$

$\sigma = 0.0031$

$T = 1972(4)\text{–}2001(1) = 114.$

The lower part of Table 9.2 contains diagnostics for the final model. In
particular, we note the insignificance of $\chi^2_{overidentification}$ (33), which shows that
the model reduction restrictions are supported by the data.

The wage growth equation implies that a one percentage point increase
in the rate of inflation raises wage growth by 0.8 percentage point. The
discretionary variables for incomes policies ($Wdum_t$) and for price controls
($Pdum_t$) are also significant. Hence, discretionary policies have clearly

succeeded in affecting consumer real wage growth over the sample period.
The equilibrium-correction term is highly significant, as expected. Finally, the
change in normal working-time Δh_t enters the wage equation with a negative
coefficient, as expected. In addition to equilibrium-correction and the dummies
representing incomes policy, price inflation is significantly influenced by wage
growth and output growth (the output gap), together with effects from import
prices and energy prices—as predicted by the theoretical model.

The question whether wage–price systems like ours imply a NAIRU property hinges on the detailed restrictions on the short-run dynamics. A necessary
condition for a NAIRU is that wage growth is homogenous with respect to the
change in producer prices, Δq_t. Using, $\Delta p_t \equiv (1 - \phi)\Delta q_t + \phi\Delta pi_t$, and since
Δpi_t does not enter the wage equation, it is clear that a homogeneity restriction
does not hold in the wage growth equation (9.5): using the maintained value
of $\phi = 0.3$ from (9.4) the implied wage elasticity with respect to the change
in producer prices, Δq_t is 0.56. The wage equation therefore implies that we
do not have a NAIRU model. Hence, the conventional Phillips curve NAIRU,
for example, does not correspond to the eventual steady-state rate of unemployment implied by the larger model obtained by grafting the wage and price
equations in a larger system of equations.

The model has constant parameters, as shown in Figure 9.3, which contains
the one-step residuals and recursive Chow-tests for the model. Finally, the lower
left panel of Figure 9.3 shows that the model parsimoniously encompasses the

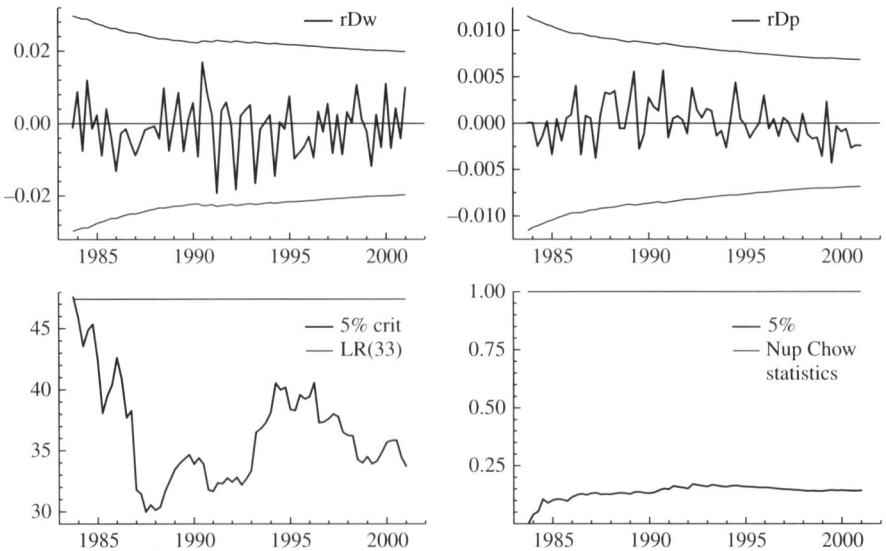

Figure 9.3. Recursive stability tests for the wage–price model. The upper
panels show recursive residuals for the model. The lower panels show recursive
encompassing tests (left) and recursive Chow tests (right)

system at every sample size. As noted in the introduction, improperly modelled expectations in the dynamic simultaneous equations model could cause the model's parameters to change when policies change, generating misleading policy simulations, as emphasised by Lucas (1976). However, as Figure 9.3 shows, there is no evidence of any mis-specified expectations mechanisms.

9.3 Closing the model: marginal models for feedback variables

We have established a wage–price model conditional upon the exchange rate v_t (which works through pi_t), GDP mainland output y_t, the rate of unemployment u_t, and average labour productivity a_t. In this section, we enlarge the model to include relationships for these four variables and functions for real credit cr_t, and two interest rates: for government bonds RBO_t and for bank loans RL_t. This serves three purposes: first, all of these variables are affected by the monetary policy instrument (represented in the model by the money market interest rate) and are therefore channels for monetary instruments to influence inflation; second, none of these variables are likely to be strongly exogenous. For example, import prices depend by definition on the nominal exchange rate. Below we report a model that links the exchange rate to the lagged real exchange rate, which in turn depends on the domestic price level; third, we make use of the marginal models to test the exogeneity assumptions that underlie the estimation strategy of the wage–price model as well as conditions for valid use of the full model for policy simulations.[6]

9.3.1 The nominal exchange rate v_t

The nominal exchange rate affects wages and prices via import prices pi_t. Let pf_t be an index of import prices in foreign currencies. Then, as a first step in the completion of the model, we make use of the identity

$$pi_t = v_t + pf_t$$

and attempt to model the (log) of the trade weighted exchange rate index v_t. In doing so, we follow Akram (2004), who models the exchange rate as equilibrium-correcting to the real exchange rate, which means that it is determined by PPP in steady state,

$$ecm_{v,t} = v_t + pw_t - p_t,$$

where pw_t is the log of a trade-weighted index of foreign consumer prices. Figure 9.4 shows the time-series properties of $ecm_{v,t}$, together with the corresponding term $ecm_{y,t}$ from the aggregate demand equation developed later.

---

[6] The marginal models reported below are estimated with OLS.

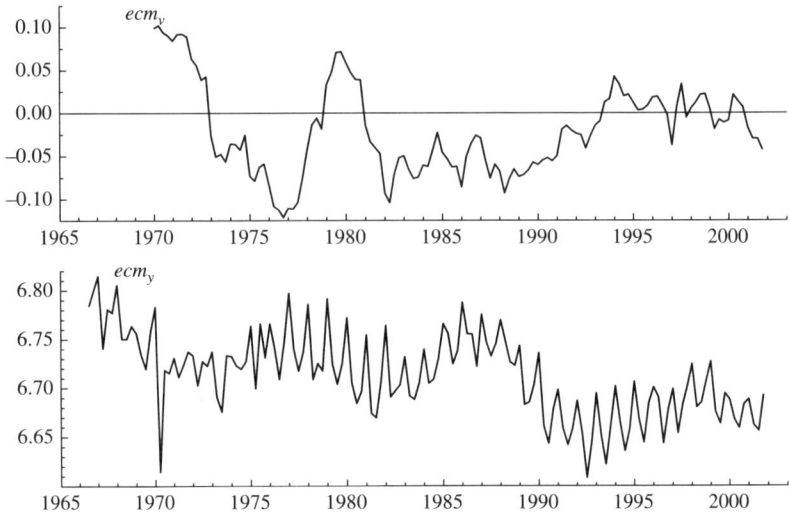

Figure 9.4. The equilibrium-correction terms of the exchange rate and the aggregate demand equations

The graphs of the *ecms* indicate stationary behaviour, corresponding to short-run deviations from steady state.

The resulting model is given as

$$\Delta v_t = -\underset{(0.08)}{0.35}\Delta RS_t - \underset{(0.19)}{0.41}\,sRISK_t + \underset{(0.04)}{0.15}\,(s \cdot \Delta(euro/dollar))_t$$
$$- \underset{(0.03)}{0.13}\,\Delta oilST_t - \underset{(0.03)}{0.06}\,(v + pw - p)_{t-2} + \underset{(0.004)}{0.04}\,Vdum_t + \underset{(0.01)}{0.02}$$

$$(9.7)$$

$$T = 1972(4)\text{--}2001(1) = 114$$
$$\hat{\sigma} = 1.24\%$$
$$F_{AR(1\text{--}5)}(5, 102) = 1.76[0.13]$$
$$\chi^2_{normality}(2) = 5.64[0.06]$$
$$F_{HET\chi^2}(12, 94) = 0.55[0.88].$$

(Reference: see Table 9.2. The numbers in [..] are *p*-values.)

Akram (2004) documents significant non-linear effects of the USD price of North Sea oil on the Norwegian exchange rate. Our model is built along the same lines and therefore features non-linear effects from oil prices (oil$_t$) in the form of a smooth transition function (see Teräsvirta 1998),

$$\Delta oilST_t = \Delta oil_t/\{1 + \exp[4(oil_t - 14.47)]\}.$$

The implication is that an oil price below 14 USD triggers depreciation of the krone.

As for the other right-hand side variables, the first term implies that there is a negative (appreciation) effect of an increase in the money market interest rate ΔRS_t. The variable $sRISK_t$ captures deviations from uncovered interest rate parity (see Rødseth 2000, p.15) after 1998(4):

$$sRISK_t = RS_{t-1} - RW_t - (\Delta v_{t-1} - 0.8v_{t-1}) \qquad \text{for } t > 1998(4)$$
$$sRISK_t = 0.0394 \qquad\qquad\qquad\qquad \text{for } t \leq 1998(4),$$

where RW_t is the three months Euro money market rate and $(\Delta v_{t-1} - 0.8v_{t-1})$ is the expected change in the nominal exhange rate, $E(\Delta v_t)$. The term $(s \cdot \Delta(euro/dollar))_t$ reflects the fact that we are modelling the trade-weighted exchange rate, which is influenced by the changes in the relative value of United States dollar to Euro (Ecu). This effect is relevant for the period after the abolition of currency controls in Norway in 1990(2), which is why we multiply with a step dummy, s_t, that is 0 before 1990(3) and 1 after.

Finally, there is a composite dummy

$$Vdum_t = [-2 \times i73q1 + i78q1 + i82q3 + i86q3 + 0.7i86q4 - 0.1s86q4\_01q4$$
$$- i97q1 + i97q2]_t$$

to take account of devaluation events. Figure 9.5 shows the sequence of 1-step residuals for the estimated Δv_t equation, together with similar graphs for the following three marginal models reported.

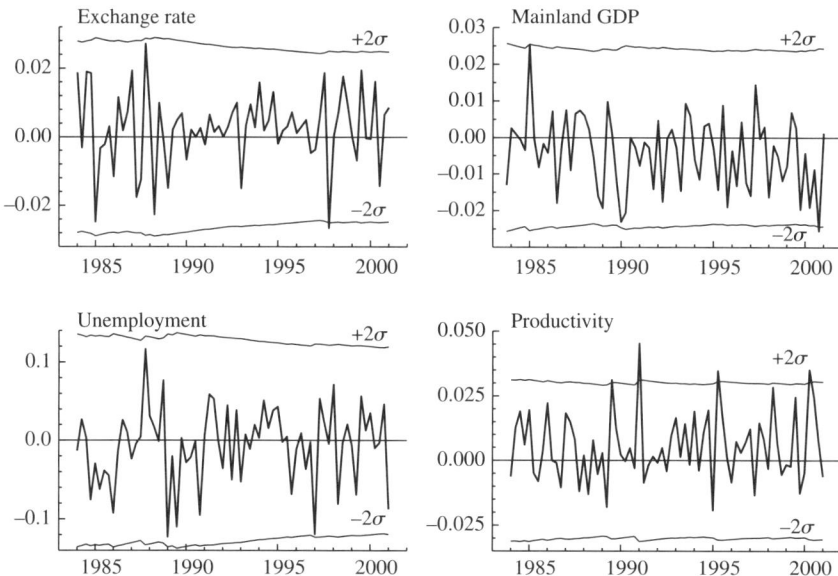

Figure 9.5. Marginal equations: recursive residuals and ± 2 standard errors (σ)

9.3.2   Mainland GDP output y_t

The model for Δy_t is adapted from the 'AD' equation in Bårdsen and Klovland (2000). The growth in output Δy_t is in the short run a function of public demand Δg_t, and growth in private demand—represented by growth in real private credit Δcr_t. Moreover, there is an effect from the change in the real exchange rate in the period after the deregulation of currency controls in Norway in 1990(2).

$$
\begin{aligned}
\Delta y_t = \; & \underset{(0.30)}{1.16} \; - \; \underset{(0.07)}{0.39 \Delta y_{t-1}} \; + \; \underset{(0.06)}{0.29 \Delta g_t} \; + \; \underset{(0.12)}{0.49 \Delta cr_{t-1}} \\
& - \; \underset{(0.05)}{0.17 \, ecm_{y,t}} \; + \; \underset{(0.12)}{0.41 \, (s \cdot \Delta(v + pw - p))_{t-2}} \; + \; \underset{(0.01)}{0.07 \mathrm{Ydum}_t} \\
& - \; \underset{(0.003)}{0.06 \, \mathrm{Seasonal}_{t-1}} \; - \; \underset{(0.005)}{0.07 \, \mathrm{Seasonal}_{t-2}} \; - \; \underset{(0.004)}{0.03 \, \mathrm{Seasonal}_{t-3}}
\end{aligned}
$$

$$ (9.8) $$

$$ T = 1972(4)\text{--}2001(1) = 114 $$
$$ \hat{\sigma} = 1.21\% $$
$$ \mathsf{F}_{\mathrm{AR}(1-5)}(5, 99) = 0.84[0.53] $$
$$ \chi^2_{\mathrm{normality}}(2) = 0.78[0.67] $$
$$ \mathsf{F}_{\mathrm{HET}x^2}(14, 89) = 0.48[0.94]. $$

(Reference: see Table 9.2. The numbers in [..] are p-values.)

The equilibrium-correction mechanism of aggregate demand, denoted $ecm_{y,t}$ is defined as:

$$ ecm_{y,t} = y_{t-1} - 0.5 y w_{t-1} - 0.5 g_{t-1} + 0.3 (\mathrm{RL} - 4\Delta p)_{t-1}, $$

where the long-run steady-state is determined by real public consumption expenditure (g_t), real foreign demand, which is proxied by the weighted GDP for trading partners (yw_t), and the real interest rate on bank loans rate $(\mathrm{RL} - 4\Delta p)_t$, where RL_t is the nominal bank loan rate. The estimated adjustment coefficient of -0.17, suggests a moderate reaction to shocks to demand. The estimated equation also includes a constant and three seasonal dummies and in addition the dummy $\mathrm{Ydum}_t = [i75q2]_t$ is required to whiten the residuals.

9.3.3   Unemployment u_t

The dynamics of unemployment Δu_t display strong hysteresis effects, with very sluggish own dynamics. Also aggregate demand shocks $\Delta_4 y_t$ and changes in the real wage $\Delta(w - p)_t$ have significant short-run effects. Moreover, there are significant effects of change in foreign demand Δyw, and the share of the workforce between 16 and 49 years old $N_{16-49,t}$.

$$
\begin{aligned}
\Delta u_t = -\ &1.23 \ + \ 0.34\Delta u_{t-1} - \ 0.06 u_{t-1} - \ 1.83\Delta_4 y_t \\
&(0.49) \quad (0.07) \qquad\quad (0.02) \qquad\quad (0.29) \\
&+ \ 1.30\Delta(w-p)_{t-1} - 2.63\Delta y w_{t-2} + 1.78 N_{16-49,t} + \ 0.22\text{Udum}_t \\
&\ \ (0.52) \qquad\qquad (1.03) \qquad\quad (0.71) \qquad\qquad (0.04) \\
&+ \ 0.41\text{Seasonal}_{t-1} + \ 0.10\text{Seasonal}_{t-2} + \ 0.29\text{Seasonal}_{t-3} \\
&\ \ (0.03) \qquad\qquad (0.02) \qquad\qquad (0.02) \\
&- \ 6.46 ch\text{Seasonal}_{t-1} - \ 7.55 ch\text{Seasonal}_{t-2} - \ 4.34 ch\text{Seasonal}_{t-3}. \\
&\ \ (0.49) \qquad\qquad (0.34) \qquad\qquad (0.39)
\end{aligned}
$$

$$(9.9)$$

$$
\begin{aligned}
T &= 1972(4)\text{--}2001(1) = 114 \\
\hat{\sigma} &= 5.97\% \\
\mathsf{F}_{\mathsf{AR}(1\text{-}5)}(5,95) &= 0.69[0.63] \\
\chi^2_{\text{normality}}(2) &= 1.91[0.38] \\
\mathsf{F}_{\mathsf{HET}\chi^2}(23,76) &= 2.21[0.005].
\end{aligned}
$$

(Reference: see Table 9.2. The numbers in [..] are p-values.)

There are two sets of seasonals in this equation. chSeasonal$_t$ is designed to capture a gradual change in seasonal pattern over the period:

$$
ch\text{Seasonal}_t = \frac{1}{1 + e^{0.5+0.35\cdot\text{Trend}_t}}\text{Seasonal}_t.
$$

Moreover, a composite dummy variable $\text{Udum}_t = [i75q1 + i75q2 - i87q2]_t$ is required to whiten the residuals.

Summing up, the unemployment equation in essence captures Okun's law. An asymptotically stable solution of the model would imply $\bar{u} = \text{const} + f(\overline{\Delta y})$, so there is a one-to-one relationship linking the equilibria for output growth and unemployment.

9.3.4 Productivity a_t

Productivity growth Δa_t is basically modelled as a moving average with declining weights

$$
\begin{aligned}
\Delta a_t = \ &0.73 \ - \ 0.76\Delta a_{t-1} - \ 0.79\Delta a_{t-2} - \ 0.48\Delta a_{t-3} \\
&(0.15) \quad (0.05) \qquad\quad (0.05) \qquad\quad (0.10) \\
&- \ 0.18 ecm_{a,t} - \ 0.06\text{Adum}_t + \ 0.08\text{Seasonal}_{t-3} \qquad (9.10) \\
&\ \ (0.04) \qquad\quad (0.02) \qquad\quad (0.01)
\end{aligned}
$$

$$T = 1972(4)–2001(1) = 114$$
$$\hat{\sigma} = 1.52\%$$
$$\mathsf{F}_{\mathsf{AR}(1–5)}(5, 102) = 0.17[0.97]$$
$$\chi^2_{\text{normality}}(2) = 1.23[0.54]$$
$$\mathsf{F}_{\mathsf{HET}x^2}(10, 96) = 0.74[0.69].$$

(Reference: see Table 9.2. The numbers in [..] are *p*-values.)

In the longer run the development is influenced by the real wage, by unemployment and by technical progress—proxied by a linear trend—as expressed by the equilibrium correction mechanism

$$ecm_{a,t} = a_{t-4} - 0.3(w - p)_{t-1} - 0.06u_{t-3} - .002\,Trend_t.$$

The dummy $Adum_t = [i86q2]_t$ picks up the effect of a lock-out in 1986(2) and helps whiten the residuals.

9.3.5　Credit expansion cr_t

The growth rate of real credit demand, Δcr_t, is sluggish, and it is also affected in the short run by income effects. In addition the equation contains a step dummy s_t for the abolition of currency controls (which again takes the value 1 after 1990(3) and (0) before) and a composite dummy variable

$$\text{CRdum}_t = [0.5i85q3 + i85q4 + 0.5i86q1 + i87q1 + Pdum]_t$$

to account for the deregulation of financial markets.

$$
\begin{aligned}
\Delta cr_t = &-\underset{(0.05)}{0.26} + \underset{(0.06)}{0.17\Delta cr_{t-1}} + \underset{(0.06)}{0.42\Delta cr_{t-2}} + \underset{(0.02)}{0.10\Delta y_t} \\
&-\underset{(0.12)}{0.27\Delta \text{RL}_{t-1}} - \underset{(0.005)}{0.026 ecm_{cr,t}} + \underset{(0.002)}{0.015\text{CRdum}_t} - \underset{(0.002)}{0.006 s_t} \ (9.11)
\end{aligned}
$$

$$T = 1972(4)–2001(1) = 114$$
$$\hat{\sigma} = 0.61\%$$
$$\mathsf{F}_{\mathsf{AR}(1-5)}(5, 101) = 0.52[0.75]$$
$$\chi^2_{\text{normality}}(2) = 0.06[0.97]$$
$$\mathsf{F}_{\mathsf{HET}x^2}(13, 92) = 0.94[0.51].$$

(Reference: see Table 9.2. The numbers in [..] are *p*-values.)

The long-run properties are those of a standard demand function—with an elasticity of 2 with respect to income and a negative effect from opportunity costs, as measured by the difference between bank loan rates RL and bond rates RBO

$$ecm_{cr,t} = cr_{t-3} - 2y_{t-1} + 2.5(\text{RL}_{t-1} - \text{RBO}_{t-1}).$$

9.3.6 Interest rates for government bonds RBO_t and bank loans RL_t

Finally, the model consists of two interest rate equations. Before the deregulation, so $s_t = 0$, changes in the bond rate RBO_t are an autoregressive process, corrected for politically induced changes modelled by a composite dummy.

$$\begin{aligned}
\Delta RBO_t = {}& \underset{(0.04)}{0.12}\Delta RBO_{t-1} + \underset{(0.03)}{0.30}s\Delta RS_t + \underset{(0.07)}{0.95}s\Delta RW_t \\
& - \underset{(0.01)}{0.02}s \cdot ecm_{RBO,t-1} + \underset{(0.001)}{0.011}RBOdum_t \qquad (9.12)
\end{aligned}$$

$$T = 1972(4)\text{--}2001(1) = 114$$
$$\hat{\sigma} = 0.18\%$$
$$\mathsf{F}_{AR(1-5)}(5, 104) = 0.83[0.53]$$
$$\chi^2_{normality}(2) = 0.46[0.80]$$
$$\mathsf{F}_{HET\chi^2}(10, 98) = 1.61[0.11]$$

(Reference: see Table 9.2. The numbers in [..] are p-values.)

where

$$\begin{aligned}
RBOdum_t = {}& [i74q2 + 0.9i77q4 - 0.6i78q1 + 0.6i79q4 + i80q1 \\
& + i81q1 + i82q1 + 0.5i86q1 - 1.2i89q1]_t.
\end{aligned}$$

After the deregulation, the bond rate reacts to the changes in the money-market rate $s\Delta RS_t$ as well as the foreign rate $s\Delta RW_t$, with the long-run effects represented by the equilibrium-correcting term:

$$ecm_{RBO,t-1} = (RBO - 0.6RS - 0.75RW)_{t-1}.$$

The equation for changes in the bank loan rate ΔRL_t is determined in the short run by changes in the bond rate, with additional effects from changes in the money-market rate $s\Delta RS_t$ after the deregulation.

$$\begin{aligned}
\Delta RL_t = {}& - \underset{(0.0002)}{0.0007} + \underset{(0.03)}{0.09}\,\Delta RL_{t-1} + \underset{(0.03)}{0.37}\,s\Delta RS_t + \underset{(0.035)}{0.11}\,\Delta RBO_{t-1} \\
& - \underset{(0.03)}{0.29}\,s \cdot ecm_{RL,t-1} + \underset{(0.0003)}{0.001}\,s66_t + \underset{(0.001)}{0.012}\,RLdum_t \qquad (9.13)
\end{aligned}$$

$$T = 1972(4)\text{--}2001(1) = 114$$
$$\hat{\sigma} = 0.15\%$$
$$\mathsf{F}_{AR(1-5)}(5, 102) = 1.01[0.42]$$
$$\chi^2_{normality}(2) = 1.04[0.59]$$
$$\mathsf{F}_{HET\chi^2}(11, 95) = 0.89[0.55].$$

(Reference: see Table 9.2. The numbers in [..] are p-values.)

Again a rather elaborated composite dummy is needed in order to obtain white noise residuals

$$RLdum_t = [i78q1 + 0.5i80q3 + 0.75i81q2 + 0.5i86q1 - 86q2$$
$$+ 0.75i86q4 - 0.5i89q1 - 89q3 - 0.67i92q4 + 2i98q3]_t.$$

In the long run, the pass-through of effects from both the money-market rate and the bond rate are considerably higher:

$$ecm_{RL,t-1} = (RL - 0.8RS - 0.5RBO)_{t-1}.$$

9.4   Testing exogeneity and invariance

Following Engle *et al.* (1983), the concepts of weak exogeneity and parameter invariance refer to different aspects of 'exogeneity', namely the question of valid conditioning in the context of estimation, and valid policy analysis, respectively. In terms of the 'road-map' of Figure 9.1, weak exogeneity of the conditional variables for the parameters of the wage–price model $D_y(y_t \mid z_t, Y_{t-1}, Z_{t-1})$ implies that these parameters are free to vary with respect to the parameters of the marginal models for output, productivity, unemployment, and exchange rates $D_{z_1}(z_{1t} \mid z_{2t}, z_{3t}, Y_{t-1}, Z_{t-1})$. Below we repeat the examination of these issues as in Bårdsen *et al.* (2003): we follow Johansen (1992) and concentrate the testing to the parameters of the cointegration vectors of the wage–price model. Valid policy analysis involves as a necessary condition that the coefficients of the wage–price model are *invariant* to the interventions occurring in the marginal models. Such invariance, together with weak exogeneity (if that holds), implies super exogeneity.

Following Johansen (1992), weak exogeneity of $z_{1,t}$ with respect to the cointegration parameters requires that the equilibrium-correction terms for wages and prices do not enter the marginal models of the conditioning levels variables. Table 9.3 shows the results of testing weak exogeneity of productivity, unemployment, and import prices[7] within the marginal system.

We observe that the weak exogeneity assumptions do not hold (at the 5% critical level) for import prices with respect to the long-run parameters, whereas those assumptions appear to be tenable for productivity and unemployment. Looking at the detailed results, we observe that it is the equilibrium correction term for the price equation $ecm_{p,t}$ that is significant for import prices (through the exchange rate equation). This means that the estimation of the long-run equations is slightly inefficient, whereas the finding of the two long-run relationships (9.3)–(9.4) is likely to be a robust result due to the superconsistency of the cointegrating equations.

---

[7] In effect we model the exchange rate, treating foreign prices as being determined by factors that are a priori unrelated to domestic conditions.

Table 9.3
Testing weak exogeneity

	$ecm_{w,t}$ and $ecm_{p,t}$
Δpi_t	$F(2, 105) = 3.67[0.03]$
Δu_t	$F(2, 98) = 1.11[0.33]$
Δa_t	$F(2, 105) = 2.45[0.09]$

To test for parameter invariance, we need the interventions occurring in the parameterisations of $D_{\mathbf{z}_1}(\mathbf{z}_{1t} \mid \mathbf{z}_{2t}, \mathbf{z}_{3t}, \mathbf{Y}_{t-1}, \mathbf{Z}_{t-1})$. Consider therefore the following stacked form of the estimated single equation marginal models (9.7)–(9.13) in Section 9.3:

$$\Delta \mathbf{z}_{1,t} = \mathbf{A}(L) \begin{pmatrix} \Delta \mathbf{Z} \\ \Delta \mathbf{Y} \end{pmatrix}_{t-1} + \mathbf{B} \cdot \text{EqCM}(\mathbf{Z}_{t-1}) + \mathbf{C} \cdot \mathbf{X}_t + \mathbf{D} \cdot \mathbf{INT}_t + \varepsilon_{\mathbf{z}_{1,t}}.$$

$$(9.14)$$

The matrix $\mathbf{B}$ contains the coefficients of the equilibrium correction terms (if any) in the marginal models (with the loadings along the diagonal). The matrix $\mathbf{C}$ contains the coefficients of the maintained exogenous variables $\mathbf{X}_t$ in the marginal models for $\mathbf{z}_{1,t}$. Intervention variables affecting the mean of the variables under investigation—significant dummies and non-linear terms— are collected in the $\mathbf{INT}_t$ matrix, with coefficients $\mathbf{D}$. By definition, the elements in $\mathbf{INT}_t$ are included because they pick up linear as well as non-linear features of $\mathbf{z}_{1,t}$ that are left unexplained by the information set underlying the wage–price model.[8]

To test for parameter invariance in the wage–price model, we test for the significance of all the intervention variables from all the marginal models (9.7)– (9.13) in Section 9.3.[9] The results from adding the set of intervention variables to the wage–price model (9.5)–(9.6) are reported in Table 9.4.

The intervention variables are jointly insignificant in the wage–price system (with p-value $= 0.32$) as is seen from Table 9.4. As a specification test, this yields support to the empirical model in (9.5)–(9.6). However, we find that three terms in the price equation are significant—the oil-price term and the dummies from the output and productivity equations. Hence, the support for super exogeneity for the conditioning variables on our sample from 1972(4)–2001(1) is weaker than in Bårdsen et al. (2003) on a sample period 1966(4)–1996(4).

[8] The idea to first let the marginal models include non-linear terms in order to obtain stability and second to use them as a convenient alternative against which to test invariance in the conditional model, was first proposed by Jansen and Teräsvirta (1996).

[9] There is no marginal model for the impact of import prices Δpi_t. Instead, we have assumed full and immediate pass-through of the exchange rate, imposing $\Delta pi_t = \Delta v_t + \Delta pw_t$ on the model. We therefore use the intervention variables of Δv_t to test for invariance of the parameters of Δpi_t.

Table 9.4
Testing invariance

$$\Delta w_t = \cdots + \underset{(0.011)}{0.005} \text{ Ydum}_t + \underset{(0.007)}{0.003} \text{ Udum}_t - \underset{(0.007)}{0.009} \text{ CRdum}_t$$
$$- \underset{(0.027)}{0.027} \Delta \text{oil}_t \times \text{oilST}_t - \underset{(0.033)}{0.043} \text{ } s\Delta(\text{euro/dollar})_t - \underset{(0.003)}{0.003} \text{ } s_t$$
$$- \underset{(0.003)}{0.003} \text{ Vdum}_t + \underset{(0.013)}{0.007} \text{ Adum}_t + \underset{(0.004)}{0.001} \text{ RBOdum}_t$$
$$+ \underset{(0.139)}{0.047} \text{ } s\Delta \text{RS}_t - \underset{(0.385)}{0.426} \text{ } s\Delta \text{RW}_t + \underset{(0.0128)}{0.0003} \text{ RLdum}_t$$

$$\Delta p_t = \cdots - \underset{(0.004)}{0.008} \text{ Ydum}_t + \underset{(0.0023)}{0.0026} \text{ Udum}_t - \underset{(0.0021)}{0.0003} \text{ CRdum}_t$$
$$+ \underset{(0.011)}{0.022} \Delta \text{oil}_t \times \text{oilST}_t - \underset{(0.0115)}{0.0014} \text{ } s\Delta(\text{euro/dollar})_t + \underset{(0.0012)}{0.0014} \text{ } s_t$$
$$+ \underset{(0.0011)}{0.0014} \text{ Vdum}_t + \underset{(0.0044)}{0.0087} \text{ Adum}_t - \underset{(0.0014)}{0.0012} \text{ RBOdum}_t$$
$$- \underset{(0.049)}{0.015} \text{ } s\Delta \text{RS}_t + \underset{(0.133)}{0.100} \text{ } s\Delta \text{RW}_t + \underset{(0.0012)}{0.0002} \text{ RLdum}_t$$

Note: Testing the invariance with respect to all interventions: $\chi^2(24) = 26.75[0.32]$.

In the same vein, we have also augmented the wage–price model (9.5)–(9.6) with all equilibrium correction terms in the marginal models (9.7)–(9.13): $ecm_{v,t}$, $ecm_{y,t}$, $ecm_{u,t}$, $ecm_{a,t}$, $ecm_{cr,t}$, $ecm_{\text{RBO},t}$, $ecm_{\text{RL},t}$. They are individually and jointly insignificant, with a joint test statistic of $\chi^2(14) = 6.82[0.94]$, providing additional support to the wage–price model specification.

9.5   Model performance

The model (9.5)–(9.13) is a small econometric model for Norway, which is characterised by the inclusion of labour market effects in addition to effects of aggregated demand and the exchange rate. The motivation for the extended model is given in the preceding chapters: in order to capture the effects of monetary policy in general and on inflation in particular, it is essential to include the workings of the labour market.

Figure 9.6 gives an overview of the transmission mechanism in the model, focusing on the relationship between interest rates and inflation. The most direct effect on inflation from a rise in the interest rate is an exchange rate appreciation which feeds into lower consumer price inflation with a time lag. This delayed 'pass-through' of exchange rates into consumer price inflation is well known in empirical work and reflects *inter alia* that price setters may find it difficult to distinguish between permanent and temporary shocks to the

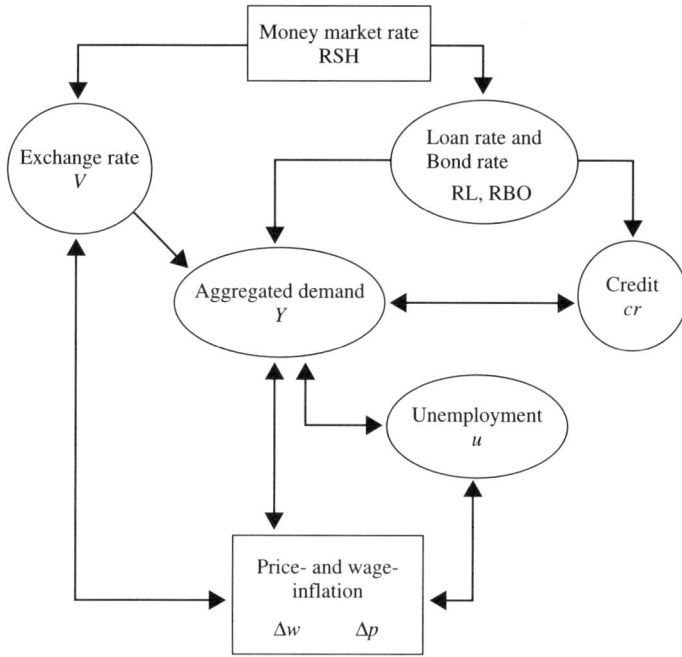

Figure 9.6. Interest rate and exchange rate channels

exchange rate. Other interest rate effects work through their effects on aggregate demand which in turn affect output growth and the rate of unemployment. Both indicators affect domestic wage and price growth and hence inflation.

There is a link between Figure 9.6 and Figures 1.1 and 1.2. The small econometric model we are studying here captures the effect of Figure 1.1 through the aggregate demand channel and of Figure 1.2 through the exchange rate channel.

In order to take account of all implied feedback links, the model is completed with the necessary set of identities for the equilibrium-correction terms, real wages, the real exchange rate, the real bond rate, and so forth. With these new equations in place it is possible to estimate the model simultaneously with full information maximum likelihood (FIML). Doing so does not change the coefficient estimates of the model much.

As it stands, the system is fundamentally driven by the following exogenous variables:

- real world trade (weighted GDP for trading partners), yw_t, and real public expenditure (g_t)
- nominal foreign prices pw_t measured as a trade-weighted index of foreign consumer prices
- the price of Brent Blend in USD (oil_t)

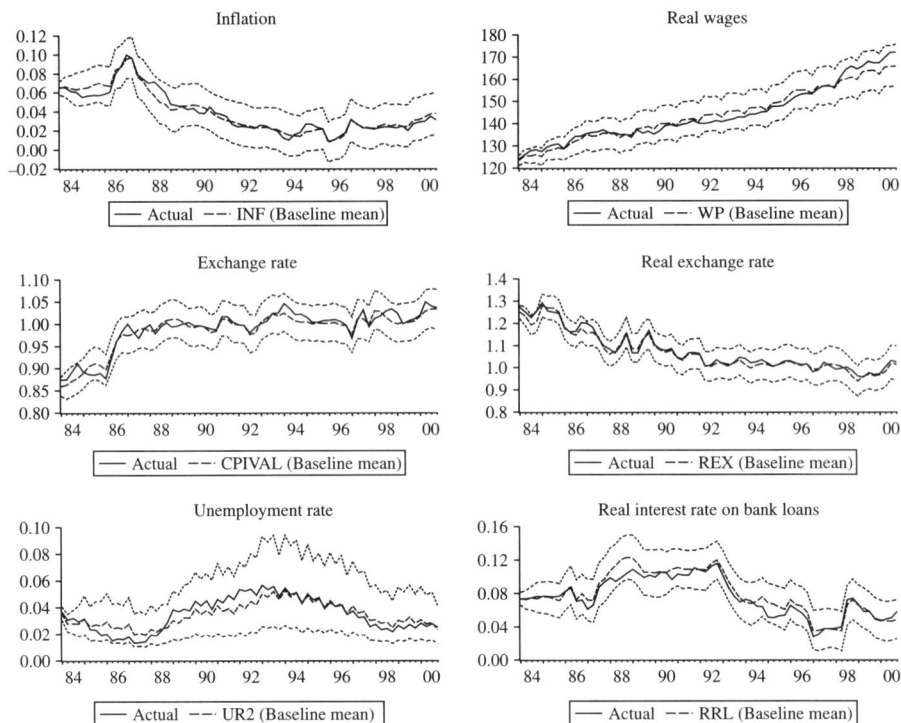

Figure 9.7. Tracking performance under dynamic simulation 1984(1)–2001(1): CPI annual inflation, real wages, loan rate, the nominal and real exchange rate, unemployment rate and real interest rate on bank loans. The dotted lines are 95% confidence intervals

- the monetary policy instrument, that is the short-term interest rate, represented through the money market interest rate (RS_t).[10]

Figure 9.7 shows the tracking performance of the model when we simulate from 1984(1) to 2001(1). The variables (listed row-wise from upper left to bottom right) are annual headline CPI inflation $((P_t/P_{t-4}) - 1)$, the real wage level (W_t/P_t), the nominal and real exchange rate V_t and $V_t(PW_t/P_t)$, respectively, unemployment rate (U_t) and real interest rate on bank loans $(RL - 4\Delta p)_t$. The dotted lines are 95% confidence intervals. The model tracks headline CPI inflation fairly well over the period, but it should be noted that dummies are used to represent active price- and wage-policies during some periods in the 1970s and 1980s.

[10] This is a convenient model simplification, implicitly treating the money market rate as if there is an instant pass-through of a change in the signal rate of the central bank.

Figure 9.8 shows the model's forecasting properties for the period 1999(1)–2001(1). The variables (listed row-wise from upper left to bottom right) are quarterly wage inflation, Δw_t, quarterly headline CPI inflation, Δp_t, deviation from PPP, $[v - (p - pw)]_t$, quarterly import price inflation, Δpi_t, annual headline CPI inflation, $\Delta_4 p_t$, unemployment, u_t, mainland output, y_t, annual output growth, $\Delta_4 y_t$, and the nominal exchange rate, v_t. The model parameters are estimated on a sample that ends in 1998(4). These dynamic forecast are conditional on the actual values of the non-modelled variables (*ex post* forecasts). However, the model has a high degree of endogeneity as all important variables describing the domestic economy are explained within the model. The model exhibits good forecasting properties and the quarterly inflation rate Δp_t is in particular accurately forecasted. However, there is a slight overprediction in each quarter, and when we look to the annual inflation $\Delta_4 p_t$ the effect accumulates over the period. The same is the case for annualised output growth $\Delta_4 y_t$ over the last 4 quarters (i.e. in 2000). The predicted nominal exchange rate is constant and tends not to capture the observed changes in v_t.

Figure 9.8 also contains the 95% prediction intervals in the form of ±2 standard errors, as a direct measure of the uncertainty of the forecasts. The prediction intervals for the annual rate of inflation are far from negligible and are growing with the length of the forecast horizon.

However, forecast uncertainty appears to be much smaller than similar results for the United Kingdom: Haldane and Salmon (1995) estimate one standard error in the range of 3 to $4\frac{1}{2}$ percentage points, while Figure 9.8 implies a standard error of 1.0 percentage points 4-periods ahead, and 1.2 percentage points 8-periods ahead. One possible explanation of this marked differences is that Figure 9.8 understates the uncertainty, since the forecast is based on the actual short-term interest rate, while Haldane and Salmon (1995) include a policy rule for interest rate.

In Bårdsen *et al.* (2003) an attempt is made to control for this difference. To make their estimate of inflation uncertainty—which is nearly of the same order of magnitude as the estimated uncertainty in Figure 9.8—comparable to Haldane and Salmon (1995), they calculated new forecasts for a model that includes an equation for the short-term interest rate as a function of the lagged rates of domestic and foreign annual inflation, of nominal exchange rate depreciation, and of the lagged output gap. The results showed a systematic bias in the inflation forecast, due to a marked bias in the forecasted interest rate, but the effect on forecast uncertainty was very small. Hence it appears that the difference in forecast uncertainty stems from the other equations in the models, not the interest rate policy rule. For example, Haldane and Salmon (1995) use a Phillips curve equation for the wage growth, and the other equations in their model are also in differences, implying non-cointegration in both labour and product markets. In contrast, Bårdsen *et al.* (1998) (see Section 6.7.2) find that a core wage–price model with equilibrium-correction terms give very similar results for Norway and the United Kingdom. Hence it is clearly possible that

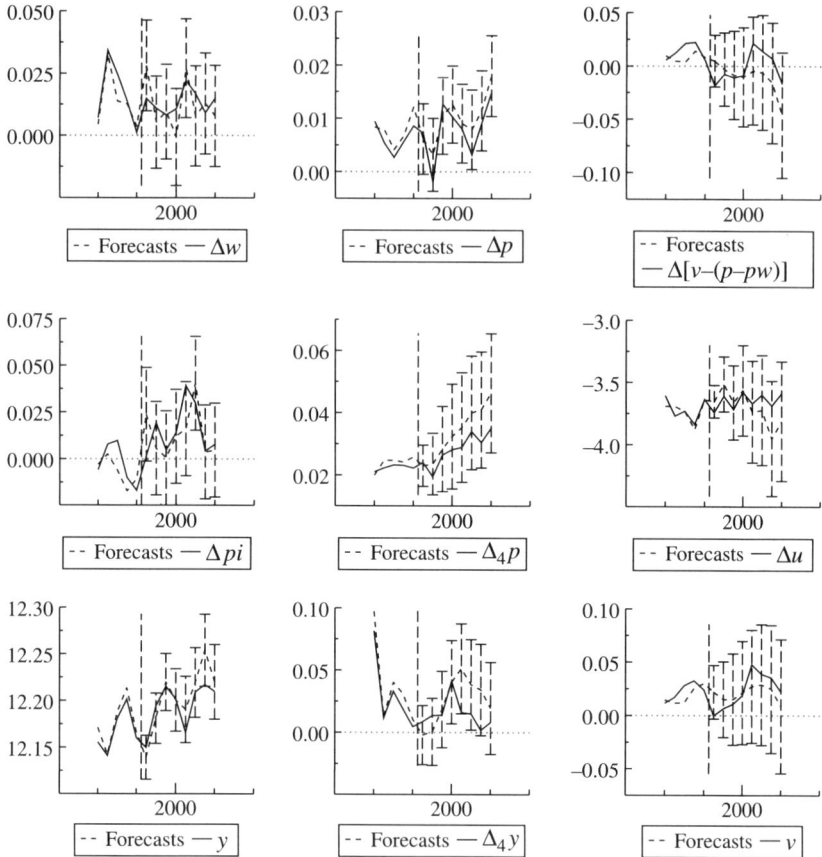

Figure 9.8. Dynamic forecasts over 1999(1)–2001(1): from top left to bottom right: quarterly wage inflation, Δw, quarterly headline CPI inflation, Δp, deviation from PPP, $[v - (p - pw)]$, quarterly import price inflation, Δpi, annual headline CPI inflation, $\Delta_4 p$, unemployment, u, mainland output, y, annual output growth, $\Delta_4 y$, and the nominal exchange rate, v. The bars show prediction intervals (± 2 standard errors)

a large fraction of the inflation forecast uncertainty in Haldane and Salmon's study is a result of model mis-specification.

9.6   Responses to a permanent shift in interest rates

In this section, we discuss the dynamic properties of the full model. In the simulations of the effects of an increase in the interest rate below we have not

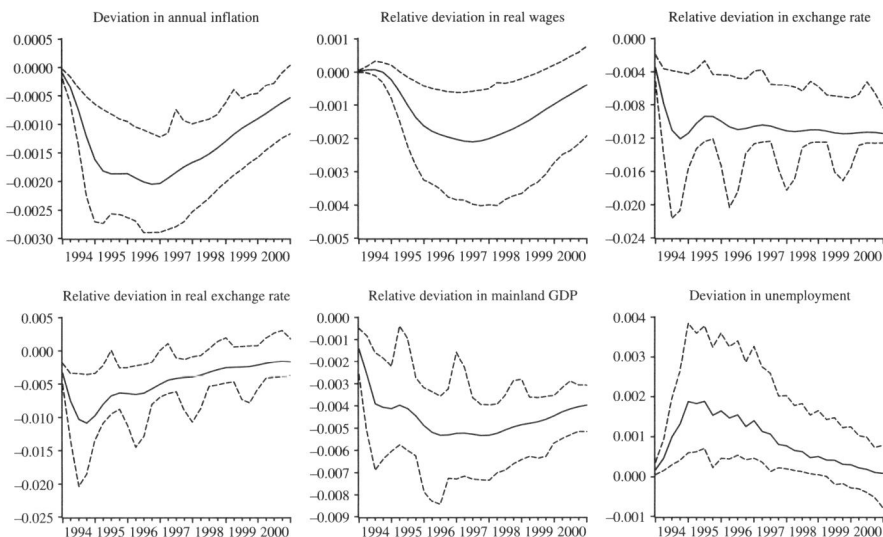

Figure 9.9. Accumulated responses of some important variables to a 1 per cent permanent increase in the interest rate RS_t

incorporated the non-linear effect in the unemployment equation. Hence the results should be interpreted as showing the impact of monetary policy when the initial level of unemployment is so far away from the threshold value that the non-linear effect will not be triggered by the change in policy.

Figure 9.9 shows the simulated responses to a permanent rise in the interest rate RS_t by 100 basis points, that is, by 0.01, as of 1994(1). This experiment is stylised in the sense that it is illuminating the dynamic properties of the model rather than representing a realistic monetary policy scenario. Notwithstanding this, we find that a permanent increase in the signal rate by 1 percentage point causes a maximal reduction in annual inflation of about 0.2% after three years.

Next, in Kolsrud and Nymoen (1998) it is shown that a main property of the competing claims model is that the system determining $(w - p)_t$ and $(pi - p)_t$ is dynamically stable. However, that prediction applied to the conditional sub-system, a priori we have no way of telling whether the same property holds for the full model, where we have taken account of the endogeneity of unemployment, productivity, the nominal exchange rate, and the output gap (via the model of GDP output). However, the upper middle and lower left graphs show that the effects of the shock on the real wage growth, $\Delta(w - p)_t$, and on the change in the real exchange rate, $\Delta(pi - p)_t$, disappear completely in the course of the 24 quarters covered by the graph, which constitute direct evidence that stability holds also for the full system. The permanent rate of appreciation is closely linked to the development of the real exchange rate $(v - p + pw)_t$: the increase in RS_t initially appreciates the krone, both in nominal and real terms.

After a couple of periods, however, the reduction in Δp_t pushes the real exchange rate back up, towards equilibrium. Because of the PPP mechanism in the nominal exchange rate equation, the new equilibrium features nominal appreciation of the krone, as Δv_t equilibrium corrects. This highlights the important role of nominal exchange rate determination—a different model, for example, one where Δv_t is not reacting to deviations from interest rate parity, would produce different responses. The two remaining graphs depict the response of the real economy. As real interest rates increase, aggregate demand falls and the unemployment rate u_t increases, which dampens wages and prices.

9.7   Conclusions

The discussion in this chapter is aimed at several ends. First, as macroeconomic models typically are built up of submodels or modules for different parts of the economy, we have emulated this procedure in the construction of a small econometric model for Norway. Second, the chapter highlights the potential usefulness of such a model for the conduct of monetary policy. More specifically, we have argued that the success of inflation targeting on the basis of conditional forecasts rests on the econometric properties of the model being used.

Inflation targeting means that the policy instrument ('the interest rate') is set with the aim of controlling the conditional forecast of inflation 2–3 years ahead. In practice, this means that central bank economists will need to form a clear opinion about (and be able to explain) how the inflation forecasts are affected by different future interest rate paths, which in turn amounts to quantitative knowledge of the transmission mechanism in the new regime. In this chapter, we show how econometrics can play a role in this process, as well as in an established regime of operational inflation targeting. In the formative period, the econometric approach will at least provide a safeguard against 'wishful thinking' among central bank economists, for example, that formally introducing an inflation target has 'changed everything' including the strength of the relationship between changes in interest rates and the overall price level. True, opting for inflation targeting is an important event in the economy, but one should take care not to overestimate its impact on the behavioural equations of a macroeconomic model that has given a realistic picture of the strength of the transmission mechanism over a sample that includes other, maybe equally substantive, changes in economic policy. Arguably, it may be a more robust procedure to regard at least the main part of the transmission mechanism as unaffected initially, and to take a practical view on the forecasting issue, that is using the model estimated on pre-inflation targeting data, and taking a practical approach to the forecasting issue, that is, using judgement and intercept corrections. Moreover, as experience with inflation targeting grows, and new data accumulate, the constancy of the model parameters becomes an obvious hypothesis to test, leading to even more information about how the economy operates under the new monetary policy regime.

We have presented a macroeconomic model for Norway, that we view both as a tool of monetary policy, and as providing a testing bed for the impact of the policy change on the economy. Conceptually, we partition the (big) simultaneous distribution function of prices, wages, output, interest rates, the exchange rate, foreign prices, and unemployment, etc. into a (much smaller) simultaneous model of wage and price-setting, and several implied marginal models of the rest of the macroeconomy. The partitioning, and the implied emphasis on the modelling of a wage-and-price block, is anything but 'theory-free', but reflect our view that inflation in Norway is rooted in this part of the economy. Moreover, previous studies—as laid out in this book—have established a certain level of consensus about how wage and price-setting can be modelled econometrically, and about how, for example, wages react to shocks to the rate of unemployment and how prices are influenced by the output gap. Thus, there is pre-existing knowledge that seems valuable to embed in the more complete model of the transmission mechanism required for inflation targeting.

In the previous study, Bårdsen *et al.* (2003), based on data for the period 1966(4)–1996(4), valid conditioning of the wage–price model was established through the estimation and testing of the marginal models for the feedback variables, and—with one exception—they found support for super exogeneity of these variables with respect to the parameters in the core model. These results does not completely carry over to our current re-estimation of the core model on a dataset covering the period 1972(4)–2001(1). While the core model sustains broad specification tests, weak exogeneity no longer holds for the exchange rate with respect to the long-run parameters of the wage–price model. This implies a loss of estimation efficiency, which is only eliminated by simultaneous estimation of the core model together with the marginal models.

When we bring together the core model with the marginal models to the small econometric model for Norway, we show that the model can be used to forecast inflation. As regards the effects of monetary policy on inflation targeting, simulations indicate that inflation can be affected by changing the short-run interest rate. A 1 percentage point permanent increase in the interest rate leads to 0.2 percentage point reduction in the annual rate of inflation. Bearing in mind that a main channel is through output growth and the level of unemployment, it is shown in Bårdsen *et al.* (2003) that interest rates can be used to counteract shocks to GDP output. Inflation impulses elsewhere in the system, for example, in wage-setting (e.g. permanently increased wage claims), can prove to be difficult to curb by anything but huge increases in the interest rate.

Thus we conclude that *econometric* inflation targeting is feasible, and we suggest it should be regarded as a possible route for inflation targeters, alongside other approaches of modern open-economy macroeconomics.

10

Evaluation of monetary policy rules

We now relax the assumption of an exogenous interest rate in order to focus on monetary policy rules. We evaluate the performance of different types of reaction functions or interest rate rules using the small econometric model we developed in Chapter 9. In addition to the standard efficiency measures, we look at the mean deviations from targets, which may be of particular interest to policy makers. Specifically, we introduce the root mean squared target error (RMSTE), which is an analogue to the well known root mean squared forecast error. Throughout we assume that the monetary policy rules aim at stabilising inflation around an inflation target and that the monetary authorities also put some weight on stabilising unemployment, output, and interest rates. Finally we conduct simulation experiments where we vary the weights in the interest rate rules as well as the weights of other variables in the loss function of a policy maker. The results are summarised by estimating response surfaces on the basis of the whole range of weights considered in the simulations.

10.1 Introduction

Taking full account of inflation targeting entails that we supplement our model description of the economy with a monetary rule in terms of an interest rate reaction function for the central bank. The monetary rule can be forecast-based or focused on contemporary values of the target variables in the reaction function. We have chosen to analyse the latter alternative, although our discussion below is related to Levin *et al.* (2003), who consider (optimised) forecast-based interest rate rules which they derive for several different models assuming that the preference function of the central bank depends on the variances of inflation and the output gap.

In this chapter, we evaluate a different, and also wider, set of interest rate rules, using the model of Chapter 9.[1] First, the choice of preference function of Levin *et al.* (2003) reflects what seems to be a consensus view, namely that inflation and output gap stabilisation are the main monetary policy objectives of a central bank. While we do not dispute the relevance of this view, there are several arguments for looking at output growth rather than the output gap. In addition to the inherent possibility of measurement error in the output gap, as emphasised by Orphanides (2003), there are also theoretical reasons why output growth might be a sensible objective. Walsh (2003) argues that changes in the output gap—growth in demand relative to growth in potential output— can lead to better outcomes of monetary policy than using the output gap. He demonstrates that such a 'speed limit policy' can induce inertia that dom- inates monetary policy based on inflation targeting and the output gap—except when inflation expectations are primarily backward-looking.[2] A policy rule with output growth and inflation is therefore used as a baseline. Second, rules based on different criteria are considered: those include criteria like simplicity, smoothness or gradualism, and fresh information, which all are considered to be important by policy makers. Finally, we also follow the common practice of central banks to adopt inflation measures that captures underlying inflation rather than the headline consumer price index (CPI) inflation.[3]

More specifically, the interest rate rules we evaluate are based on

- output growth and inflation—as a baseline
- interest rate smoothing
- open economy information: exchange rates
- real-time information on the state of the economy: unemployment, wage growth, and credit growth.[4]

The third item is particularly relevant to the small open economy—and that perspective has not previously been emphasised either in the theoretical or the empirical literature.

The different interest rules are presented in Section 10.2. Section 10.3 gives an overview of the basis of three different sets of evaluation criteria. We evalu- ate the rules along the dimensions fit, relative losses, and optimality, all derived from the counterfactual simulations. The fit is evaluated on standard efficiency measures as well as using a new measure called root mean squared target errors

---

[1] This chapter draws on Akram *et al.* (2003).

[2] Walsh's results are based on simulations from a calibrated stylised New-Keynesian model. The forecasting properties of the New Keynesian Phillips curve are compared with those of alternative inflation models (on data for Norway and for the Euro area) in Chapter 8.

[3] The model of Chapter 9 is therefore supplemented with a technical equation linking headline inflation (Δp_t) and underlying inflation ($\Delta p u_t$), which is the inflation measure entering the reaction functions of this chapter. $\Delta p u_t$ measures inflation net of changes in energy prices and indirect taxes.

[4] These are 'real-time' variables in the sense that reliable current-quarter information is either available or arrives with only a short time lag; see Orphanides (2001).

(RMSTEs), which takes into account both the bias (i.e. the average deviation from target) and the variability of selected report variables, such as alternative measures of inflation (e.g. headline CPI inflation, underlying inflation), and output growth etc. Relative losses summarise the performance of any given rule relative to a benchmark rule as we vary the monetary authorities' weight on output variability and interest rate variability. Finally, in Section 10.3.4 we trace out optimal rules using an estimated response surface based on counterfactual simulations over a grid range of weights in the instrument rule and with varying parameters in the loss function.

10.2 Four groups of interest rate rules

The rules we consider are of the type

$$\mathrm{RS}_t = \omega_r \mathrm{RS}_{t-1} + (1 - \omega_r)(\pi^* + \mathrm{RR}^*) + \omega_\pi(\hat{\pi}_{t+\theta} - \pi^*)$$
$$+ \omega_y(\widehat{\Delta_4 y}_{t+\kappa} - g_y^*) + \omega_{\mathrm{real}}\mathbf{z}_{\mathrm{real},t} + \omega_{\mathrm{open}}\mathbf{z}_{\mathrm{open},t},$$

where RS_t denotes the short-term nominal interest rate, RR_t^* is the equilibrium real interest rate, $\hat{\pi}_{t+\theta}$ is a model-based forecast of inflation (i.e. $\Delta_4 pu$) θ periods ahead, π^* is the inflation target for $\Delta_4 pu_t$, $\widehat{\Delta_4 y}_{t+\kappa}$ is a model-based forecast of output growth κ periods ahead, g_y^* is the target output growth rate, $\mathbf{z}_{\mathrm{real},t}$ denote real-time variables and $\mathbf{z}_{\mathrm{open},t}$ denotes open economy variables (typically the real exchange rate). When the target horizons θ and κ are set to zero, the rules are based on contemporary values of output and inflation. In Section 10.3.4 the optimality of the different rules are evaluated in terms of welfare losses based on minimising the loss function

$$\mathcal{L}(\lambda) = V[\pi_t] + \lambda V[\Delta_4 y_t],$$

where $V[\cdot]$ denotes the unconditional variance and π_t is the inflation measure, in our case $\Delta_4 pu_t$. A large number of possible variations over this theme obtains by combining different rules and loss functions, cf. the survey in Taylor (1999).[5] The interest rate rules we consider are specified in Table 10.1 and they fall

---

[5] A recent example is Levin *et al.* (2003). In their study of the United States economy they consider (optimised) forecast-based interest rate rules of the type
$$\mathrm{RS}_t = \omega_r \mathrm{RS}_{t-1} + (1 - \omega_r)(\mathrm{RR}_t^* + \hat{\pi}_{t+\theta}) + \omega_p(\hat{\pi}_t - \pi^*) + \omega_y \hat{y}_{t+\kappa}^{\mathrm{gap}},$$
where $\hat{y}_{t+\kappa}^{\mathrm{gap}}$ is a model-based forecast of the output level κ periods ahead and all other symbols are as defined in the main text.
    For any given values of (RR^*, π^*) each rule is fully described by the triplet $(\omega_r, \omega_p, \omega_y)$, and Levin *et al.* (2003) derive the parameters of such interest rate rules for five different models under the assumption that the Central Bank's preference function is given by
$$\mathcal{L}(\lambda) = V[\pi_t] + \lambda V[y_t^{\mathrm{gap}}], \qquad \text{subject to } V[\Delta \mathrm{RS}_t] \leq \bar{\sigma}_{\Delta \mathrm{RS}}^2, \quad \lambda \in (0, 1/3, 1, 3).$$
This loss function is then minimised subject to an upper bound on the volatility of the interest rate, $\bar{\sigma}_{\Delta rs}^2$.

Table 10.1

Interest rate rules used in the counterfactual simulations,
as defined in equation (10.1)

Variables		RS_{t-1}	$\Delta_4 pu_t$	$\Delta_4 y_t$	vr_t	u	$\Delta_4 w$	$\Delta_4 cr$
Target/trigger		$\pi^* + \text{RR}^*$	π^*	g_y^*	vr^*	u^*	g_w^*	g_{cr}^*
Trigger value			0.06	0.025	0.025	0	0.04	0.045	0.05
Weights		ω_r	ω_π	ω_y	ω_{vr}	ω_u	ω_w	ω_{cr}
Flexible	FLX		1.5	0.5				
Strict	ST		1.5					
Smoothing	SM	0.75	1.5	0.5				
Real exchange rate	RX		1.5	0.5	0.33			
Unemployment	UR		1.5			−1.00		
Wage growth	WF		1.5				1.00	
Credit growth	CR		1.5					0.20

$$\text{RS}_t = \omega_r \text{RS}_{t-1} + (1 - \omega_r)(\pi^* + \text{RR}^*) + \omega_\pi(\Delta_4 pu_t - \pi^*) + \omega_y(\Delta_4 y_t - g_y^*)$$
$$+ \omega_{vr}(vr_t - vr^*) + \omega_u(u_t - u^*) + \omega_w(\Delta_4 w_t - g_w^*) + \omega_{cr}(\Delta_4 cr_t - g_{cr}^*).$$

into four categories. The first category has two members: (1) a variant of the standard Taylor rule for a closed economy ('flexible' rule) where interest rates respond to inflation and output (FLX in the table), and (2) a strict inflation targeting rule where all weight is put on inflation (ST). The next class of rules introduces interest rate smoothing ('smoothing' rule), where we also include the lagged interest rate (SM), and the third category contains an 'open economy' rule, in which the interest rate responds to the real exchange rate, vr_t (RX). Similar rules have previously been used in, for example, Ball (1999) and Batini *et al.* (2001). The fourth category includes real-time variables, where we use unemployment (UR), wage growth (WF), and credit growth (CR) as alternative indicators for the state of the real economy. The motivation for using real-time variables is well known. As discussed in the introduction, the output gap is vulnerable to severe measurement problems, partly due to a lack of consensus about how to measure potential output, motivating our choice of output growth, following Walsh (2003). However, another source of uncertainty is data revisions. In practice, statistical revisions of output would also render output growth subject to this source of uncertainty, so using output growth rates does not necessarily remove the measurement problem in real time. The alternative 'real-time' interest rate rules use variables, observed with greater timeliness, which are less vulnerable to later data revisions.

The first lines of Table 10.1 contain the different variables (x, say), their associated target parameters (h^*) and the assumptions about the target parameter's trigger values. Each rule correspond to a line in Table 10.1 and the weights attached to the different variables are shown in the columns.[6] In Table 10.1 g_w and g_{cr} are the target growth rates for wages and credit.

All the interest rate rules considered can be written as a special case of equation (10.1). The first line in equation (10.1) defines the standard FLX and SM rules. The second line defines the rule which responds directly to the real exchange rate (rule RX). And finally, in line three, we include the different 'real-time' variables which are potential candidates to replace output growth in the interest rate rule—registered unemployment (UR), annual wage growth (WF), or annual credit growth (CR)—cf. Table 10.1.

$$
\begin{aligned}
RS_t = {}& \omega_r RS_{t-1} + (1 - \omega_r)(\pi^* + RR^*) + \omega_\pi(\Delta_4 pu_t - \pi^*) + \omega_y(\Delta_4 y_t - g_y^*) \\
& + \omega_{vr}(vr_t - vr^*) \\
& + \omega_u(u_t - u^*) + \omega_w(\Delta_4 w_t - g_w^*) + \omega_{cr}(\Delta_4 cr_t - g_{cr}^*)
\end{aligned}
\tag{10.1}
$$

In order to facilitate the comparison between the different interest rate rules we maintain the weights on inflation ($\omega_\pi = 1.5$) and output growth ($\omega_y = 0.5$) in all rules where applicable in Table 10.1. Note that these values alone define the interest rate rule denoted FLX. Hence, the FLX rule serves as a benchmark for comparison with all other rules in Table 10.1.

10.2.1 Revisions of output data: a case for real-time variables?

A first version of the quarterly national accounts (QNA) data is published by Statistics Norway shortly after the end of each quarter, based on a limited information set. As more information accrues, the data are revised and the final figures appear with a 18-months lag. Often there are substantial discrepancies between the first and the final quarterly data. The Norwegian QNA show that on average for the period 1995–99 growth in GDP for Mainland Norway was revised up by almost 1 percentage point per year, and, for example, the output growth for 1999 was adjusted from 1.1% to 2.7%. In Figure 10.1(a) we plot the growth rates for output according to the two sources together. The graphs reveal substantial revisions of output growth in the Norwegian mainland economy. The estimated change in interest rates according to the standard Taylor rule in Table 10.1 (FLX) is shown in Figure 10.1(b). Since the data revisions alone may induce up to 50 basis points change in the interest rate, there is a clear case for using interest rate rules with real-time variables.

---

[6] For the real exchange rate vr the trigger value of the target is 0. Hence $(vr - vr^*)$ is equivalent to deviations from purchasing power parity (PPP), $(v + pw - p)$, cf. Section 9.3.1.

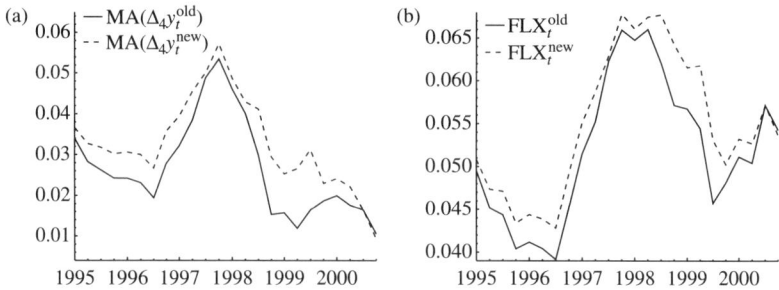

Figure 10.1. Old and revised data for output in the mainland economy and corresponding Taylor-rates, 1990(1)–2000(4). (a) Old and revised data for output growth in the Norwegian mainland economy, five quarters centred moving average. (b) Standard Taylor-rule interest rates from using five quarters centred moving averages of old and revised data for output growth

10.2.2   Data input for interest rate rules

Figure 10.2 shows the variation in the variables we use in the different interest rate rules over the period 1995(1) to 2000(4). Underlying inflation $\Delta_4 pu_t$ is headline inflation corrected for changes in excise duties and energy prices, and is clearly less volatile than headline CPI inflation during the 1990s, cf. Figure 10.2(a). Output growth picked up towards the end of the 1990s, and during 1997–98 we see from Figure 10.2(b) that the four-quarter output growth rate shifts rather abruptly. Figure 10.2(c) shows the development in three variables used in the 'real-time' rules, that is, the rate of unemployment, u_t, annual wage growth, $\Delta_4 w_t$, and annual growth in nominal domestic credit $\Delta_4 ncr_t$. The potential volatility in the interest rate implied by a real-time rule based on wage growth can be anticipated by the hike in wages in 1998. Finally, Figure 10.2(d) shows the deviations from PPP, $v_t - (p_t - pw_t)$, which we use in the 'open economy' interest rate rules.

10.2.3   *Ex post* calculated interest rate rules

To get a feel for the properties and implications of the different monetary policy rules in Table 10.1, we have calculated *ex post interest rates* corresponding to the different rules, by inserting the actual outcomes of the variables into the various versions of equation (10.1). The results are shown in the four charts in Figure 10.3. The upper left panel shows the realised interest rate together with the implied interest rate of following the flexible rule FLX. Following the rule would have meant a much higher interest rate during 1997, as a consequence of the spurt in output growth, shown in Figure 10.2(b). The strict rule ST of the upper right panel is basically reflecting the development of underlying inflation of Figure 10.2(a), while the smoothing rule SM appears more

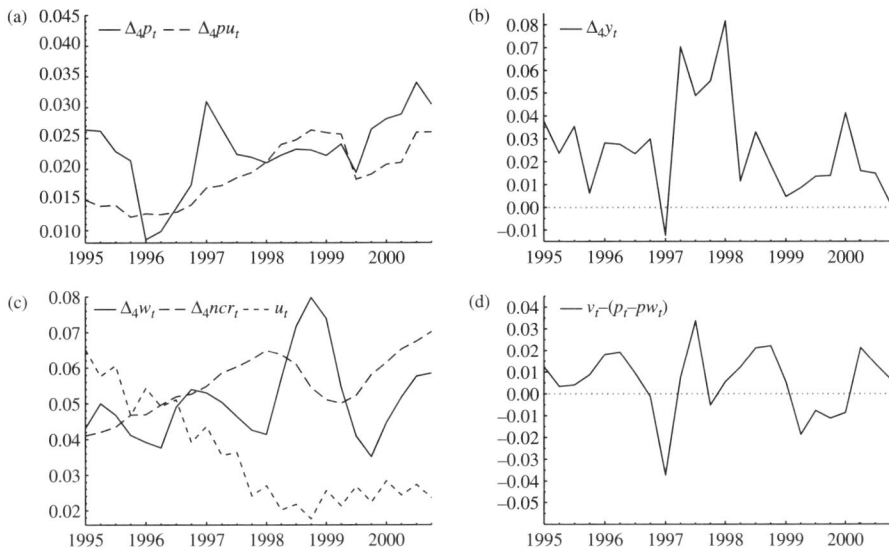

Figure 10.2. Data series for the variables which are used in the Taylor rules, 'real time'-rules and open economy-rules respectively, over the period 1995(1)–2000(4). (a) Taylor rules: headline inflation, $\Delta_4 p_t$, and underlying inflation, $\Delta_4 pu_t$. (b) Taylor rules: output growth, $\Delta_4 y_t$. (c) 'Real time' rules: unemployment, u_t, wage growth, $\Delta_4 w_t$, and credit growth $\Delta_4 ncr_t$. (d) Open economy rules: deviations from PPP, $v_t - (p_t - pw_t)$

volatile, again reflecting the output growth volatility—even though the rule implies considerable interest rate smoothing. Of the real-time rules, unemployment and credit growth appear the smoothest, while the wage growth rule WF would imply a very contractive response to the wage hikes. Finally, the real exchange rate rule RX implies quite volatile interest rate responses. Of course, these responses are only indicatory, as there is no feedback onto the variables entering the different rules. In later sections of this chapter we will investigate the properties of these rules in counterfactual model simulations, where we allow the economy to react to changes in monetary policy according to the prescribed interest rate rules, and the changed outcome for the set of variables in each rule will feed back and change the interest rate according to the rule.

10.3 Evaluation of interest rate rules

10.3.1 A new measure—RMSTEs

Since we set the monetary policy instrument RS_t in order to make a target variable x_t stay close to its target level x^*, it makes sense to evaluate the rules according to how well they achieve their objective. In the theoretical literature,

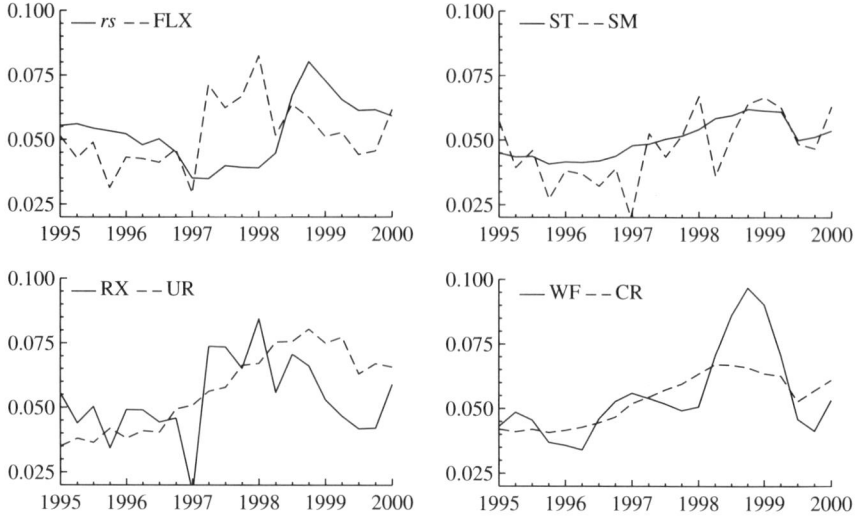

Figure 10.3. *Ex post* calculations of the implied interest rates from different interest rate rules over the period 1995(1)–2000(4).

however, policy evaluation is often based on the unconditional variance of x_t, denoted $V[x]$. An alternative measure which puts an equally large weight on the bias of the outcome, that is, on how close the expected value of x_t is to the target x^*, is the RMSTE. Since the bias could differ considerably between different monetary policy rules, it is of interest to investigate its effect in small samples. If we estimate the expected level $E[x]$ by its sample mean $\bar{x}$, the measure can be written as

$$\text{RMSTE}(x) = \sqrt{\frac{1}{T} \sum_{t=1}^{T} (x_t - x^*)^2} = \sqrt{\widehat{V[x]} + (\bar{x} - x^*)^2},$$

which is the form we will adopt in the following sections.

10.3.2   RMSTEs and their decomposition

Table 10.2 shows the results from a series of counterfactual model simulations. For each interest rate rule we show the bias, standard deviation, and RMSTE measured relative to a baseline scenario. The baseline is the results we obtain for the variables from a model simulation where the interest rate is kept equal to actual sample values.[7]

---

[7] In the baseline simulation the model residuals have been calibrated such that the actual values of the data are reproduced exactly when we simulate the model with historical values for the short-run interest rate, RS_t. For each of counterfactual simulations with the different

Flexible and strict rules The least volatile development in interest rates is seen to follow from the strict targeting rule (ST). The sharp rise in output growth in 1997 is reflected in the volatility of the interest rates implied by the flexible rule (FLX) and the smoothing rule (SM). The FLX rule puts three times more weight on inflation than on output growth. Table 10.2 shows that the FLX rule gives a slightly more expansive monetary policy compared with the baseline over the period 1995(1)–2000(4): a lower interest rate and weaker exchange rate give rise to somewhat higher output growth (relative bias greater than one) and higher inflation growth (relative bias less than one). The explanation is that while average output growth in the baseline scenario is higher than the target growth of 2.5%, average headline and underlying inflation is lower. Thus the relative bias from a more expansionary monetary policy will become larger than one for output growth (moving output growth further away from the target) and smaller than one for inflation (moving inflation closer to the target). The relative variability of underlying inflation and output growth is 11% lower than in the baseline, while interest rates and exchange rates show greater variability.

The strict targeting rule ST leads to less variability in interest rates since the weight on output growth is reduced to zero. The exchange rate is somewhat weaker. This contributes to reducing the bias in underlying inflation compared to the FLX scenario.

Smoothing Giving the lagged interest rate in the (smoothing) rule SM a positive weight $\omega_r = 0.75$, gives rise to a considerably more expansionary monetary policy. This reduces the bias for underlying inflation and gives a negative bias for headline inflation, which means that inflation on average is above the target of 2.5% in the SM scenario.

Open economy rules The RX rule puts some weight on the real exchange rate, vr_t, such that a weaker real exchange rate leads to a tightening of monetary policy. In addition to its direct contractionary effect, the increase in interest rates also partly counteracts the weakening of the exchange rate and dampen the expansionary effects initially working through the exchange rate channel. In our simulation the RX scenario leads on average to a less expansionary monetary policy than the baseline scenario, and a relative bias larger than one for headline as well as underlying inflation. The exchange rate is more stable exchange rate (less variability in v_t) at a cost of higher variability in interest rate changes.

interest rate rules we maintain these add factors over the simulation period. Thus, we isolate the partial effect from changing the interest rate rule while maintaining a meaningful comparison with the historical sample values. In the counterfactual simulation we also make the usual assumption that the models' parameters are invariant to the proposed changes in the interest rate rule across the period from 1995(1) to 2000(4).

Table 10.2

Counterfactual simulations 1995(1)–2000(4)

		$\Delta_4 p_t$	$\Delta_4 pu_t$	$\Delta_4 y_t$	u_t	$\Delta_4 cr_t$	v_t	ΔRS_t	RS_t
		Simstart 1995(1), evaluation over 1995(1)–2000(4)							
Policy rule	*Target/ trigger*	0.025	0.025	0.025	0.040	0.050	0	0	0.06
Baseline	Mean	0.023	0.019	0.027	0.032	0.074	0.013	0.000	0.055
(no rule)	bias	−0.002	−0.006	0.002	−0.008	0.024	0.013	0.000	−0.005
	sdev	0.006	0.005	0.023	0.009	0.023	0.017	0.007	0.013
	RMSTE	0.007	0.008	0.023	0.012	0.033	0.022	0.007	0.014
Flexible rule	Mean	0.023	0.019	0.028	0.032	0.075	0.016	−0.001	0.052
FLX	Rel. bias	0.94	0.99	1.24	1.03	1.03	1.19	−2.13	1.63
	Rel. sdev	1.06	0.89	0.83	0.95	1.02	1.35	1.86	0.76
	Rel. RMSTE	1.05	0.95	0.83	0.99	1.03	1.30	1.86	0.91
Strict rule	Mean	0.024	0.020	0.027	0.032	0.076	0.017	0.000	0.052
ST	Rel. bias	0.80	0.95	1.21	1.04	1.06	1.28	−0.72	1.78
	Rel. sdev	1.01	0.90	0.91	0.95	0.96	1.22	0.76	0.52
	Rel. RMSTE	0.99	0.93	0.91	0.99	1.01	1.25	0.76	0.78
Smoothing	Mean	0.025	0.021	0.028	0.031	0.081	0.028	−0.001	0.042
rule SM	Rel. bias	0.04	0.75	1.63	1.14	1.26	2.08	−3.70	3.88
	Rel. sdev	1.02	0.88	0.83	0.93	1.12	1.38	1.46	1.12
	Rel. RMSTE	0.99	0.81	0.84	1.03	1.19	1.68	1.47	1.69
Real exchange	Mean	0.023	0.019	0.027	0.032	0.074	0.014	0.000	0.054
rate RX	Rel. bias	1.05	1.02	1.13	1.01	1.00	1.05	−1.61	1.24
	Rel. sdev	1.06	0.92	0.85	0.97	1.02	1.20	2.13	0.87
	Rel. RMSTE	1.06	0.97	0.85	0.99	1.01	1.15	2.13	0.92
Unemployment	Mean	0.022	0.018	0.026	0.033	0.069	0.006	0.000	0.062
UR	Rel. bias	1.75	1.20	0.62	0.93	0.80	0.46	1.06	−0.34
	Rel. sdev	0.97	0.75	0.90	0.91	0.87	1.17	0.93	0.80
	Rel. RMSTE	1.05	1.02	0.90	0.91	0.83	0.97	0.93	0.76
Wage growth	Mean	0.023	0.019	0.027	0.032	0.073	0.012	0.000	0.057
WF	Rel. bias	1.26	1.07	1.01	0.99	0.94	0.89	0.01	0.73
	Rel. sdev	1.06	0.85	0.90	0.93	1.03	1.27	2.09	1.08
	Rel. RMSTE	1.07	0.98	0.90	0.96	0.99	1.15	2.09	1.05
Credit growth	Mean	0.023	0.019	0.027	0.032	0.073	0.012	0.000	0.056
CR	Rel. bias	1.19	1.05	0.94	1.00	0.96	0.94	−0.60	0.94
	Rel. sdev	1.03	0.83	0.84	0.93	0.96	1.32	1.92	0.86
	Rel. RMSTE	1.05	0.96	0.84	0.96	0.96	1.19	1.92	0.87

Note: RMSTE and its decomposition in bias, standard deviations and RMSTE of the different interest rate rules, relative to the baseline scenario (with interest rates kept equal to actual sample values).

Real-time interest rate rules When the interest rate rule responds to changes in unemployment we observe an early contraction of monetary policy compared with the FLX rule. This is due to the fact that in the model the unemployment rate follows Okun's law when demand changes, and thus shows substantial persistence. Hence we observe a gradual tightening of monetary policy under the UR scenario over the simulation period as unemployment falls under the trigger level (of 4%) and on average we observe that this rule has the highest average interest rate level across all alternatives. This runs together with the lowest relative bias in output growth and unemployment and the highest relative bias in inflation. In the two alternative real-time rules the interest rate responds to wage growth $\Delta_4 w_t$ (WF rule) and credit growth $\Delta_4 cr_t$ (CR rule), respectively. The WF rule gives rise to more volatile interest rates than the FLX rule and also to a slightly more contractive monetary policy over the simulation period. The observed volatility in inflation is, however, at the same level as for the FLX rule. The credit growth based rule CR shares many of the characteristics observed for the flexible rule FLX, except that the interest rate is higher in particular towards the end of the simulation period.

Comparing the rules The main features of the counterfactual simulations can be seen in Figure 10.4. For each monetary policy rule the figure shows the deviations from the baseline scenario (with 'exogenous' short-term interest rates). Figure 10.4 shows that most of the rules give a more expansive monetary policy with lower interest rates in the first two years, compared with the baseline scenario. The initial easing averages around 2 percentage points (pp) and it is followed by a tightening of more than 3 pp. It is hard to evaluate details on the individual rules from the figure although we see that the smoothing rule SM appears to give rise to the most expansionary monetary policy over the simulation period.

When we evaluate the implications for inflation, output and unemployment, we see from Figure 10.4 that the SM scenario and the UR scenario form the boundaries of a corridor for the relative responses for each rule compared with the data. For inflation the width of this corridor is about plus/minus 0.5 pp relative to actual inflation. Output growth deviates from actual growth with about plus/minus 2 pp, and unemployment deviates from actual with about plus/minus 0.7 pp. The width of the corridor would be considerably smaller if we take out the SM scenario. Note, however, that the parameters in the monetary policy rules were chosen to illustrate some main features of each rule, and are not necessarily optimising the rule.

It is also of interest to compare the counterfactual simulations with the actual data as shown in Figure 10.2. Output growth increases sharply during 1997 to levels above their assumed steady-state growth rate of 2.5%, and this is a driving force behind the tightening of monetary policy during 1997 and 1998 in the counterfactual simulations in Figure 10.4, where interest rates

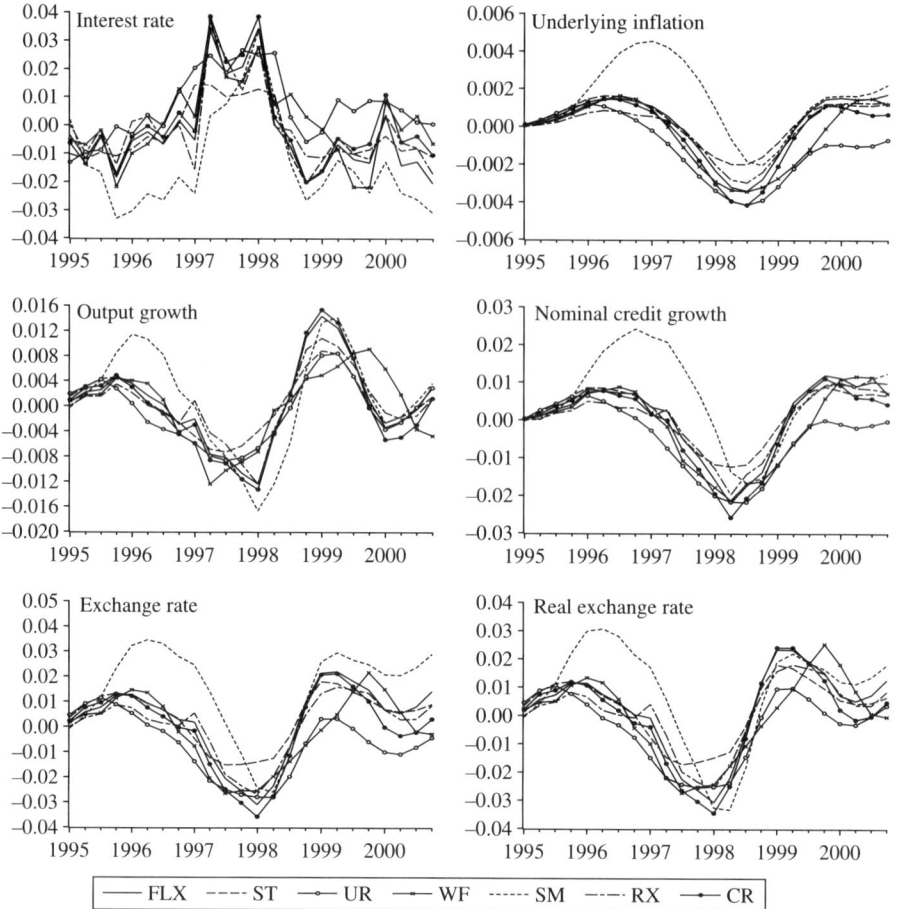

Figure 10.4. Counterfactual simulations 1995(1)–2000(4) for each of the interest rate rules in Table 10.1. The variables are measured as deviations from the baseline scenario

on average rise towards a peak level of 8%. Interestingly, as can be seen from Figure 10.3, there was a marked tightening of actual monetary policy, but this happened one year later when interest rates were increased sharply under the fixed exchange rate regime in an attempt to resist speculative attacks at the Norwegian krone. The actual monetary policy was eased later in 1998 and at this point we note in Figure 10.4 that there are considerable differences between the interest rates implied by the different interest rate rules. This motivates us to make a further assessments of the rules by turning to their implications for the variables in the monetary authorities' loss function.

10.3.3 Relative loss calculations

So far we have summarised the counterfactual results through the effects on the mean and variability of a number of key variables. In the following we will investigate how the interest rate rules in Table 10.1 perform when we select different weights λ, ϕ in the monetary authorities' loss function. We write the loss function as a linear combination of the unconditional variances of output growth $\Delta_4 y_t$ and underlying inflation, $\Delta_4 p u_t$, with a possible extension in terms of the variance of interest rate changes $\Delta \mathrm{RS}_t$.

$$\pounds(\lambda, \phi) = V[\Delta_4 p u_t] + \lambda V[\Delta_4 y_t] + \phi V[\Delta \mathrm{RS}_t]. \tag{10.2}$$

In Table 10.3 we report the square root of the loss according to equation (10.2) for different values of central bank preference parameters (i.e. the weights λ and ϕ). The loss calculations reported in the upper part of Table 10.3 are calculated on the basis of the pure measures of volatility (sdev's). The lower part of Table 10.3 reports similar loss calculations based on the RMSTEs. The first column in the table shows the results from the flexible rule FLX relative to the baseline scenario where we assume that interest rates are kept at their historical values.

Let us first consider losses based on sdev's for the FLX rule. When we assume that the central bank pays no regard to interest rate variability (i.e. when $\phi = 0$) we find that the loss is reduced by 11–17% under the FLX rule depending on λ. This is because both underlying inflation and output growth show less variability under the FLX rule (cf. columns two and three in Table 10.2) compared to under the baseline alternative. The loss reduction grows larger with increased weight λ on output. If λ is set to 0 the loss reductions is 11%, whereas $\lambda = 2$, leads to a loss reduction. This is because the FLX rule gives rise to a larger relative reduction in variability for output than for underlying inflation. As we increase the weight ϕ on interest rate variability from 0 to 1, we find that relative losses increase from 0.89 to 1.62 when $\lambda = 0$, since the variability in interest rate changes is 86% higher under the FLX rule compared with the baseline. As more weight is put on the variability of output, the partial effect from interest rate variability counts less and we find that when $\lambda = 2$, relative losses only increase from 0.83 to 0.91 as we increase ϕ from 0 to 1. We find qualitatively similar results when we apply RMSTEs but since the bias for underlying inflation is relatively larger compared with that for output growth, we find the largest differences between losses based on RMSTE compared with those based on sdev for small values of λ.

The strict rule ST puts zero weight on output growth and gives rise to considerably less variation in interest rate changes compared with the FLX rule and also compared with the baseline scenario. This puts the ST rule at an advantage as we increase the weight on interest rate variability ϕ from 0 to 1. When we span the relative loss measures in 3-dimensional plots in Figures 10.5(a) and (b) we note that the ST rule in both cases gives rise to a relatively flat surface

Table 10.3

Counterfactual simulations 1995(1)–2000(4). Loss function evaluation based on relative sdev (upper half) and relative RMSTE (lower half)—relative to the baseline scenario (actual observations of interest rates)

Central Bank preferences		FLX	ST	SM	RX	UR	WF	CR
λ	ϕ							
Loss based on relative sdev								
0	0	0.894	0.895	0.876	0.915	0.752	0.851	0.834
0	0.1	1.125	0.873	1.003	1.218	0.785	1.167	1.104
0	0.5	1.475	0.829	1.212	1.654	0.846	1.612	1.495
0	1.0	1.617	0.807	1.301	1.828	0.873	1.788	1.652
0.5	0	0.836	0.908	0.836	0.856	0.885	0.891	0.835
0.5	0.1	0.867	0.906	0.852	0.898	0.885	0.928	0.868
0.5	0.5	0.972	0.896	0.908	1.034	0.888	1.054	0.981
0.5	1.0	1.073	0.886	0.965	1.162	0.891	1.173	1.088
1	0	0.833	0.909	0.834	0.853	0.891	0.893	0.835
1	0.1	0.850	0.908	0.842	0.876	0.891	0.913	0.853
1	0.5	0.910	0.903	0.874	0.955	0.892	0.985	0.918
1	1.0	0.975	0.897	0.910	1.039	0.894	1.062	0.987
2	0	0.832	0.909	0.832	0.852	0.894	0.894	0.835
2	0.1	0.840	0.909	0.837	0.863	0.894	0.904	0.844
2	0.5	0.873	0.906	0.854	0.907	0.894	0.943	0.879
2	1.0	0.911	0.903	0.874	0.956	0.895	0.988	0.919
Loss based on relative RMSTE								
0	0	0.946	0.926	0.807	0.972	1.024	0.979	0.961
0	0.1	1.057	0.913	0.883	1.120	1.016	1.120	1.078
0	0.5	1.312	0.877	1.061	1.447	0.993	1.433	1.346
0	1.0	1.465	0.850	1.171	1.638	0.977	1.617	1.505
0.5	0	0.856	0.915	0.835	0.876	0.920	0.912	0.860
0.5	0.1	0.883	0.913	0.849	0.912	0.920	0.944	0.889
0.5	0.5	0.975	0.904	0.901	1.033	0.920	1.055	0.987
0.5	1.0	1.066	0.894	0.953	1.150	0.921	1.163	1.084
1	0	0.846	0.914	0.837	0.866	0.909	0.905	0.849
1	0.1	0.861	0.913	0.845	0.886	0.909	0.923	0.866
1	0.5	0.917	0.908	0.875	0.960	0.910	0.990	0.925
1	1.0	0.978	0.902	0.909	1.038	0.910	1.062	0.990
2	0	0.840	0.913	0.839	0.860	0.902	0.901	0.843
2	0.1	0.849	0.913	0.843	0.870	0.902	0.910	0.852
2	0.5	0.880	0.910	0.859	0.912	0.903	0.948	0.885
2	1.0	0.916	0.907	0.879	0.959	0.903	0.990	0.924

$$\mathcal{L}(\lambda, \theta) = m[\Delta_4 pu_t] + \lambda m[\Delta_4 y_t] + \phi m[\Delta RS_t]$$

for $\lambda \in (0, 0.5, 1, 2)$, $\phi \in (0, 0.1, 0.5, 1)$, $m = (\text{sdev}, \text{RMSTE})$.

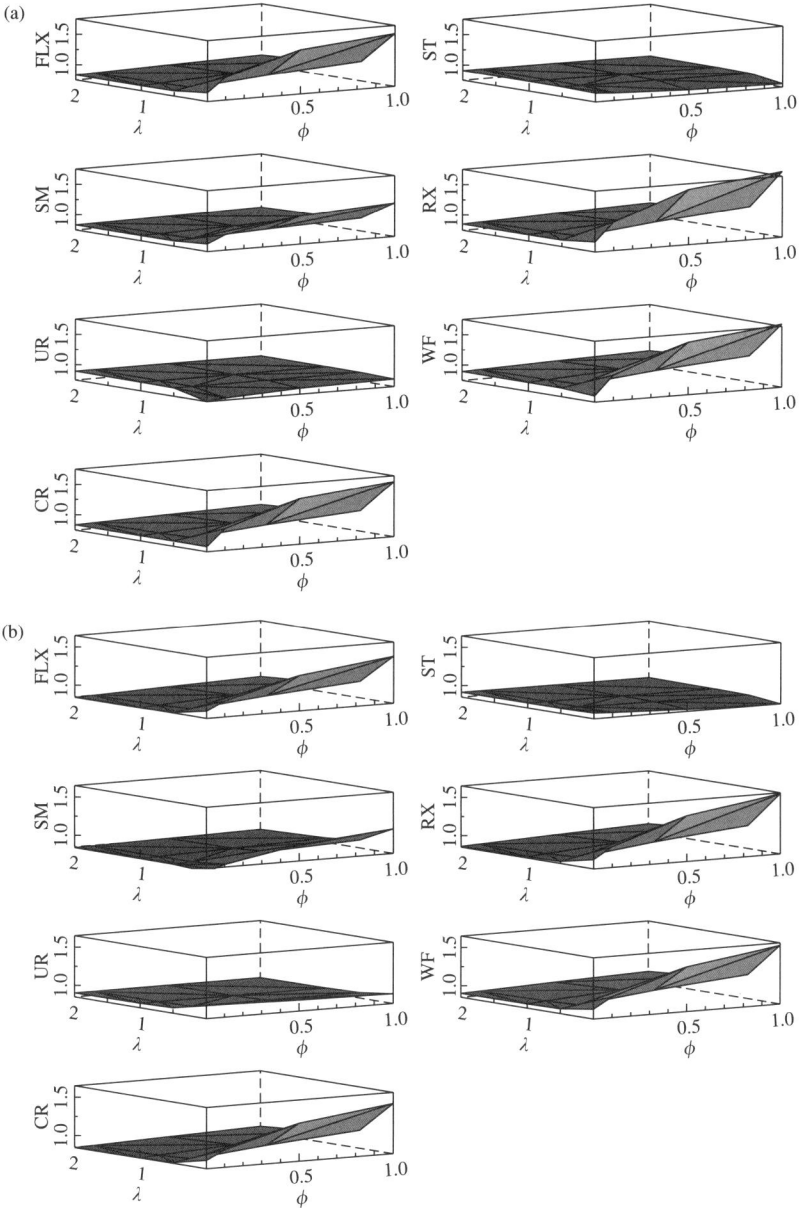

Figure 10.5. Counterfactual simulations 1995(1)–2000(4). (a) Loss function evaluation based on relative sdev (relative to the baseline scenario). (b) Loss function evaluation based on relative RMSTE (relative to the baseline scenario).

$$\mathcal{L}(\lambda, \phi) = m[\pi_t] + \lambda m[\Delta y_t] + \phi m[\Delta r_t]$$
$$\text{for } \lambda \in (0, 0.5, 1, 2), \quad \phi \in (0, 0.1, 0.5, 1),$$
$$m = (\text{sdev}, \text{RMSTE}).$$

compared with the other rules, which means that relative losses by adopting this rule are constant across values assigned to the central bank preferences parameters.

The smoothing rule SM gives rise to a more expansive monetary policy, with higher output growth and inflation. This entails an increase in the bias of output growth and a decrease in the bias for inflation, which we would expect to give different results depending on whether we calculate losses based on sdev's or RMSTEs. As we increase ϕ the relative loss increases less sharply than the FLX rule due to the smaller volatility in interest rate changes under smoothing. Figures 10.5(a) and (b) show that SM does well compared to many of the other rules although the surface is far from being as flat as, for example, the strict ST rule.

The real exchange rate based RX rule gives increased interest rate volatility, and as we increase ϕ this translates into the largest relative loss compared with the other rules. The RX rule gives a slightly more contractive monetary policy compared with FLX, and the relative loss increases for all values of λ (no matter which measure we base the loss calculations on). The RX stands out in Figures 10.5(a) and (b) showing the largest relative loss as we increase ϕ from 0 to 1. For large values of λ this 'open economy' rule performs as well as or even better than many of the other rules.

Finally, we compare the results for the 'real-time' rules where output growth is replaced by either unemployment, wage growth, or credit growth. The unemployment based rule, UR, shows a remarkably flat surface in Figures 10.5(a) and (b). This is due to the fact that interest rate volatility is almost as low as in the case with the strict rule ST. The UR rule gives rise to the most contractionary monetary policy, and this is why that rule has a markedly different impact depending on whether the loss function is based on sdev's or RMSTEs. This is mainly due to the increase in the inflation bias under a more contractionary monetary policy.

When we use wage growth or credit growth as basis for the 'real-time' rule we find higher interest rate volatility and this translates into a rising surface in Figures 10.5(a) and (b) as we increase ϕ from 0 to 1. Again these rules are (on average) more contractionary than the FLX rule and the increase in the inflation bias makes these rules score less well with the RMSTE based losses. For large values of λ or for small values of ϕ the WF and CR rules stand out as superior to the other rules.

10.3.4 Welfare losses evaluated by response surface estimation

Taylor (1979*a*) argues that the tradeoff between inflation variability and output variability can be illustrated by the convex relationship in Figure 10.6. In point A monetary policy is used actively in order to keep inflation close to its target, at the expense of somewhat larger variability in output. Point C

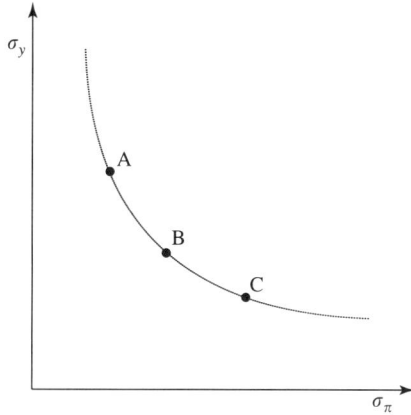

Figure 10.6. The Taylor curve

illustrates a situation in which monetary policy responds less actively to keep the variability of inflation low, and we have smaller output variability and larger inflation variability. Point B illustrates a situation with a flexible inflation target, and we obtain a compromise between the two other points. The downward sloping curve illustrates a frontier along which the variability of output can only be brought down at the cost of increasing the variability of inflation. The preferred allocation along the Taylor curve depend on the monetary authorities' loss function. It is, however, pointed out, for example, in Chatterjee (2002), that the Taylor curve in itself does not resolve the decision problem on which monetary policy should be adopted, and that further analysis on the welfare consequences for households of different combinations of variability of inflation and unemployment rates along the Taylor curve is required.

In the following, we will investigate how different interest rate rules behave under different choices of weights $(\omega_\pi, \omega_y, \omega_r)$, and under different weights λ in the monetary authorities' loss function, which we assume can be written as a linear combination of the unconditional variances of output growth $\Delta_4 y_t$ and underlying inflation, $\Delta_4 pu_t$.

$$\mathcal{L}(\lambda) = V[\Delta_4 pu_t] + \lambda V[\Delta_4 y_t].$$

For given levels of target inflation, π^*, target output growth rate g_y^* and equilibrium real interest rate RR*, the interest rate reaction function is described by the triplet $(\omega_\pi, \omega_y, \omega_r)$.[8] We have designed a simulation experiment in order to uncover the properties of different interest rate rules across a range of different values of these coefficients. The experiment constitutes a simple grid search across $\Omega_p \times \Omega_y \times \Omega_r$ under different interest rate rules.

---

[8] It follows that the experiment is particularly relevant for the first three types of rules in Table 10.1 (FLX, ST, and SM).

For each simulation the variance of underlying inflation, $V[\Delta_4 pu_t]$, and output growth, $V[\Delta_4 y_t]$ is calculated over the period 1995(1)–2000(4).

To summarise the different outcomes we have used the loss function $\mathcal{L}(\lambda) = V[\Delta_4 pu_t] + \lambda V[\Delta_4 y_t]$ for $\lambda \in (0, \ldots, 4)$ (11 different values).

The inflation coefficient is varied across $w_\pi \in (0, 0.5, \ldots, 4) (\Rightarrow 9$ values), the output growth coefficient is varied across $w_y \in (0, 0.5, \ldots, 4) (\Rightarrow 9$ values), and the smoothing coefficient is varied across $w_r \in (0, 0.1, \ldots, 1) (\Rightarrow 11$ values). This makes a total of $9 \times 9 \times 11 = 891$ simulations and 9801 loss evaluations for each type of rule/horizon.

In order to analyse such large amounts of data we need some efficient way to obtain a data reduction. We suggest to analyse the performance of the different interest rate rules by estimating a response surface for the loss function $\mathcal{L}(\lambda)$ across different weights of the loss function $\lambda \in (0, 0.1, \ldots, 1)$.

We consider a second-order Taylor expansion around some values $\bar{w}_\pi, \bar{w}_y, \bar{w}_r$, and we have chosen the standard Taylor rule $(0.5, 0.5, 0)$ as our preferred choice.

$$\mathcal{L}(\lambda) \simeq \alpha_0 + \alpha_1 w'_\pi + \alpha_2 w'_y + \alpha_3 w'_r + \beta_{12} w'_\pi w'_y + \beta_{13} w'_\pi w'_r + \beta_{23} w'_y w'_r$$
$$+ \beta_1 w'^2_\pi + \beta_2 w'^2_y + \beta_3 w'^2_r + \text{error}$$
$$w'_\pi = w_\pi - \bar{w}_\pi, \quad w'_y = w_y - \bar{w}_y, \quad w'_r = w_r - \bar{w}_r.$$

α's and β's are estimated by OLS for each choice of weights in the loss function $\lambda \in (0, 0.1, \ldots, 1)$. We minimise the estimated approximation to this loss function with respect to the three weights (w_π, w_y, w_r), and apply the first-order conditions to solve for these weights as functions of λ in the loss function, as linear combinations of the estimated α's and β's.

$$\underset{w}{\text{Min }} \mathcal{L}(\lambda) \Rightarrow \begin{array}{l} \dfrac{\partial \mathcal{L}(\lambda)}{\partial w_p} = 0 \\[2mm] \dfrac{\partial \mathcal{L}(\lambda)}{\partial w_y} = 0 \\[2mm] \dfrac{\partial \mathcal{L}(\lambda)}{\partial w_r} = 0 \end{array} \Rightarrow \begin{bmatrix} -2\hat{\beta}_1 & -\hat{\beta}_{12} & -\hat{\beta}_{13} \\ -\hat{\beta}_{12} & -2\hat{\beta}_2 & -\hat{\beta}_{23} \\ -\hat{\beta}_{13} & -\hat{\beta}_{23} & -2\hat{\beta}_3 \end{bmatrix}_\lambda \begin{bmatrix} w'_\pi \\ w'_y \\ w'_r \end{bmatrix}_\lambda = \begin{bmatrix} \hat{\alpha}_1 \\ \hat{\alpha}_2 \\ \hat{\alpha}_3 \end{bmatrix}_\lambda$$

$$\Rightarrow \begin{bmatrix} \tilde{w}_\pi \\ \tilde{w}_y \\ \tilde{w}_r \end{bmatrix}_\lambda = \begin{bmatrix} \bar{w}_\pi \\ \bar{w}_y \\ \bar{w}_r \end{bmatrix}_\lambda + \begin{bmatrix} -2\hat{\beta}_1 & -\hat{\beta}_{12} & -\hat{\beta}_{13} \\ -\hat{\beta}_{12} & -2\hat{\beta}_2 & -\hat{\beta}_{23} \\ -\hat{\beta}_{13} & -\hat{\beta}_{23} & -2\hat{\beta}_3 \end{bmatrix}_\lambda^{-1} \begin{bmatrix} \hat{\alpha}_1 \\ \hat{\alpha}_2 \\ \hat{\alpha}_3 \end{bmatrix}_\lambda$$

The optimal reaction function according to this minimisation is shown in Figure 10.7, where the optimal weights $\tilde{w}_\pi, \tilde{w}_y, \tilde{w}_r$ are plotted as function of the relative weight λ assigned to output variability in the loss function. The main findings are that there is a tradeoff between variability in inflation and variability in output, irrespective of the degree of smoothing. The inflation coefficient drops as we increase the output growth weight λ. Interest rate smoothing increases variability in the inflation rate without any substantial reduction in

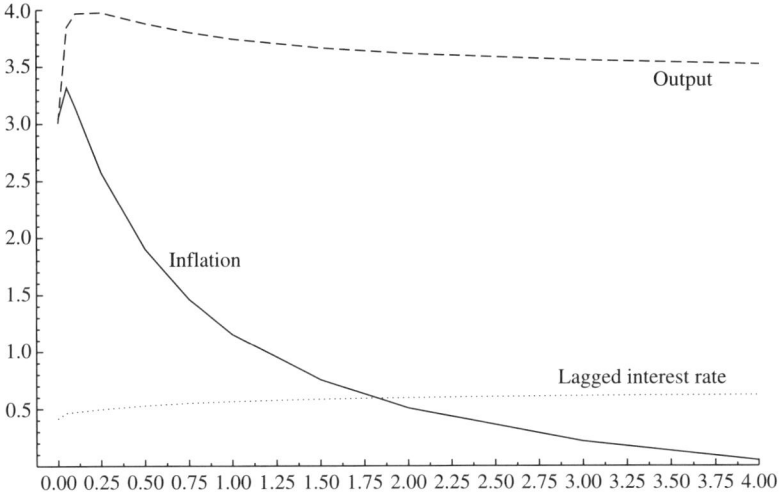

Figure 10.7. Estimated weights $\tilde{\omega}_\pi, \tilde{\omega}_y, \tilde{\omega}_r$ as a function of λ, the weight of output growth in the loss function. The weights are based on an estimated response surface for a Taylor approximation of the loss function.

$$\mathcal{L}(\lambda) = V[\Delta_4 pu_t] + \lambda V[\Delta_4 y_t], \qquad \text{for } \lambda \in (0, \dots, 4)$$

output variability. This may explain the relatively low weight on interest rate smoothing suggested by the plot of the smoothing coefficient as a function of λ.

10.4 Conclusions

The results from the counterfactual simulations indicate that a standard Taylor rule does quite well, across different values of the central bank preference parameters in a loss function, even in the case of a small open economy like Norway. 'Open economy' rules that respond to exchange rate misalignments, are shown to perform slightly worse than the Taylor rule. These rules contribute towards lower exchange rate variability without increasing interest rate variability, but at a cost of raising the variability in other target variables like headline and underlying inflation, output growth, and unemployment. Rules which respond to volatile variables like output growth produce higher interest rate volatility as a consequence. The counterfactual simulations illustrate substantial differences in the bias across the different interest rate rules, which are picked up by the RMSTE. The derivation of weights in the interest rate rules from estimated response surfaces indicate a tradeoff between variability in inflation and variability in output, irrespective of the degree of interest rate smoothing. In contrast with many other studies, interest rate smoothing seems to increase variability in the inflation rate without any substantial reduction in output variability. We conclude from this observation that statements about the optimal degree of interest rate smoothing appear to be non-robust or—to put it differently—that they are model dependent. In a situation with such conflicting evidence, the central bank should evaluate the empirical relevance and realism in the underlying models and base its decisions on the one with the highest degree of congruence.

11

Forecasting using econometric models

The non-stationary nature of many economic time series has a bearing on virtually all aspects of econometrics, including forecasting. Recent developments in forecasting theory have taken this into account, and provide a framework for understanding typical findings in forecast evaluations: for example, why certain types of models are more prone to forecast failure than others. In this chapter we discuss the sources of forecast failure most likely to occur in practice, and we compare the forecasts of a large econometric forecasting model with the forecasts stemming from simpler forecasting systems, such as dVARs. The large scale model holds its ground in our experiment, but the theoretical discussion about vulnerability to deterministic shifts is very relevant for our understanding of the instances where a dVAR does better. We also analyse the theoretical and empirical forecast properties of two wage–price models that have appeared earlier in the book: the dynamic incomplete competition model (ICM) and the Phillips curve, PCM. The analysis shows that although the PCM shares some of the robustness of dVARs, it also embodies equilibrium correction, in the form of natural rate dynamics. Since that form of correction mechanism is rejected empirically, the PCM forecasts are harmed both by excessive uncertainty (from its dVAR aspect), and by their econometric mis-specification of the equilibrium-correction mechanism in wage formation.

11.1 Introduction

Economic forecasts are statements about the future which are generated with a range of methods, ranging from wholly informal ('gut feeling') to sophisticated statistical techniques and the use of econometric models. However, professional forecasters never stick to only one method of forecasting, so formal and informal forecasting methods both have an impact on the final (published) forecast. The use of judgemental correction of forecasts from econometric models is one example.

It is fair to say that the combined use of different forecasting methods reflects how practitioners have discovered that there is no undisputed and overall 'best' way of constructing forecasts. A related observation, brought into the literature already in Bates and Granger (1969), is that a combination of forecasts of an economic variable often turn out to be more accurate than the individual projections that are combined together.

Nevertheless, intercept correction and pooling are still looked upon with suspicion in wide circles. Hence, being open-minded about intercept correction often has a cost in terms of credibility loss. For example, the forecaster will often find herself accused of an inconsistency (i.e. 'if you believe in the model, why do you overrule its forecasts?'), or the model can be denounced on the logic that 'if intercept correction is needed, why use a model in the first place?'.

It is probable that such reactions are based on an unrealistic description of the forecasting situation, namely that the econometric model in question is correctly specified simplification of the data generation process, which in turn is assumed to be without regime shifts in the forecasting period. Realistically however, there is genuine uncertainty about how good a model is, even within the sample. Moreover, since the economy is evolving, we can take it for granted that the data generation process will change in the forecast period, causing any model of it to become mis-specified over that period, and this is eventually the main problem in economic forecasting. The inevitable conclusion is that there is no way of knowing *ex ante* the degree of mis-specification of an econometric model over the forecast period. The implication is that all measures of forecast uncertainty based on within-sample model fit are underestimating the true forecast uncertainty. Sometimes, when regimes shifts affect parameters like growth rates and the means and coefficients of cointegration relationships, one is going to experience forecast failure, that is, *ex post* forecast errors are systematically larger than indicated by within-sample fit.

On the basis of a realistic description of the forecasting problem it thus becomes clear that intercept corrections have a pivotal role in robustifying the forecasts from econometric models, when the forecaster has other information which indicate that structural changes are 'immanent'; see Hendry (2001*a*). Moreover, correcting a model's forecast through intercept correction does not incriminate the use of that model for policy analysis. That issue hinges more precisely on which parameters of the model are affected by the regime shift. Clearly, if the regime shift entails significant changes in the parameters that determine the (dynamic) multipliers, then the continued use of the model is untenable. However, if it is the intercepts and long-run means of cointegrating relationships which are affected by a regime shift, the model may still be valid for policy analysis, in spite of forecast failure. Clements and Hendry (1999*a*) provide a comprehensive exposition of the theory of forecasting non-stationary time-series under realistic assumptions about regime shifts, and Hendry and Mizon (2000) specifically discuss the consequences of forecast failure for policy analysis. Ericsson and Hendry (2001) is a non-technical presentation of recent developments in economic forecasting theory and practice.

A simple example may be helpful in defining the main issues at stake. Let M1 in equation (11.1) represent a model of the rate of inflation π_t (i.e. denoted Δp_t in the earlier chapters). In equation (11.1) μ denotes the unconditional mean of the rate of inflation, while z_t denotes an exogenous variable, whose change affects the rate of inflation with semi-elasticity given by γ (hence γ is the derivative coefficient of this model). Assume next that M1 corresponds to the data generation process over the sample period $t = 1, 2, \ldots, T$, hence as in earlier chapters, ε_t denote a white noise innovation (with respect to π_{t-1} and Δz_t), and follows a normal distribution with zero mean and constant variance.

$$\text{M1:} \quad \Delta \pi_t = \delta - \alpha(\pi_{t-1} - \mu) + \gamma \Delta z_t + \varepsilon_t. \tag{11.1}$$

By definition, any alternative model is mis-specified over the sample period. M2 in equation (11.2) is an example of a simple model in differenced form, a dVAR, often used as a benchmark in forecast comparisons since it produces the naive forecasts that tomorrow's rate of inflation is identical to today's rate. The M2-disturbance is clearly not an innovation, but is instead given by the equation below M2.

$$\text{M2:} \quad \Delta \Delta \pi_t = \nu_t, \tag{11.2}$$
$$\nu_t = -\alpha \Delta \pi_{t-1} + \gamma \Delta^2 z_t + \varepsilon_t - \varepsilon_{t-1}.$$

As noted above, M2 is by definition inferior to M1 when viewed as an alternative model of the rate of inflation. However, our concern now is a different one: if we use M1 and M2 to forecast inflation over H periods $T+1, T+2, \ldots, T+H$, which set of forecasts is the best or most accurate? In other words: which of M1 and M2 provides the best *forecast mechanism*? It is perhaps surprising that the answer depends on which other additional assumptions we make about the forecasting situation. Take for instance the exogenous variable z_t in (11.1): only if the forecasting agency also controls the process governing this variable in the forecast period can we assume that the conditional inflation forecast based on M1 is based on the correct sequence of exogenous variables $(z_{T+1}, z_{T+2}, \ldots, z_{T+H})$. Thus, as is well documented, errors in the projections of the exogenous variables are important contributors to forecast errors. Nevertheless, for the purpose of example, we shall assume that the future z's are correctly forecasted. Another simplifying assumption is to abstract from estimation uncertainty, that is, we evaluate the properties of forecasting mechanism M1 *as if* the coefficients δ, α, and γ are known coefficients. Intuitively, given the first assumption that M1 corresponds to the data generating process, the assumption about no parameter uncertainty is of second- or third-order importance.

Given this description of the forecasting situation, we can concentrate on the impact of deterministic non-stationarities, or structural change, on the forecasts of M1 and M2. Assume first that there is no structural change. In this case, M1 delivers the predictor with the minimum mean squared forecast error (MMSFE): see, for example, Clements and Hendry (1998: ch. 2.7). Evidently, the imputed forecast errors from M2, and hence the conventional 95% predictions intervals, are too large (notably by 100% for the $T+1$ forecast).

However, if there is a structural change in the long-run mean of the rate of inflation, μ, it is no longer obvious that M1 is the winning forecasting mechanism. Exactly when μ shifts to its new value, μ^*, before or after the preparation of the forecast in period T, is important, as shown by the biases of the two 1-step ahead forecasts:

$$E[\pi_{T+1} - \hat{\pi}_{M1,T+1} \mid \mathcal{I}_T] = \alpha(\mu - \mu^*), \qquad \text{if } \mu \to \mu^* \text{ (no matter when)}$$
$$E[\pi_{T+1} - \hat{\pi}_{M2,T+1} \mid \mathcal{I}_T] = -\alpha\Delta\pi_T + \gamma\Delta^2 z_{T+1}, \qquad \text{if } \mu \to \mu^*, \text{ before } T$$
$$E[\pi_{T+1} - \hat{\pi}_{M2,T+1} \mid \mathcal{I}_T] = \alpha(\mu - \mu^*), \qquad \text{if } \mu \to \mu^*, \text{ after } T$$

demonstrating that

- The forecast mechanism corresponding to M2 'error corrects' to the structural change occurring before the forecast period. Hence, M2 *post-break* forecasts are robust.
- M1 produces forecast failure, also when M2-forecasts do not break down, that is, M1 *post-break* forecasts are not robust.
- Both forecasts are damaged if the regime shift occurs after the forecast is made (i.e. in the forecast period). In fact, M1 and M2 share a common bias in this *pre-break* case (see first and third line).

Thus, apart from the special case where the econometric model corresponds to the true mechanism in the forecast period, it is impossible to prove that it provides the best forecasting mechanism, establishing the role of supplementary forecasting mechanisms in economic forecasting. Moreover, in this example, forecast failure of M1 is due to a change in a parameter which does *not* affect the dynamic multipliers (the relevant parameters being α and γ in this example). Thus, forecast failure per se does not entail that the model is invalid for policy analysis.

M1-regime shifts that occur prior to the forecast period are detectable, in principle, and the forecaster therefore has an opportunity to avoid forecast failure by intercept correction. In comparison, it is seen that the simple forecast mechanism M2 has built in intercept correction: its forecast is back on track in the first period after the break. Intriguingly, M2 has this capability almost by virtue of being a wrong model of the economy.

In the rest of this chapter, we investigate the relevance of these insights for macroeconometric forecasting. Section 11.2 contains a broader discussion of the relative merits of equilibrium-correction models (EqCMs) and differenced VARs (dVARs) in macroeconometric forecasting. This is done by first giving an extended algebraic example, in Section 11.2.1. In Section 11.2.2, we turn to the theory's practical relevance for understanding the forecasts of the Norwegian economy in the 1990s. The model that takes the role of the EqCM is the macroeconometric model RIMINI. The rival forecasting systems are dVARs derived from the full scale model as well as univariate autoregressive models.

So far we have discussed forecasting mechanisms as if the choice of forecasting method is clear-cut and between using a statistical model, M2, and a well-defined econometric model, M1. In practice, the forecaster has not one but many econometric models to choose from. In earlier chapters of this book, we have seen that different dynamic model specifications can be compatible with the same theoretical framework. We showed, for example, that the open economy Phillips curve model with a constant NAIRU, can be seen as an EqCM version of a bargaining model (see Chapter 4), but also that there was an alternative EqCM which did not imply a supply-side natural rate (in Chapter 6 we referred to it as the dynamic incomplete competition model). In Section 11.3 we discuss the forecasting properties of the two contending specifications of inflation, theoretically and empirically.

11.2   EqCMs vs. dVARs in macroeconometric forecasting

The development of macroeconometric models in the course of the 1980s and 1990s, with more emphasis on dynamic specification and on model evaluation, meant that the models became less exposed to the critique against earlier generations of models, namely that models that largely ignore dynamics and temporal properties of the data, will necessarily produce suboptimal forecasts; see, for example, Granger and Newbold (1986: ch. 6). At the same time, other model features also changed in response to developments in the real economy, for example, the more detailed and careful modelling of the supply-side factors and the transmission mechanism between the real and financial sectors of the economy; see Wallis (1989) for an overview. Given these developments, macroeconomic model builders and forecasters may be justified in claiming that modern models of the EqCM type, would forecast better than models that only use differenced data, dVARs. Forecast competitions between models of these two types have been reported in Eitrheim *et al.* (1999, 2002*a*). This chapter draws on these results, and extends the horse race competition between the different inflation models reported in Chapter 8 (Section 8.7.6).

As noted above, Michael Clements and David Hendry have re-examined several issues in macroeconometric forecasting, including the relative merits of dVARs and EqCMs (see, for example, Clements and Hendry 1995*a,b*, 1996, 1998). Assuming constant parameters in the forecast period, the dVAR is mis-specified relative to a correctly specified EqCM, and dVAR forecasts will therefore be suboptimal. However, if parameters change after the forecast is made, then the EqCM is also mis-specified in the forecast period. Clements and Hendry have shown that forecasts from a dVAR are *robust* with respect to certain classes of parameter changes. Hence, in practice, EqCM forecasts may turn out to be less accurate than forecasts derived from a dVAR. Put differently, the 'best model' in terms of economic interpretation and econometrics, may not be the best model for forecasts. At first sight, this is paradoxical, since any dVAR can be viewed as a special case of an EqCM, since it imposes additional unit root restrictions on the system. However, if the parameters of the levels variables that are excluded from the dVAR change in the forecast period, this in turn makes also the EqCM mis-specified. Hence, the outcome of a horse race is no longer given, since both forecasting models are mis-specified relative to the generating mechanism that prevails in the period we are trying to forecast.

11.2.1   Forecast errors of bivariate EqCMs and dVARs

In this section, we illustrate how the forecast errors of an EqCM and the corresponding dVAR are affected differently by structural breaks. Practical forecasting models are typically open systems, with exogenous variables. Although the model that we study in this section is of the simple kind, its properties will prove helpful in interpreting the forecasts errors of the large systems in Section 11.2.3.

A simple DGP   This book has taken as a premise that macroeconomic time-series can be usefully viewed as integrated of order one, I(1), and that they also frequently include deterministic terms allowing for a linear trend. The following simple bivariate system (a first-order VAR) can serve as an example:

$$y_t = \kappa + \lambda_1 y_{t-1} + \lambda_2 x_{t-1} + e_{y,t}, \tag{11.3}$$
$$x_t = \varphi + x_{t-1} + e_{x,t}, \tag{11.4}$$

where the disturbances $e_{y,t}$ and $e_{x,t}$ have a jointly normal distribution. Their variances are σ_y^2 and σ_x^2 respectively, and the correlation coefficient is denoted by $\rho_{y,x}$. The openness of practical forecasting models is captured by x_t which is (strongly) exogenous. x_t is integrated of order one, denoted I(1), and contains a linear deterministic trend if $\varphi \neq 0$. We will assume that (11.3) and (11.4) constitute a small cointegrated system such that y_t is also I(1) but cointegrated with x_t. This entails that $0 < \lambda_1 < 1$ and $\lambda_2 \neq 0$. With a change in notation, the DGP can be written as

$$\Delta y_t = -\alpha[y_{t-1} - \beta x_{t-1} - \zeta] + e_{y,t}, \qquad 0 < \alpha < 1, \tag{11.5}$$
$$\Delta x_t = \varphi + e_{x,t}, \tag{11.6}$$

where $\alpha = (1 - \lambda_1)$, $\beta = \lambda_2/\alpha$, and $\zeta = \kappa/\alpha$. In equation (11.5), α is the equilibrium-correction coefficient and β is the derivative coefficient of the cointegrating relationship.

The system can be re-written in 'model form' as a conditional equilibrium-correcting model for y_t and a marginal model for x_t.

$$\Delta y_t = \gamma + \pi \Delta x_t - \alpha[y_{t-1} - \beta x_{t-1} - \zeta] + \varepsilon_{y,t}, \qquad (11.7)$$
$$\Delta x_t = \varphi + e_{x,t}, \qquad (11.8)$$

where

$$\pi = \rho_{y,x} \frac{\sigma_y}{\sigma_x},$$
$$\gamma = -\varphi \pi,$$
$$\varepsilon_{y,t} = e_{y,t} - \pi e_{x,t}$$

from the properties of the bivariate normal distribution.

We define two parameters, μ and η, such that $E[y_t - \beta x_t] = \mu$ and $E[\Delta y_t] = \eta$. By taking expectations in (11.6) we see that $E[\Delta x_t] = \varphi$. Similarly, by taking expectations in (11.5) and substituting for these definitions, noting that $\eta = \beta \varphi$, we find the following relationship between these parameters:

$$\beta \varphi = \alpha(\zeta - \mu). \qquad (11.9)$$

Solving with respect to μ yields

$$\mu = \zeta - \frac{\beta \varphi}{\alpha} = \frac{\kappa - \beta \varphi}{\alpha}. \qquad (11.10)$$

In the case when $\varphi \neq 0$, both series contain a deterministic trend which stems from the x_t-process and conversely, if $\varphi = 0$ there is no deterministic growth in either of the variables. In the latter case we see from (11.10) that $\mu = \zeta$.

The case with a linear deterministic trend is relevant for many variables of interest for forecasters. In the empirical part of this chapter, Section 11.2.3, we will show examples of both cases. Typical examples of exogenous variables associated with positive drift are indicators of foreign demand, foreign price indices, and average labour productivity, while the zero drift assumption is the most appealing one for variables like, for example, oil prices and monetary policy instruments, that is, money market interest rates and exchange rates.

EqCM and dVAR models of the DGP   The purpose of this section is to trace the impact of parameter changes in the DGP on the forecasts of two *models* of the DGP. First, the equilibrium correction model, EqCM, which coincides with the DGP within sample, that is, there is no initial mis-specification, and second, the dVAR.

The EqCM is made up of equations (11.7) and (11.8). Equation (11.7) is the conditional model of y_t (see, for example, Hendry 1995a: ch. 7), which has many counterparts in practical forecasting models, following the impact of econometric methodology and cointegration theory on applied work. Equation (11.8) is the marginal equation for the explanatory variable x_t. The dVAR model of y_t and x_t (wrongly) imposes one restriction, namely that $\alpha = 0$, hence the dVAR model consists of

$$\Delta y_t = \gamma + \pi \Delta x_t + \epsilon_{y,t}, \tag{11.11}$$
$$\Delta x_t = \varphi + e_{x,t}. \tag{11.12}$$

Note that the error process in the dVAR model, $\epsilon_{y,t}$ $(=\varepsilon_{y,t}-\alpha[y_{t-1}-\beta x_{t-1}-\zeta])$, will in general be autocorrelated provided there is some autocorrelation in the omitted disequilibrium term (for $0 < \alpha < 1$).

We further assume that

- parameters are known;
- in the forecasts, $\Delta x_{T+j} = \varphi$ $(j = 1, \ldots, h)$;
- forecasts for the periods $T + 1, T + 2, \ldots, T + h$, are made in period T.

The first assumption abstracts from small sample biases in the EqCM and inconsistently estimated parameters in the dVAR case. The second assumption rules out one source of forecast failure that is probably an important one in practice, namely that non-modelled or exogenous variables are poorly forecasted. In our framework systematic forecast errors in Δx_{T+j} are tantamount to a change in φ.

Although all other coefficients may change in the forecast period, the most relevant coefficients in our context are α, β, and ζ, that is, the coefficients that are present in the EqCM but not in the dVAR. Among these, we concentrate on α and ζ, since β represents partial structure by virtue of being a cointegration parameter; see Doornik and Hendry (1997b) and Hendry (1998) for an analysis of the importance and detectability of shifts.

In the following two sections we derive the biases for the forecasts of EqCM and dVAR, when both models are mis-specified in the forecast period. We distinguish between the case where the parameter change occurs *after* the forecast is made (post-forecast break) and a shift that takes place *before* the forecast period (pre-forecast break).

Parameter change after the forecast is prepared   We first assume that the intercept ζ in (11.5) changes from its initial level to a new level, that is, $\zeta \rightarrow \zeta^*$, *after* the forecast is made in period T. Since we maintain a constant α in this section, the shift in ζ is fundamentally the product of a change in κ, the intercept in equation (11.3). In equilibrium correction form, the DGP in the forecast period is therefore

$$\Delta y_{T+h} = \gamma + \pi \Delta x_{T+h} - \alpha [y_{T+h-1} - \beta x_{T+h-1} - \zeta^*] + \varepsilon_{y,T+h},$$
$$\Delta x_{T+h} = \varphi + e_{x,T+h},$$

where $h = 1, \ldots, H$. The 1-period forecast errors for the EqCM and the dVAR models can be written:

$$y_{T+1} - \hat{y}_{T+1,\text{EqCM}} = -\alpha [\zeta - \zeta^*] + e_{y,T+1}, \qquad (11.13)$$
$$y_{T+1} - \hat{y}_{T+1,\text{dVAR}} = -\alpha [y_T - \beta x_T - \zeta^*] + e_{y,T+1}. \qquad (11.14)$$

In the following, we focus on the bias of the forecast errors. The 1-step biases are defined by the conditional expectation (on $\mathcal{I}_T$) of the forecast errors and are denoted $\text{bias}_{T+1,\text{EqCM}}$ and $\text{bias}_{T+1,\text{dVAR}}$ respectively:

$$\text{bias}_{T+1,\text{EqCM}} = -\alpha [\zeta - \zeta^*], \qquad (11.15)$$
$$\text{bias}_{T+1,\text{dVAR}} = -\alpha [y_T - \beta x_T - \zeta^*]. \qquad (11.16)$$

Let x_t°, denote the steady-state values of the x_t-process. The corresponding steady-state values of the y_t-process, denoted y_t°, are then given by

$$y_t^\circ = \mu + \beta x_t^\circ. \qquad (11.17)$$

Using this definition and (11.15), the dVAR forecast error (11.16) can be rewritten as

$$\text{bias}_{T+1,\text{dVAR}} = -\alpha \left[(y_T - y_T^\circ) - \beta(x_T - x_T^\circ) - \frac{\beta\varphi}{\alpha} + (\zeta - \zeta^*) \right]$$
$$= -\alpha \left[(y_T - y_T^\circ) - \beta(x_T - x_T^\circ) - \frac{\beta\varphi}{\alpha} \right] + \text{bias}_{T+1,\text{EqCM}}. $$
$$(11.18)$$

Note that both EqCM and dVAR forecasts are harmed by the parameter shift from ζ to ζ^*; see Clements and Hendry (1996). Assuming that the initial values' deviations from steady state are negligible, that is, $x_T \approx x_T^\circ$ and $y_T \approx y_T^\circ$, we can simplify the expression into

$$\text{bias}_{T+1,\text{dVAR}} = \beta\varphi + \text{bias}_{T+1,\text{EqCM}}. \qquad (11.19)$$

The two models' 1-step forecast error biases are identical if y_T equals its long-run mean $\bar{y}_T$. An example of such a case will be ordinary least squares (OLS)-estimated unrestricted dVAR (see Clements and Hendry 1998: ch. 5.4).

For comparison we also write down the biases of the 2-period forecast errors (maintaining the steady-state assumption).

$$\text{bias}_{T+2,\text{EqCM}} = -\alpha\delta_{(1)}[\zeta - \zeta^*], \tag{11.20}$$

$$\text{bias}_{T+2,\text{dVAR}} = \beta\varphi\alpha - \alpha\delta_{(1)}\left[(y_T - y_T^\circ) - \beta(x_T - x_T^\circ) - \frac{\beta\varphi}{\alpha} + (\zeta - \zeta^*)\right] \tag{11.21}$$

$$\approx \beta\varphi(\alpha + \delta_{(1)}) + \text{bias}_{T+2,\text{EqCM}}$$

$$= 2\beta\varphi + \text{bias}_{T+2,\text{EqCM}},$$

where $\delta_{(1)} = 1 + (1 - \alpha)$.

More generally, for h-period forecasts we obtain the following expressions

$$\text{bias}_{T+h,\text{EqCM}} = -\alpha\delta_{(h-1)}[\zeta - \zeta^*], \tag{11.22}$$

$$\text{bias}_{T+h,\text{dVAR}} = \beta\varphi(\alpha\psi_{(h-2)} - \delta_{(h-1)}) - \alpha\delta_{(h-1)}[(y_T - y_T^\circ)$$
$$- \beta(x_T - x_T^\circ) + (\zeta - \zeta^*)] \tag{11.23}$$

for forecast horizons $h = 2, 3, \ldots$, where δ_{h-1} and ψ_{h-2} are given by

$$\delta_{(h-1)} = 1 + \sum_{j=1}^{h-1}(1 - \alpha)^j, \qquad \delta_{(0)} = 1 \tag{11.24}$$

$$= 1 + (1 - \alpha)\delta_{(h-2)},$$

$$\psi_{(h-2)} = 1 + \sum_{j=1}^{h-2}\delta_{(j)}, \qquad \psi_{(0)} = 1, \quad \psi_{(-1)} = 0 \tag{11.25}$$

$$= (h - 1) + (1 - \alpha)\psi_{(h-3)}$$

and we have again used (11.17). As the forecast horizon h increases to infinity, $\delta_{(h-1)} \rightarrow 1/\alpha$, hence the EqCM-bias approaches asymptotically the size of the shift itself, that is, $\text{bias}_{T+h,\text{EqCM}} \rightarrow \zeta^* - \zeta$.

Assuming that $x_T \approx x_T^\circ$ and $y_T \approx y_T^\circ$, we can simplify the expression and the dVAR forecast errors are seen to contain a bias term that is due to the growth in x_t and which is not present in the EqCM forecast bias, cf. the term $\beta\varphi(\alpha\psi_{(h-2)} + \delta_{(h-1)})$ in (11.23). We can simplify this expression, since the term in square brackets containing the recursive formulae $\delta_{(h-1)}$ and $\psi_{(h-2)}$ can be rewritten as $[\alpha\psi_{(h-2)} + \delta_{(h-1)}] = h$, and we end up with a simple linear trend in the h-step ahead dVAR forecast error bias in the case when $\varphi \neq 0$, thus generalising the 1-step and 2-step results[1]:

$$\text{bias}_{T+h,\text{dVAR}} = \beta\varphi h - \alpha\delta_{(h-1)}[(y_T - y_T^\circ) - \beta(x_T - x_T^\circ)] + \text{bias}_{T+h,\text{EqCM}}.$$
(11.26)

We note furthermore that the two models' forecast error biases are identical if there is no autonomous growth in x_t ($\varphi = 0$), and y_T and x_T equal their steady-state values. In the case with positive deterministic growth in x_t ($\varphi > 0$), while maintaining the steady-state assumption, the dVAR bias will dominate the EqCM bias in the long run due to the trend term in the dVAR bias.

Change in the equilibrium-correction coefficient α Next, we consider the situation where the adjustment coefficient α changes to a new value, α^*, after the forecast for $T+1, T+2, \ldots, T+h$ have been prepared. Conditional on $\mathcal{I}_T$, the 1-step biases for the two models' forecasts are:

$$\text{bias}_{T+1,\text{EqCM}} = -(\alpha^* - \alpha)[y_T - \beta x_T - \zeta], \tag{11.27}$$
$$\text{bias}_{T+1,\text{dVAR}} = -\alpha^*[y_T - \beta x_T - \zeta]. \tag{11.28}$$

Using the steady-state expression (11.17), we obtain

$$\text{bias}_{T+1,\text{EqCM}} = -(\alpha^* - \alpha)\left[(y_T - y_T^\circ) - \beta(x_T - x_T^\circ) - \frac{\beta\varphi}{\alpha}\right], \tag{11.29}$$

$$\text{bias}_{T+1,\text{dVAR}} = -\alpha^*\left[(y_T - y_T^\circ) - \beta(x_T - x_T^\circ) - \frac{\beta\varphi}{\alpha}\right]. \tag{11.30}$$

In general, the EqCM bias is proportional to the size of the shift, while the dVAR bias is proportional to the magnitude of the level of the new equilibrium-correction coefficient itself. Assuming that $x_T \approx x_T^\circ$ and $y_T \approx y_T^\circ$, we can simplify the expression into

$$\text{bias}_{T+1,\text{dVAR}} = \beta\varphi + \text{bias}_{T+1,\text{EqCM}}. \tag{11.31}$$

Hence, the difference between the dVAR and EqCM 1-step forecast error biases is identical to (11.19). For the multi-period forecasts, the EqCM and dVAR

---

[1] From the definition of $\psi_{(h-2)}$ in (11.25) it follows that $\psi_{(h-3)} = \psi_{(h-2)} - \delta_{(h-2)}$. Inserting this in the recursive formula for $\psi_{(h-3)}$ and rearranging terms yields $\alpha\psi_{(h-2)} = (h-1) - (1-\alpha)\delta_{(h-2)}$. Finally, when we add $\delta_{(h-1)}$ on both sides of this equality and apply the recursive formula for $\delta_{(h-1)}$ in (11.25), the expression simplifies to $(h-1) + 1 = h$.

forecast error biases are

$$\text{bias}_{T+h,\text{EqCM}} = \beta\varphi(\alpha^*\psi^*_{(h-2)} - \alpha\psi_{(h-2)}) - (\alpha^*\delta^*_{(h-1)} - \alpha\delta_{(h-1)})$$
$$\times \left[(y_T - y_T^\circ) - \beta(x_T - x_T^\circ) - \frac{\beta\varphi}{\alpha} \right], \qquad (11.32)$$

$$\text{bias}_{T+h,\text{dVAR}} = \beta\varphi\alpha^*\psi^*_{(h-2)}$$
$$- \alpha^*\delta^*_{(h-1)} \left[(y_T - y_T^\circ) - \beta(x_T - x_T^\circ) - \frac{\beta\varphi}{\alpha} \right] \qquad (11.33)$$

$h = 2, 3, \ldots$, where y_T° is defined in (11.17), $\delta_{(h-1)}$ in (11.24), $\psi_{(h-2)}$ in (11.25). $\delta^*_{(h-1)}$ and $\psi^*_{(h-2)}$ are given by

$$\delta^*_{(h-1)} = 1 + \sum_{j=1}^{h-1}(1-\alpha^*)^j, \qquad \delta^*_{(0)} = 1,$$

$$\psi^*_{(h-2)} = 1 + \sum_{j=1}^{h-2}\delta^*_{(j)}, \qquad \psi^*_{(0)} = 1, \quad \psi^*_{(-1)} = 0.$$

To facilitate comparison we again assume that $x_T \approx x_T^\circ$ and $y_T \approx y_T^\circ$, and insert (11.33) in (11.32). Using a similar manipulation as when deriving (11.26), we arrive at the following $\text{bias}_{T+h,\text{dVAR}}$-expression:

$$\text{bias}_{T+h,\text{dVAR}} = \beta\varphi h + \text{bias}_{T+h,\text{EqCM}}.$$

We see that under the simplifying steady-state assumption, the difference between dVAR and EqCM h-step forecast error biases is identical to (11.26). Hence there will be a linear trend in the difference between the dVAR and EqCM forecast error biases due to the mis-representation of the growth in x_t in the dVAR.

Parameter change before the forecast is made   This situation is illustrated by considering how the forecasts for $T + 2, T + 3, \ldots, T + h + 1$ are updated conditional on outcomes for period $T + 1$. Remember that the shift $\zeta \to \zeta^*$ first affects outcomes in period $T + 1$. When the forecasts for $T + 2, T + 3, \ldots$ are updated in period $T + 1$, information about parameter non-constancies will therefore be reflected in the starting value y_{T+1}.

Change in the intercept ζ   Given that ζ changes to ζ^* in period $T + 1$, the (updated) forecast for y_{T+2}, conditional on y_{T+1} yields the following forecast error biases for the EqCM and dVAR models:

$$\text{bias}_{T+2,\text{EqCM}} \mid \mathcal{I}_{T+1} = -\alpha[(\zeta - \zeta^*)], \qquad (11.34)$$
$$\text{bias}_{T+2,\text{dVAR}} \mid \mathcal{I}_{T+1} = -\alpha[y_{T+1} - \beta x_{T+1} - \zeta^*]. \qquad (11.35)$$

Equation (11.34) shows that the EqCM forecast error is affected by the parameter change in exactly the same manner as before, cf. (11.15), despite the fact that in this case the effect of the shift is incorporated in the initial value y_{T+1}. Manifestly, the EqCM forecasts do not correct to events that have occurred prior to the preparation of the forecast. Indeed, unless the forecasters detect the parameter change and take appropriate action by (manual) intercept correction, the effect of a parameter shift prior to the forecast period will bias the forecasts 'forever'. The situation is different for the dVAR.

Using the fact that

$$y^{\circ}_{T+1} = \mu^* + \beta x^{\circ}_{T+1},$$

where

$$\mu^* = \zeta^* + \frac{\beta\varphi}{\alpha}. \tag{11.36}$$

Equation (11.35) can be expressed as

$$\text{bias}_{T+2,\text{dVAR}} \mid \mathcal{I}_{T+1} = -\alpha \left[(y_{T+1} - y^{\circ}_{T+1}) - \beta(x_{T+1} - x^{\circ}_{T+1}) - \frac{\beta\varphi}{\alpha} \right]$$
$$\approx \beta\varphi \tag{11.37}$$

under the steady-state assumption. We see that if there is no deterministic growth in the DGP, that is, $\varphi = 0$, the dVAR will be immune with respect to the parameter change. In this important sense, there is an element of inherent 'intercept correction' built into the dVAR forecasts, while the parameter change that occurred before the start of the forecast period will produce a bias in the 1-step EqCM forecast. A non-zero drift in the x_t-process will, however, produce a bias in the 1-step dVAR forecast as well, and the relative forecast accuracy between the dVAR model and the EqCM will depend on the size of the drift relative to the size of the shift.

The expression for the h-period forecast biases, conditional on $\mathcal{I}_{T+1}$, takes the form:

$$\text{bias}_{T+(h+1),\text{EqCM}} \mid \mathcal{I}_{T+1} = -\alpha\delta_{(h-1)}[\zeta - \zeta^*] \tag{11.38}$$
$$\text{bias}_{T+(h+1),\text{dVAR}} \mid \mathcal{I}_{T+1} = \beta\varphi h - \alpha\delta_{(h-1)}[(y_{T+1} - y^{\circ}_{T+1}) - \beta(x_{T+1} - x^{\circ}_{T+1})] \tag{11.39}$$

for $h = 1, 2, \ldots$. This shows that the EqCM forecast remains biased also for long forecast horizons. The forecast does 'equilibrium correct', but unfortunately towards the old (and irrelevant) 'equilibrium'. For really long (infinite) forecast horizons the EqCM bias approaches the size of the shift $[(\zeta^* - \zeta)]$ just as in the case where the parameter changed before the preparation of the forecast and therefore was undetectable.

For the dVAR forecast there is once again a trend in the bias term that is due to the growth in x_t. In the case with no deterministic growth in the DGP, the dVAR forecasts are unbiased for all h.

Change in the equilibrium-correction coefficient α Just as with the long-run mean, the EqCM forecast do not adjust automatically when the change $\alpha \to \alpha^*$ occurs prior to the preparation of the forecasts (in period $T+1$). The biases for period $T+2$, conditional on $\mathcal{I}_{T+1}$, take the form

$$\text{bias}\,_{T+2,\text{EqCM}}|\mathcal{I}_{T+1} = -(\alpha^* - \alpha)\left[(y_{T+1} - y^\circ_{T+1}) - \beta(x_{T+1} - x^\circ_{T+1}) - \frac{\beta\varphi}{\alpha}\right]$$

$$(11.40)$$

$$\text{bias}\,_{T+2,\text{dVAR}}|\mathcal{I}_{T+1} = -\alpha^*\left[(y_{T+1} - y^\circ_{T+1}) - \beta(x_{T+1} - x^\circ_{T+1}) - \frac{\beta\varphi}{\alpha}\right]$$

$$(11.41)$$

where we have used (11.17).

So neither of the two forecasts 'intercept correct' automatically to parameter changes occurring prior to the preparation of the forecast. For that reason, the 1-step biases are functionally similar to the formulae for the case where α change to α^* after the forecast has been prepared. The generalisation to multi-step forecast error biases is similar to previous derivations.

Estimated parameters In practice both EqCM and the dVAR forecasting models use estimated parameters. Since the dVAR is mis-specified relative to the DGP (and the EqCM), estimates of the parameters of (11.11) will in general be inconsistent. Ignoring estimated parameter uncertainty, the dVAR model will be

$$\Delta y_t = \gamma^* + \pi^* \Delta x_t + \epsilon^*_{y,t}, \qquad\qquad (11.42)$$

$$\Delta x_t = \varphi + e_{x,t}, \qquad\qquad (11.43)$$

where γ^* and π^* denote the probability limits of the parameter estimates. In the forecast period $\gamma^* + \pi^* \Delta x_{T+h} = g \neq 0$, hence the dVAR forecast of y_{T+h} will include an additional deterministic trend (due to estimation bias) which does not necessarily correspond to the trend in the DGP (which is inherited from the x_t-process).

The parameter bias may be small numerically (e.g. if differenced terms are close to orthogonal to the omitted equilibrium correction term), but can nonetheless accumulate to a dominating linear trend in the dVAR forecast error bias.

One of the dVAR-type models we consider in the empirical section, denoted dRIM, is a counterpart to (11.42). The empirical section shows examples of how dVAR-type models can be successfully robustified against trend-misrepresentation.

Discussion Although we have looked at the simplest of forecasting systems, the results have several traits that one might expect to be able to recover from the forecast errors of full sized macroeconomic models that we consider in Section 11.2.2.

The analysis above shows that neither the EqCM nor the dVAR protect against *post-forecast* breaks. In the case we have focused upon, where the dVAR model excludes growth when it is present in the DGP, the dVAR forecast error biases contain a trend component. Even in this case, depending on initial conditions, the dVAR may compete favourably with the EqCM over forecast horizons of moderate length.

We have seen that the dVAR does offer protection against *pre-forecast* shifts in the long-run mean, which reiterates a main point made by Hendry and Clements. While the dVAR automatically intercept corrects to the pre-forecast break, the EqCM will deliver inferior forecasts unless model users are able to detect the break and correct the forecast by intercept correction. Experience tells us that this is not always achieved in practice: in a large model, a structural break in one or more equations might pass unnoticed, or it might be (mis)interpreted as 'temporary' or as only seemingly a breakdown because the data available for model evaluation are preliminary and susceptible to future revision.[2]

One suggestion is that the relative merits of EqCMs and dVARs for forecasting depends on

- the 'mix' of pre- and post-forecast parameter changes
- the length of the forecast horizon.

In the next section we use this perspective to interpret the forecast outcomes from a large-scale model of the Norwegian economy.

11.2.2 A large-scale EqCM model and four dVAR type forecasting systems based on differenced data

Section 11.2.1 brought out that even for very simple systems, it is in general difficult to predict which version of the model is going to have the smallest forecast error, the EqCM or the dVAR. While the forecast errors of the dVAR are robust to changes in the adjustment coefficient α and the long-run mean ζ, the dVAR forecast error may still turn out to be larger than the EqCM forecast error. Typically, this is the case if the parameter change (included in the EqCM) is small relative to the contribution of the equilibrium-correcting term (which is omitted in the dVAR) at the start of the forecast period.

---

[2] The underprediction of consumption expenditures in Norway during the mid-1980s, which marred Norwegian forecasters for several consecutive forecasting rounds at that time, is a relevant example; see Brodin and Nymoen (1989, 1992). Eitrheim *et al.* (2002*b*) give a detailed analysis of the breakdown and reconstruction of the Norwegian consumption function that took place in the wake of these forecast failures, and show that what happened can be explained in the light of forecasting theory, see Section 2.4.2.

In the following, we generate multi-period forecasts from the econometric model RIMINI, and compare these to the forecasts from models based on differenced data. In order to provide some background to those simulations, this section first describes the main features of the incumbent EqCM and then explains how we have designed the dVAR forecasting systems.

The incumbent EqCM model—eRIM   The quarterly macroeconometric model RIMINI has 205 equations[3] which can be divided into three categories:

- 146 definitional equations, for example, national accounting identities, composition of the work-force, etc;
- 33 estimated 'technical' equations, for example, price indices with different base years and equations that serve special reporting purposes (with no feedback to the rest of the model);
- 26 estimated stochastic equations, representing economic behaviour.

The two first groups of equations are identical in RIMINI and the dVAR versions of the model. It is the specifications of the 26 econometric equations that distinguish the models. Together they contain putative quantitative knowledge about behaviour relating to aggregate outcomes, for example, consumption, savings, and household wealth; labour demand and unemployment; wage and price interactions (inflation); capital formation; foreign trade. Seasonally unadjusted data are used for the estimation of the equations. To a large extent, macroeconomic interdependencies are contained in the dynamics of the model. For example, prices and wages are Granger-causing output, trade and employment and likewise the level of the real activity feeds back on to wage-price inflation. The model is an open system: examples of important non-modelled variables are the level of economic activity by trading partners, as well as inflation and wage-costs in those countries. Indicators of economic policy (the level of government expenditure, the short-term interest rate, and the exchange rate), are also non-modelled and the forecasts are therefore conditional on a particular scenario for these variables. In the following, we refer to the incumbent version of RIMINI as eRIM.

Two full scale dVAR models—dRIM and dRIMc   Because all the stochastic equations in RIMINI are in equilibrium correction form, a simple dVAR version of the model, dRIM, can be obtained by omitting the equilibrium correcting terms from the equation and re-estimating the coefficients of the remaining (differenced variables). Omission of significant equilibrium-correcting

---

[3] See Section 1.4, in this application we have used Version 2.9 of the model. A large share of the 205 endogenous variables are accounting identities or technical relationships creating links between variables; see Eitrheim and Nymoen (1991) for a brief documentation of an earlier version of the model.

terms means that the resulting differenced equations become mis-specified, with autocorrelated and heteroskedastic residuals. From one perspective, this is not a big problem: the main thrust of the theoretical discussion is that the dVAR is indeed mis-specified within sample, cf. that the error-term $\epsilon_{y,t}$ in the dVAR equation (11.11) is autocorrelated provided that there is some autocorrelation in the disequilibrium term in (11.7). The dVAR might still forecast better than the EqCM, if the coefficients relating to the equilibrium-correcting terms change in the forecast period. That said, having a mis-specified dVAR does put that model at a disadvantage compared to the EqCM. Section 11.2.1 suggests that simply omitting the levels term while retaining the intercept may seriously damage the dVAR forecasts. Hence we decided to re-model all the affected equations, in terms of differences alone, in order to make the residuals of the dVAR-equations empirically white noise. The intercept was only retained for levels variables. This constitutes the backbone of the dRIMc model.

Two univariate models—dAR and dARr All three model versions considered so far are 'system of equations' forecasting models. For comparison, we have also prepared single equation forecasts for each variable. The first set of single equation forecasts is dubbed dAR, and is based on unrestricted estimation of AR(4) models. Finally, we generate forecasts from a completely restricted fourth-order autoregressive model, hence forecasts are generated from $\Delta_4 \Delta \ln X_t = 0$, for a variable X_t that is among the endogenous variables in the original model. This set of forecasts is called dARr, where the r is a reminder that the forecasts are based on (heavily) restricted AR(4) processes. Both dAR and dARr are specified without drift terms, hence their forecasts are protected against trend-misrepresentation. Thus, we will compare forecast errors from five forecasting systems.

Table 11.1 summarises the five models in terms of the incumbent 'baseline' EqCM model and the four 'rival' dVAR type models.

Relative forecast performance 1992(1)–1994(4) All models that enter this exercise were estimated on a sample ending in 1991(4). The period 1992(1)–1994(4) is used for forecast comparisons. That period saw the start of a marked upswing in the Norwegian economy. Hence, several of the model-endogenous variables change substantially over the 12 quarter forecast period.

In this paragraph, we first use graphs to illustrate how the eRIM forecast the interest rate level (RLB), housing price growth ($\Delta_4 ph$), the rate of inflation ($\Delta_4 cpi$), and the level of unemployment (UTOT) compared to the four dVARs: dRIM, dRIMc, dAR, and dARr. We evaluate three dynamic forecasts, distinguished by the start period: the first forecast is for the whole 12 quarter horizon, so the first period being forecasted is 1992(1). The second simulation starts in 1993(1) and the third in 1994(1). Furthermore, all forecasts are conditional on the actual values of the models' exogenous variables and the initial conditions,

Table 11.1
The models used in the forecasts

Model	Name	Description
Baseline	eRIM	26 Behavioural equations, equilibrium-correcting equations 33 + 146 Technical and definitional equations
1. Rival	dRIM	26 Behavioural equations, re-estimated after omitting level terms 33 + 146 Technical and definitional equations
2. Rival	dRIMc	26 Behavioural equations, remodelled without levels-information 33 + 146 Technical and definitional equations
3. Rival	dAR	71 equations modelled as 4.order AR models
4. Rival	dARr	71 equations modelled as restricted 4.order AR models

which of course change accordingly when we initialise the forecasts in different start periods.

The results are summarised in Figures 11.1–11.3. Figure 11.1 shows actual and forecasted values from the 12-quarter dynamic simulation. Looking at the graph for the interest rate first, the poor forecast from the dRIM model is immediately evident. Remember that this model was set up by deleting all the levels term in the individual EqCM equations, and then re-estimating these mis-specified equations on the same sample as in eRIM. Hence, dRIM imposes a large number of units roots while retaining the intercepts, and there is no attempt to patch-up the resulting mis-specification. Not surprisingly, dRIM is a clear loser on all the four variables in Figure 11.1. This turns out to be a typical result, it is very seldom that a variable is forecasted more accurately with dRIM than with dRIMc, the re-modelled dVAR version of eRIM.

Turning to dRIMc vs. eRIM, one sees that for the 12-quarter dynamic forecasts in Figure 11.1, the incumbent equilibrium-correcting model seems to outperform dRIMc for interest rates, growth in housing prices, and the inflation rate. However, dRIMc beats the EqCM when it comes to forecasting the rate of unemployment.

One might wonder how it is possible for dRIMc to be accurate about unemployment in spite of the poor inflation forecasts. The explanation is found by considering eRIM, where the level of unemployment affects inflation, but where there is very little feedback from inflation per se on economic activity. In eRIM, the level of unemployment only reacts to inflation to the extent that inflation accrues to changes in level variables, such as the effective real exchange rates or real household wealth. Hence, if eRIM generated inflation forecast errors of the same size that we observe for dRIMc, that would be quite damaging for the unemployment forecasts of that model as well. However, this mechanism is not

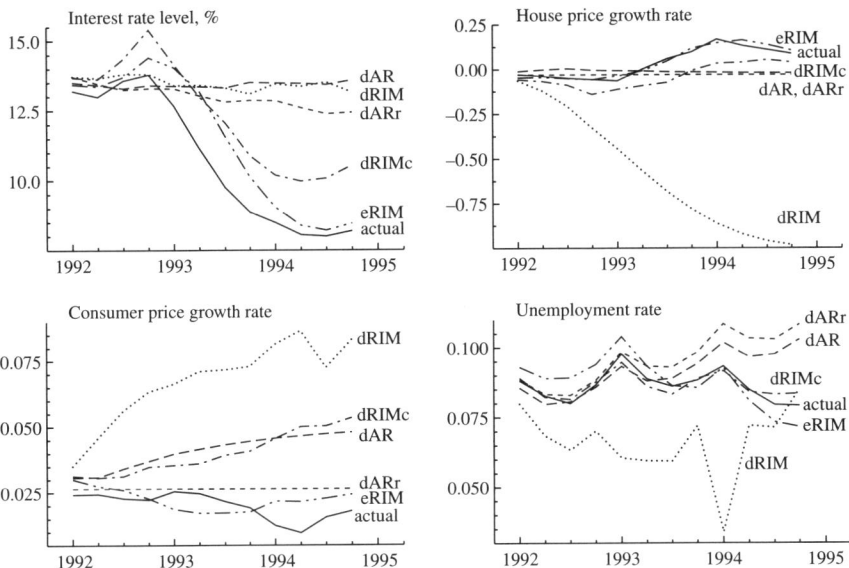

Figure 11.1. The period 1992(1)–1994(4) forecasts and actual values for the interest rate level (RLB), housing price growth ($\Delta_4 ph$), the rate of inflation ($\Delta_4 cpi$), and the level of unemployment (UTOT)

present in dRIMc, since all levels terms have been omitted. Hence, the unemployment forecasts of the dVAR versions of RIMINI are effectively insulated from the errors in the inflation forecast. In fact, the figures confirm the empirical relevance of Hendry's (1997a) claim that when the data generating mechanism is unknown and non-constant, models with less causal content (dRIMc) may still outperform the model that contains a closer representation of the underlying mechanism (eRIM). The univariate forecasts, dAR and dARr, are also way off the mark for the interest rate and for the unemployment rate. However, the forecast rule $\Delta_4 \Delta cpi_t = 0$, in dARc, predicts a constant inflation rate that yields a quite good forecast for inflation in this period; see Figure 11.1.

Figure 11.2 shows the dynamics forecast for the same selection of variables, but now the first forecast period is 1993(1). For the interest rate, the ranking of dRIMc and eRIM forecasts is reversed from Figure 11.1: dRIMc is spot on for most of the forecast-horizon, while eRIM consistently overpredicts. Evidently, dRIMc uses the information embodied in the actual development in 1992 much more efficiently than eRIM. The result is a good example of the intercept-correction provided by differencing. Equations (11.34) and (11.35) show that if the parameters of the EqCM change prior to the start of the forecast (i.e. in 1992 in the present case), then the dVAR might constitute the better forecasting model. Since the loan interest rate is a major explanatory variable for housing price growth (in both eRIM and dRIMc), it is not surprising that

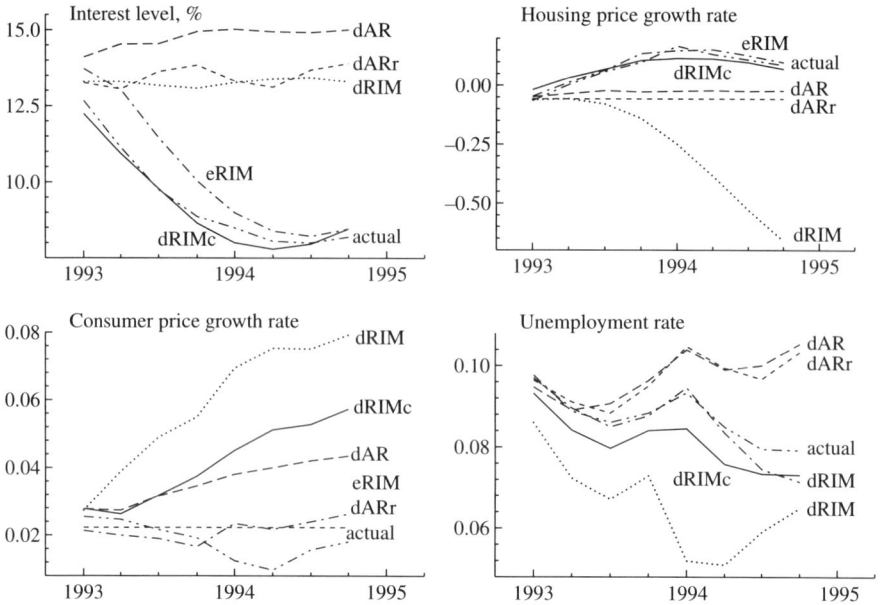

Figure 11.2. The period 1993(1)–1994(4) forecasts and actual values for the interest rate level (RLB), housing price growth ($\Delta_4 ph$), the rate of inflation ($\Delta_4 cpi$), and the level of unemployment (UTOT)

the housing price forecasts of the **dRIMc** are much better than in Figure 11.1. That said, we note that, with the exception of 1993(4) and 1994(2), **eRIM** forecasts housing prices better than **dRIMc**, which is evidence of countervailing forces in the forecasts for housing prices. The impression of the inflation forecasts are virtually the same as in the previous figure, while the graph of actual and forecasted unemployment shows that **eRIM** wins on this forecast horizon.

The 4-period forecasts are shown in Figure 11.3, where simulation starts in 1994(1). Interestingly, also the **eRIM** interest rate forecasts have now adjusted. This indicates that the parameter instability that damaged the forecasts that started in 1993(1) turned out to be a transitory shift. **dRIMc** now outperforms the housing price forecasts of **eRIM**. The improved accuracy of **dARr** as the forecast period is moved forward in time is very clear. It is only for the interest rate that the **dARr** is still very badly off target. The explanation is probably that using $\Delta_4 \Delta x_t = 0$ to generate forecasts works reasonably well for series with a clear seasonal pattern, but not for interest rates. This is supported by noting the better interest rate forecast of **dAR**, the unrestricted AR(4) model.

The relative accuracy of the **eRIM** forecasts, might be confined to the four variables covered by Figures 11.1–11.3. In Eitrheim *et al.* (1999) we therefore compare the forecasting properties of the five different models on a larger

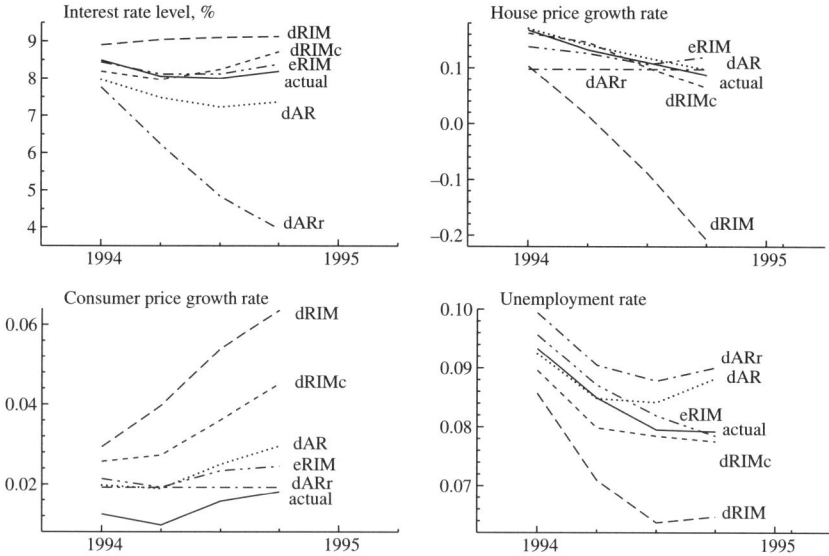

Figure 11.3. The period 1994(1)–1994(4) forecasts and actual values for the interest rate level (RLB), housing price growth ($\Delta_4 ph$), the rate of inflation ($\Delta_4 cpi$), and the level of unemployment (UTOT)

(sub)set of 43 macroeconomic variables). The list includes most of the variables that are regularly forecasted, such as GDP growth, the trade balance, wages, and productivity.

Eitrheim *et al.* (1999) follow convention and use the empirical root mean square forecast errors (RMSFE). The theoretical rationale for RMSFE is the mean squared forecast error (MSFE)

$$\text{MSFE}_{\text{mod}} = \mathsf{E}[y_{T+h} - \hat{y}_{T+h,\text{mod}} \mid \mathcal{I}_T]^2,$$

where $\hat{y}_{T+h,\text{mod}} = E[y_{T+h} \mid \mathcal{I}_T]$ and mod is either dVAR or EqCM. The MSFE can be rewritten as

$$\text{MSFE}_{\text{mod}} = \text{bias}^2_{T+1,\text{mod}} + \text{Var}[y_{T+h} \mid \mathcal{I}_T].$$

Conditional on the same information set $\mathcal{I}_T$, the model with the largest squared bias has also the highest MSFE, and consequently the highest squared RMSFE.[4]

---

[4] Abstracting from the problem that the information sets differ across the models considered, and apart from the fact that we use the empirical RMSFE (rather than the theoretical), ranking of the models according to RMSFE is the same as ranking by the squared bias. For a more comprehensive analysis of the use of RMSFEs for model comparisons and the potential pitfalls involved, see for example, Ericsson (1992), Clements and Hendry (1993).

Table 11.2 shows the placements of the five models in the 43 horse races. The incumbent model has the lowest RMSFE for 24 out of the 43 variables, and also has 13 second places. Hence eRIM comes out best or second best for 86% of the horse races, and seems to be a clear winner on this score. The two 'difference' versions of the large econometric model (dRIMc and dRIM) have very different fates. dRIMc, the version where each behavioural equation is carefully re-modelled in terms of differences is a clear second best, while dRIM is just as clear a loser, with 27 bottom positions. Comparing the two sets of univariate forecasts, it seems like the restricted version ($\Delta_4 \Delta x_t$) behaves better than the unrestricted AR model. Finding that the very simple forecasting rule in dARr outperforms the full model in 6 instances (and is runner-up in another 8), in itself suggests that it can be useful as a baseline and yardstick for the model-based forecasts.

Table 11.2
Results of 43 RMSFE forecast contests

Place #	eRIM	dRIMc	dRIM	dAR	dARr
(a) 12 period forecasts, 1992(1)–1994(4)					
1	24	13	1	1	6
2	13	11	4	5	8
3	2	6	5	7	13
4	2	12	6	15	10
5	2	1	27	7	6
(b) 4 period forecasts, 1992(1)–1992(4)					
1	7	8	10	6	12
2	17	13	3	4	6
3	13	7	8	10	7
4	3	11	2	17	9
5	3	4	20	6	9
(c) 4 period forecasts, 1993(1)–1993(4)					
1	17	9	7	1	11
2	16	13	7	2	3
3	3	12	11	12	5
4	3	9	2	17	12
5	4	0	16	11	12
(d) 4 period forecasts, 1994(1)—1994(4)					
1	13	4	5	5	16
2	11	17	1	9	6
3	7	8	11	9	7
4	7	8	13	9	6
5	5	6	13	11	8

Parts (b)–(d) in Table 11.2 collect the results of three 4-quarter forecast contests. Interestingly, several facets of the picture drawn from the 12-quarter forecasts and the graphs in Figures 11.1–11.3 appear to be modified. Although the incumbent eRIM model collects a majority of first and second places, it is beaten by the double difference model $\Delta_4\Delta x_t = 0$, dARr, in terms of first places in two of the three contests. This shows that the impression from the 'headline' graphs, namely that dARr works much better for the 1994(1)–1994(4) forecast, than for the forecast that starts in 1992, carries over to the larger set of variables covered by Table 11.2. In this way, our result shows in practice what the theoretical discussion foreshadowed, namely that forecasting systems that are blatantly mis-specified econometrically, nevertheless can forecast better than the econometric model with a higher causal content.

The results seem to corroborate the analytical results above. For short forecast horizons like, for example, 4-quarters, simple univariate dARr models offer much more protection against pre-forecast breaks compared with the other models, and their forecast errors are also insulated from forecast errors elsewhere in a larger system. However, the dARr model seems to lose this advantage relative to the other models as we increase the forecast horizon. The autonomous growth bias in dVAR type models tend to multiply as we increase the forecast horizon, causing the forecast error variance to 'explode'. Over long forecast horizons we would then typically see huge dVAR biases relative to the EqCM forecast bias. Finally, neither of the models protect against breaks that occur after the forecast is made.

11.3 Model specification and forecast accuracy

Forecasters and policy decision-makers often have to choose a model to use from a whole range of different models, all claiming to represent the economy (or the part of it that is the focal point of the forecasting exercise). The current range of wage and price models that can be used for inflation forecasting provides an example. As we have discussed earlier, in Chapter 9, inflation targeting implies that the central bank's conditional forecast 1–2 years ahead becomes the intermediate target of monetary policy. Consequently, there is a strong linkage between model choice, forecasting, and policy analysis in this case.

The statistical foundation for a conditional forecast as an operational target is that forecasts calculated as the conditional mean are unbiased and no other predictor (conditional on the same information set) has smaller MSFE, provided the first two moments exist. However, as discussed earlier in this chapter, the practical relevance of the result is reduced by the implicit assumption that the model corresponds to the data generating process (DGP), and that the DGP is constant over the forecast horizon. Credible forecasting methods must take into account that neither condition is likely to be fulfilled in reality. However, the specific inflation models have one important trait in common: they explain inflation—a growth rate—by not only other growth rates but also cointegrating combinations of levels variables. Thus, they are explicitly or implicitly EqCMs.

Implications for inflation models' forecasting properties between models with and without equilibrium-correcting mechanisms have been analysed in Bårdsen *et al.* (2002*a*). This chapter draws on their results, and extends the analysis of different inflation models reported in earlier chapters. Specifically, we consider the two most popular inflation models, namely Phillips curves and dynamic wage curve specifications (or dynamic ICMs). These models were dealt with extensively in Chapters 4–6. The standard Phillips Curve Model (denoted PCM), is formally an EqCM, the cointegrating term being the output gap or, alternatively, the difference between the rate of unemployment and the natural rate, that is, $u_t - u^{\text{phil}}$ in the notation of Chapter 4. An alternative to the PCM, consistent with the concept of a wage curve, was discussed extensively in Chapters 5 and 6, where it was dubbed the Imperfect Competition Model (ICM) because of the role played by bargaining and of imperfect competition. Since wage-curve models are EqCM specifications, they are vulnerable to regime shifts, for example, changes in equilibrium means.

The existing empirical evidence is mixed. Although varieties of Phillips curves appear to hold their ground when tested on United States data—see Fuhrer (1995), Gordon (1997), Blanchard and Katz (1999), Galí and Gertler (1999)—studies from Europe usually conclude that ICMs are preferable, see, for example, Drèze and Bean (1990, table 1.4), Wallis (1993), OECD (1997*b*, table 1.A.1), and Nymoen and Rødseth (2003). In Chapters 4 and 6 we presented both models (PCM and ICM) for Norway. In those models, and unlike most of the papers cited above, which focus only on wage formation or inflation, the rate of unemployment was modelled as part of the system. It transpired that the speed of adjustment in the PCM was so slow that little practical relevance could be attached to the formal dynamic stability of the PCM. No such inconsistency existed for the ICM, where the adjustment speed was fast, supporting strong dynamic stability.

Inflation targeting central banks seem to prefer the PCM, because it represents the consensus model, and it provides a simple way of incorporating the thesis about no long-run tradeoff between inflation and the activity level, which is seen as the backbone of inflation targeting (see, for example, King 1998).

In Section 11.3.1, we discuss the algebra of inflation forecasts based on the competing models. Section 11.3.3 evaluates the forecasting properties of the two models for Norwegian inflation.

11.3.1 Forecast errors of stylised inflation models

We formulate a simple DGP to investigate the theoretical forecasting capabilities of the ICM and the PCM, thus providing a background for the interpretation of the actual forecast errors in Section 11.3.3. The variable symbols take the same meaning as in the earlier chapters on wage–price modelling (see Chapter 6), hence (in logs) w_t is the wage rate, p_t is the consumer price index, pi_t denotes import prices, and u_t is the rate of unemployment.

In order to obtain an analytically tractable distillation of the models, we introduce simplifying assumptions. For example, we retain only one cointegrating relationship, the 'wage-curve', and we also abstract from productivity.[5] Thus (11.44) is a simplified version of the dynamic wage equation of Chapter 6:

$$\Delta(w - p)_t = \kappa - \pi_w[(w - p)_{t-1} + \lambda u_{t-1} - \mu] + \epsilon_{w,t}, \qquad \pi_w > 0, \quad \lambda > 0.$$
$$(11.44)$$

The wage-curve is the term in square brackets. The parameter μ denotes the mean of the long-run relationship for real wages, that is, $\mathsf{E}[(w - p)_{t-1} - \lambda u_{t-1} - \mu] = 0$. Since we abstract from the cointegration relationship for consumer prices, the simultaneous equation representation of the inflation equation is simply that Δp_t is a linear function of $\Delta p i_t$ and Δw_t, and the reduced form equation for Δp_t is:

$$\Delta p_t = \phi_p + \varphi_{pi}\Delta p i_t - \pi_p[(w - p)_{t-1} + \lambda u_{t-1} - \mu] + \epsilon_{p,t},$$
$$\varphi_{pi} \geq 0, \quad \pi_p \geq 0. \qquad\qquad (11.45)$$

Multi-step (dynamic) forecasts of the rate of inflation require that also import price growth and the rate of unemployment are forecasted. In order to simplify as much as possible, we let $\Delta p i_t$ and u_t follow exogenous stationary processes:

$$\Delta p i_t = \phi_{pi} + \epsilon_{pi,t}, \qquad\qquad (11.46)$$
$$\Delta u_t = \phi_u - \pi_u u_{t-1} + \epsilon_{u,t}, \qquad \pi_u > 0. \qquad (11.47)$$

$\mathcal{I}_T$ denotes the information set available in period T. The four disturbances $(\epsilon_{w,t}, \epsilon_{p,t}, \epsilon_{pi,t}, \epsilon_{u,t})$ are innovations relative to $\mathcal{I}_T$, with contemporaneous covariance matrix Ω. Thus, the system (11.44)–(11.47) represents a simple DGP for inflation, the real wage, import price growth, and the rate of unemployment. The forecasting rule

$$\widehat{\Delta p}_{T+h} = \mathsf{E}[\Delta p_{T+h} \mid \mathcal{I}_T] = a_0 + a_1\delta_{pi} + a_2\mathsf{E}[(w - p)_{T+h-1} \mid \mathcal{I}_T]$$
$$+ a_3\mathsf{E}[u_{T+h-1} \mid \mathcal{I}_T], \qquad h = 1, 2, \dots, H$$
$$(11.48)$$

with coefficients

$$a_0 = \phi_p + \pi_p\mu,$$
$$a_1 = \varphi_{pi},$$
$$a_2 = -\pi_p,$$
$$a_3 = -\pi_p\lambda$$

is the minimum MSFE predictor of Δp_{T+h}, by virtue of being the conditional expectation.

---

[5] Compared to the algebraic sections of Chapter 6, we omit productivity. Naturally, it is included as a non-modelled explanatory variable in the empirical models.

First, we abstract from estimation uncertainty and assume that the parameters are known. The dynamic ICM forecast errors have the following means and variances:

$$E[\Delta p_{T+h} - \widehat{\Delta p}_{T+h,\text{ICM}} \mid \mathcal{I}_T] = 0, \tag{11.49}$$

$$\text{Var}[\Delta p_{T+h} - \widehat{\Delta p}_{T+h,\text{ICM}} \mid \mathcal{I}_T] = \sigma_p^2 + \sigma_{pi}^2 + a_2^2 \sum_{i=1}^{h-1}(1-\pi_w)^{2(h-1-i)}\sigma_w^2$$

$$+ a_2^2(\pi_w\lambda)^2 \sum_{i=1}^{h-1}(1-\pi_w)^{2(h-1-i)}$$

$$\times \sum_{j=1}^{i}(1-\pi_u)^{2(i-j)}\sigma_u^2$$

$$+ a_3^2 \sum_{i=1}^{h-1}(1-\pi_u)^{2(h-1-i)}\sigma_u^2. \tag{11.50}$$

The first two terms on the right-hand side of (11.50) are due to $\epsilon_{p,T+h}$ and $\epsilon_{pi,T+h}$. The other terms on the right-hand side of (11.50) are only relevant for $h = 2, 3, 4, \ldots, H$. The third and fourth terms stem from $(w-p)_{T+h-1}$—it is a composite of both wage and unemployment innovation variances. The last line contains the direct effect of $\text{Var}[u_{T+h-1}]$ on the variance of the inflation forecast. In addition, off-diagonal terms in Ω might enter.

We next consider the case where a forecaster imposes the PCM restriction $\pi_w = 0$ (implying $\pi_p = 0$ as well). The 'Phillips curve' inflation equation is then given by:

$$\Delta p_t = \tilde{a}_0 + \tilde{a}_1 \Delta p i_t + \tilde{a}_3 u_{t-1} + \tilde{\epsilon}_{p,t}, \tag{11.51}$$

with

$$\tilde{a}_0 = a_0 + a_2\lambda E[u_{t-1}] + a_2\mu \quad \text{and} \quad \tilde{\epsilon}_{p,t} = \epsilon_{p,t} + a_2[(w-p)_{t-1} - \lambda u_{t-1} - \mu].$$

This definition ensures a zero-mean disturbance $E[\tilde{\epsilon}_{p,t} \mid \mathcal{I}_T] = 0$. Note also that $\text{Var}[\tilde{\epsilon}_{p,t} \mid \mathcal{I}_{t-1}] = \sigma_p^2$, that is, the same innovation variance as in the ICM-case. The PCM forecast rule becomes

$$\widehat{\Delta p}_{T+h,\text{PCM}} = E[\Delta p_{T+h,\text{PCM}} \mid \mathcal{I}_T] = \tilde{a}_0 + \tilde{a}_1\delta_{pi} + \tilde{a}_4\hat{u}_{T+h-1}.$$

The mean and variance of the 1-step forecast error are

$$E[\Delta p_{T+1} - \widehat{\Delta p}_{T+1,\text{PCM}} \mid \mathcal{I}_T] = (a_1 - \tilde{a}_1)\delta_{pb} + u_T(a_3 - \tilde{a}_3)u_T$$
$$+ a_2\{(w-p)_T - \lambda E[u_t] - \mu\},$$

$$\text{Var}[\Delta p_{T+1} - \widehat{\Delta p}_{T+1,\text{PCM}} \mid \mathcal{I}_T] = \sigma_p^2 + \sigma_{pi}^2.$$

The 1-step ahead prediction error variance conditional on $\mathcal{I}_T$ is identical to the ICM-case. However, there is a bias in the 1-step PCM forecast arising from two sources: first, omitted variables bias implies that $a_1 \neq \tilde{a}_1$ and/or $a_3 \neq \tilde{a}_3$, in general. Second,

$$(w - p)_T - \lambda \mathsf{E}[u_t] - \mu \neq 0$$

unless $(w - p)_T = \mathsf{E}[(w - p)_t]$, that is, the initial real wage is equal to the long-run mean of the real-wage process.

For dynamic h period ahead forecasts, the PCM prediction error becomes

$$\Delta p_{T+h} - \widehat{\Delta p}_{T+h,\text{PCM}} = (a_1 - \tilde{a}_1)\delta_{pb} + (a_3 - \tilde{a}_3)\hat{u}_{T+h-1}$$

$$+ a_3 \sum_{i=1}^{h-1} (1 - \pi_u)^{h-1-i} \epsilon_{u,T+i} + \epsilon_{pi,T+h} + \epsilon_{p,T+h}$$

$$+ a_2(w - p)_{T+h-1} - a_2(\lambda \mathsf{E}[u_t] - \mu).$$

Taking expectation and variance of this expression gives:

$$\mathsf{E}[\Delta p_{T+h} - \widehat{\Delta p}_{T+h,\text{PCM}} \mid \mathcal{I}_T] = (a_1 - \tilde{a}_1)\delta_{pi} + (a_4 - \tilde{a}_4)\hat{u}_{T+h-1}$$

$$+ a_2\{\mathsf{E}[(w - p)_{T+h-1} \mid \mathcal{I}_T] - \lambda \mathsf{E}[u_t] - \mu\},$$

$$\mathsf{Var}[\Delta p_{T+h} - \widehat{\Delta p}_{T+h,\text{PCM}} \mid \mathcal{I}_T] = \mathsf{Var}[\Delta p_{T+h} - \widehat{\Delta p}_{T+h,\text{ICM}} \mid \mathcal{I}_T],$$

$$\text{for } h = 2, 3, \ldots, H.$$

Hence systematic forecast error is again due to omitted variables bias and the fact that the conditional mean of real wages $h - 1$ periods ahead, departs from its (unconditional) long-run mean. However, for long forecast horizons, large H, the bias expression can be simplified to become

$$\mathsf{E}[\Delta p_{T+H} - \widehat{\Delta p}_{T+H,\text{PCM}} \mid \mathcal{I}_T] \approx (a_1 - \tilde{a}_1)\delta_{pi} + (a_4 - \tilde{a}_4)\frac{\varphi_u}{\pi_u}$$

since the conditional forecast of the real wage and of the rate of unemployment approach their respective long-run means.

Thus far we have considered a constant parameter framework: the parameters of the model in equations (11.44)–(11.47) remain constant not only in the sample period ($t = 1, \ldots, T$) but also in the forecast period ($t = T+1, \ldots, T+h$). However, as discussed, a primary source of forecast failure is structural breaks, especially shifts in the long-run means of cointegrating relationships and in parameters of steady-state trend growth. Moreover, given the occurrence of deterministic shifts, it no longer follows that the 'best' econometric model over the sample period also gives rise to the minimum MSFE; see, for example, Section 11.2.

That a tradeoff between close modelling and robustness in forecasting also applies to wage–price dynamics is illustrated by the following example: assume that the long-run mean μ of the wage-equation changes from its initial level to a new level, that is, $\mu \to \mu^*$, *before* the forecast is made in period T, but that the change is undetected by the forecaster. There is now a bias in the (1-step) ICM real-wage forecast:

$$\mathsf{E}[(w - p)_{T+1} - \widehat{(w - p)}_{T+1,\text{ICM}} \mid \mathcal{I}_T] = -\pi_w[\mu - \mu^*], \quad (11.52)$$

which in turn produces a non-zero mean in the period 2 inflation forecast error:

$$\mathsf{E}[\Delta p_{T+2} - \widehat{\Delta p}_{T+2,\text{ICM}} \mid \mathcal{I}_T] = -a_2\pi_w[\mu - \mu^*]. \quad (11.53)$$

The PCM-forecast on the other hand, is insulated from the parameter change in wage formation, since $\widehat{(w-p)}_{T+h-1}$ does not enter the predictor—the forecast error is unchanged from the constant parameter case. Consequently, both sets of forecasts for Δp_{T+2+h} are biased, but for different reasons, and there is no logical reason why the PCM forecast could not outperform the ICM forecast on a comparison of biases. In terms of forecast properties, the PCM, despite the inclusion of the rate of unemployment, behaves *as if* it was a dVAR, since there is no feedback from wages and inflation to the rate of unemployment in the example DGP.

Finally, consider the consequences of using estimated parameters in the two forecasting models. This does not change the results about the forecast biases. However, the conclusion about the equality of forecast error variances of the ICM and PCM is changed. Specifically, with estimated parameters, the two models do not share the same underlying innovation errors. In order to see this, consider again the case where the ICM corresponds to the DGP. Then a user of a PCM does not know the true composition of the disturbance $\tilde{\epsilon}_{p,t}$ in (11.51), and the estimated PCM will have an estimated residual variance that is larger than its ICM counterpart, since it is influenced by the omitted wage-curve term. In turn, the PCM prediction errors will overstate the degree of uncertainty in inflation forecasting. We may write this as

$$\mathsf{Var}[\tilde{\epsilon}_{p,t} \mid \mathcal{I}_T, \mathrm{PCM}] > \mathsf{Var}[\epsilon_{p,t} \mid \mathcal{I}_T, \mathrm{ICM}]$$

to make explicit that the conditioning is with respect to the two models (the DGP being unknown). From equation (11.51) it is seen that the size of the difference between the two models' residual variances depends on (1) the strength of equilibrium correction and (2) the variance of the long-run wage curve.

The main results of this section can be summarised in three points:

1. With constant parameters in the DGP, forecasting using the PCM will bias the forecasts and overstate the degree of uncertainty (i.e. if the PCM involves invalid parameter restrictions relative to the DGP).
2. PCM forecasts are however *robust* to changes in means of (omitted) long-run relationships.
3. Thus PCM shares some of the robustness of dVARs, but also some of their drawbacks (specifically, excess inflation uncertainty).

In sum, the outcome of a forecast comparison is not a given thing, since in practice we must allow for the possibility that both forecasting models are mis-specified relative to the generating mechanism that prevails in the period we are trying to forecast. A priori we cannot tell which of the two models will forecast best. Hence, there is a case for comparing the two models' forecasts directly, even though the econometric evidence presented in earlier chapters has gone in favour of the ICM as the best model.

11.3.2 Revisiting empirical models of Norwegian inflation

The definitions of the variables are in line with those we presented for the ICM in Chapter 9, but the sample is different and covers the period 1966(4)–1996(4). The wage variable w_t is average hourly wages in the *mainland economy*, excluding the North Sea oil-producing sector and international shipping. The productivity variable a_t is defined accordingly. The price index p_t is measured by the official consumer price index. The import prices index pi_t is a weighted average of import price indices from trading countries. The unemployment variable u_t is defined as a 'total' unemployment rate, including labour market programmes. The tax-rates $t1_t$ and $t3_t$ are rates of payroll tax and indirect tax, respectively.[6]

The output gap variable gap_t is measured as deviations from the trend obtained by the Hodrick–Prescott (HP) filter. The other non-modelled variables contain first the length of the working day Δh_t, which captures wage compensation for reductions in the length of the working day—see Nymoen (1989b). Second, incomes policies and direct price controls have been in operation on several occasions in the sample period; see, for example, Bowitz and Cappelen (2001). The intervention variables Wdum and Pdum, and one impulse dummy $i80q2$, are used to capture the impact of these policies. Finally, $i70q1$ is a VAT dummy.

The dynamic ICM As in the earlier chapters we have two simultaneous equations for Δw_t and Δp_t, with separate and identified equilibrium correction equations terms. Estimation is by full information maximum likelihood (FIML), and the coefficients and diagnostics of the final ICM for our current sample are shown in (11.54) and in Table 11.3.

$$\widehat{\Delta w_t} = \Delta p_t - 0.4 \times 0.36 \Delta pi_t - \Delta t1_{t-2} - \underset{(0.08)}{0.36} \ \Delta t3_{t-2} - \underset{(0.11)}{0.3} \ \Delta h_t$$

$$- \underset{(0.01)}{(0.08)}[w_{t-2} - p_{t-2} - a_{t-1} + 0.1u_{t-2}] + \text{dummies}$$

$$\hat{\sigma}_{\Delta w} = 1.02\%$$

$$\widehat{\Delta p_t} = \underset{(0.05)}{0.12} \ (\Delta w_t + \Delta t1_{t-2}) + \underset{(0.02)}{0.05} \ gap_{t-1} + 0.4 \times 0.07 \Delta pi_t - \underset{(0.03)}{0.07} \ \Delta t3_{t-2}$$

$$- \underset{(0.01)}{0.08} \ [p_{t-3} - 0.6(w_{t-1} - a_{t-1} + t1_{t-1}) - 0.4pi_{t-1} + t3_{t-3}] + \text{dummies}$$

$$\hat{\sigma}_{\Delta p} = 0.41\%. \tag{11.54}$$

---

[6] An income tax rate could appear as well. It is omitted from the empirical model, since it is insignificant. This is in accordance with previous studies of aggregate wage formation, see, for example, Calmfors and Nymoen (1990) and Nymoen and Rødseth (2003), where no convincing evidence of important effects from the average income tax rate on wage growth could be found.

Table 11.3
Diagnostic tests for the dynamic ICM[a]

$$\hat{\sigma}_{\Delta w} = 1.02\%$$
$$\hat{\sigma}_{\Delta p} = 0.41\%$$
Correlation of residuals $= -0.4$
$$\chi^2_{\text{overidentification}}(9) = 9.23[0.42]$$
$$F^v_{\text{AR(1-5)}}(20, 176) = 1.02[0.31]$$
$$\chi^{2,v}_{\text{normality}}(4) = 6.23[0.18]$$
$$F^v_{\text{HET}\chi^2}(102, 186) = 0.88[0.76]$$

[a]The sample is 1966(4)–1994(4), 113 observations.

Table 11.4
Diagnostic tests for the PCM[b]

$$\hat{\sigma}_{\Delta w} = 1.07\%$$
$$\hat{\sigma}_{\Delta p} = 0.47\%$$
Correlation of residuals $= -0.6$
$$\chi^2_{\text{overidentification}}(16) = 25.13[0.07]$$
$$F^v_{\text{AR(1-5)}}(20, 176) = 1.02[0.44]$$
$$\chi^{2,v}_{\text{normality}}(4) = 6.23[0.18]$$
$$F^v_{\text{HET}\chi^2}(102, 257) = 0.81[0.84]$$

[b]The sample is 1967(1)–1994(4), 112 observations.

The PCM   When estimating a PCM, we start out from the same information set as for the ICM, but with more lags in the dynamics, to make sure we
end up with a data-congruent specification. This is to secure that the forecast
comparison below is not harmed by econometric mis-specification. It is not
implied that the resulting model, given in (11.55), would be seen as the preferred
choice if one started out (possibly from another information set) with the aim
of finding the best PCM, also in terms of economic interpretation.[7] As the
diagnostic tests in Table 11.4 show, the model encompasses its reduced form
and shows no sign of mis-specification. The estimated standard errors, however,
are for both equations higher than the corresponding ones found in the ICM.

[7] Dynamic price homogeneity in the wage Phillips curve cannot be rejected statistically,
and is therefore imposed.

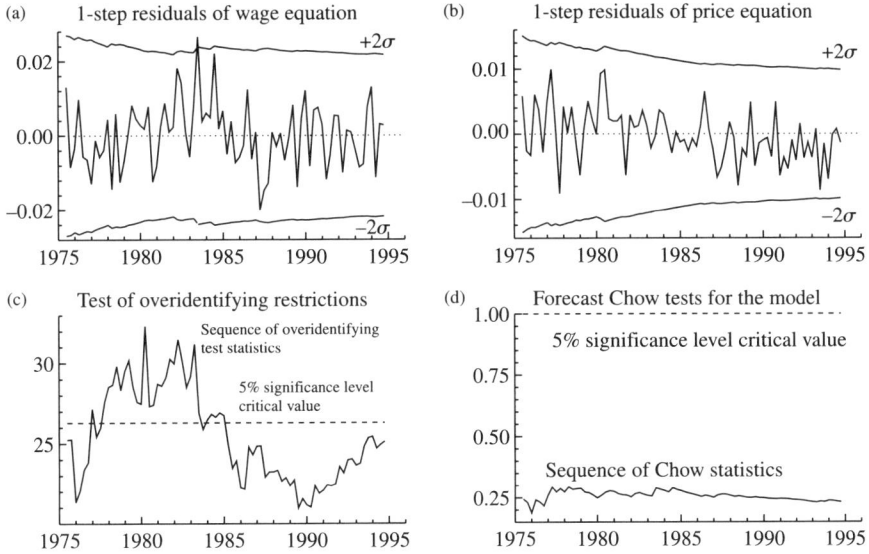

Figure 11.4. Recursive stability tests for the PCM

$$\widehat{\Delta w_t} = \underset{(0.04)}{1.11} \; \Delta p_t - 0.11\Delta pi_t - \underset{(0.22)}{0.65} \; \Delta t1_t - \underset{(0.21)}{0.41} \; \Delta t1_{t-2} - \underset{(0.005)}{0.01} \; \Delta u_{t-3}$$

$$- \underset{(0.001)}{0.006} \; u_{t-1} - \underset{(0.09)}{0.16} \; \Delta t3_{t-1} - \underset{(0.09)}{0.34} \; \Delta t3_{t-2} - \underset{(0.11)}{0.30} \; \Delta h_t + \text{dummies}$$

$$\hat{\sigma}_{\Delta w} = 1.07\%$$

$$\widehat{\Delta p_t} = \underset{(0.03)}{0.14} \; \Delta w_t + \underset{(0.02)}{0.07} \; \Delta w_{t-3} + \underset{(0.05)}{0.17} \; \Delta p_{t-1} + \underset{(0.05)}{0.27} \; \Delta p_{t-2} + \underset{(0.02)}{0.05} \; \Delta pi_t$$

$$- \underset{(0.006)}{0.03}\Delta a_{t-1} + \underset{(0.01)}{0.05} \; \text{gap}_{t-1} + \text{dummies}$$

$$\hat{\sigma}_{\Delta p} = 0.47\%. \tag{11.55}$$

Parameter constancy of the PCM is demonstrated graphically in Figure 11.4. The two 1-step residuals with their ±2 estimated residual standard errors (±2σ in the graphs) are in the upper panels, while the lower right panel shows the sequence of recursive forecast Chow tests together with their one-off 5% critical level. The lower left panel shows that the model encompasses the unrestricted reduced form as the sample size increases (i.e. the end point of the graph corresponds to Overidentification $\chi^2(16)$ in Table 11.4).

Hence, using these conventional design criteria, the PCM seems passable, and it is attractive as a forecasting model since it is simpler than the ICM.

11.3.3   Forecast comparisons

Both models condition upon the rate of unemployment u_t, average labour productivity a_t, import prices pi_t, and GDP mainland output y_t. In order to investigate the dynamic forecasting properties we enlarge both models with relationships for these four variables, in the same manner as in Chapter 9.

Figure 11.5 illustrates how the ICM-based model forecast the growth rates of wages and prices, Δw_t and Δp_t. It is also instructive to consider the forecasts for the change in the real wage $\Delta(w-p)_t$ and the annual rate of inflation, $\Delta_4 p_t$. The forecast period is from 1995(1) to 1996(4). The model parameters are estimated on a sample which ends in 1994(4). These dynamic forecasts are conditional on the actual values of the non-modelled variables (*ex post* forecasts). The quarterly inflation rate Δp_t only has one significant bias, in 1996(1). In that quarter there was a reduction in the excises on cars that explains around 40% of this particular overprediction. In the graphs of the annual rate of inflation $\Delta_4 p_t$ this effect is naturally somewhat mitigated. The quarterly change in the wage rate Δw_t is very accurately forecasted, so the only forecast error of any importance for the change in real wages $\Delta(w-p)_t$ also occurs in 1996(1). The forecasts for the rate of unemployment are very accurate for the first 5 quarters, but the reduction in unemployment in the last 3 quarters does not appear to be predictable with the aid of this model.

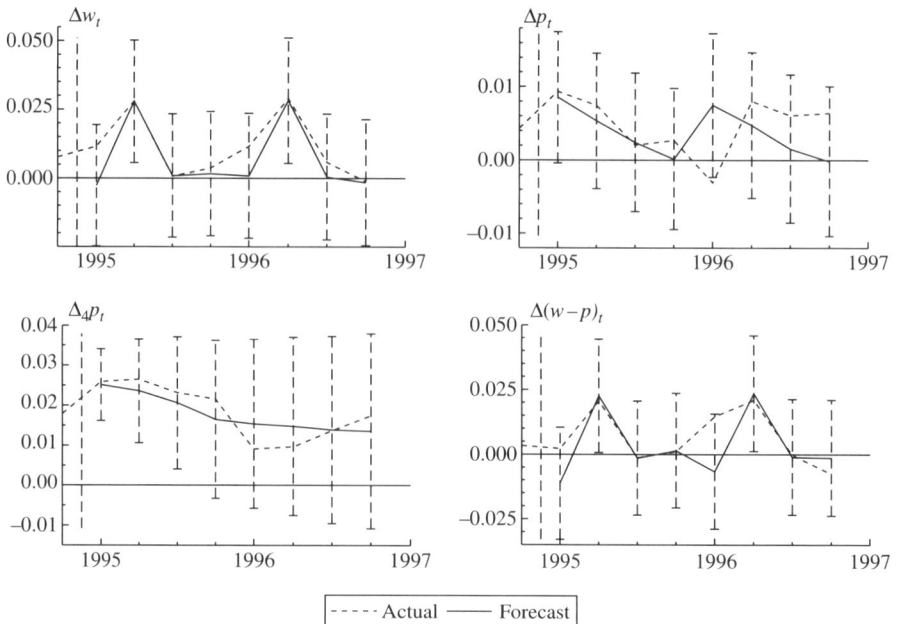

Figure 11.5. The 8-step dynamic forecasts for the period 1995(1)–1996(4), with 95% prediction bands of the ICM

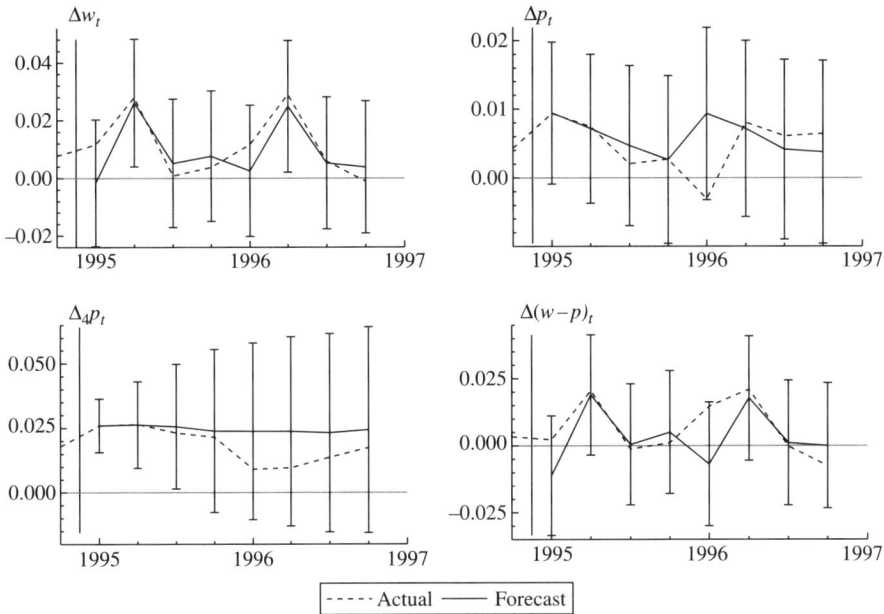

Figure 11.6. The 8-step dynamic forecasts for the period 1995(1)–1996(4), with 95% prediction bands of the PCM

Figure 11.5 also contains the 95% prediction intervals in the form of ± 2 standard errors, as a direct measure of the uncertainty of the forecasts. The prediction intervals for the annual rate of inflation are far from negligible and are growing with the length of the forecast horizon.

Next, Figure 11.6 illustrates how the model based on the Phillips curve forecast the same variables over the same period from 1995(1) to 1996(4). For most variables the differences are negligible. For the quarterly inflation rate Δp_t in particular, the Phillips curve specification seems to be no worse than the ICM as regards the point forecasts, although the prediction intervals are somewhat wider, due to the larger residual variances in wage- and price-setting.

However, in the graphs of the annual rate of inflation $\Delta_4 p_t$ there is after all a clear difference between the predictions on this one-off comparison. $\Delta_4 \hat{p}_{T+h,\text{mod}}$ is simply a 4-quarter moving average of the quarterly rates, and the same is true for the prediction errors, thus

$$\Delta_4 p_{T+h} - \Delta_4 \hat{p}_{T+h,\text{mod}} = \sum_{i=0}^{3} (\Delta p_{T+h-i} - \Delta \hat{p}_{T+h-i,\text{mod}}),$$

$$\text{mod} = \text{ICM}, \text{PCM}. \tag{11.56}$$

Until 1995(4) there is zero bias in $\Delta_4\hat{p}_{T+h,\text{PCM}}$ because all the preceding quarterly forecasts are so accurate. However, $\Delta_4\hat{p}_{T+h,\text{PCM}}$ becomes biased from 1996(1) and onwards because, after the overprediction of the quarterly rate in 1996(1), there is no compensating underprediction later in 1996. The ICM forecasts on the other hand achieve exactly that correction, and do not systematically overpredict inflation.

For the annualised inflation rate the uncertainty increases quite rapidly for both models, but markedly more so for the Phillips curve forecast. Indeed, by the end of the two-year period, the forecast uncertainty of the Phillips curve is about twice as big as the dynamic ICM. This effect is clearly seen when the annual inflation forecasts from the two models are shown in the same graph (Figure 11.7). The dotted lines denote the point forecasts and the 95% prediction bands of the dynamic ICM, while the solid lines depict the corresponding results from the forecasts of the Phillips curve specification. At each point of the forecast the uncertainty of the Phillips curve is bigger than for the ICM. Indeed, while the ICM has a standard error of 0.9 percentage points 4 periods ahead, and 1.2 percentage points 8 periods ahead, the Phillips curve standard errors are 1.6 and 2 percentage points, respectively. Considering equation (11.56) it transpires that the explanation is not only that each $\text{Var}[\Delta p_{T+h} - \Delta\hat{p}_{T+h,\text{PCM}}] > \text{Var}[\Delta p_{T+h} - \Delta\hat{p}_{T+h,\text{ICM}}]$, but also that the PCM quarterly prediction errors are more strongly positively autocorrelated than the ICM counterparts.

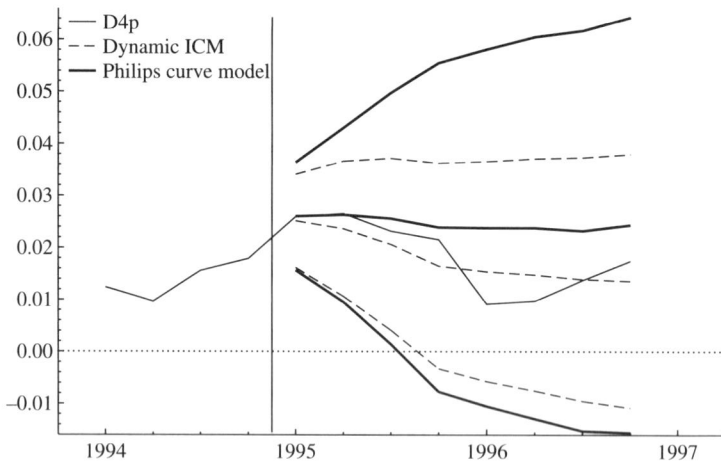

Figure 11.7. Comparing the annual inflation forecasts of the two models. The thin line is actual annual inflation in Norway. The dashed lines denote the point forecasts and the 95% prediction error bands of the ICM model, while the solid lines depict the corresponding results from the forecasts of the PCM in (11.55)

11.4   Summary and conclusions

The dominance of EqCMs over systems consisting of relationships between differenced variables (dVARs) relies on the assumption that the EqCM model coincides with the underlying data generating mechanism. However, that assumption is too strong to form the basis of practical forecasting. First, parameter non-constancies, somewhere in the system, are almost certain to arise in the forecast period. The example in Section 11.2.1 demonstrated how allowance for non-constancies in the intercept of the cointegrating relations, or in the adjustment coefficients, make it impossible to assert the dominance of the EqCM over a dVAR. Second, the forecasts of a simple EqCM were shown to be incapable of correcting for parameter changes that happen *prior to* the start of the forecast, whereas the dVAR is capable of utilising the information about the parameter shift embodied in the initial conditions. Third, operational macroeconometric models that are used for forecasting are bound to be mis-specified to some degree, for example, because of limited information in the data set, measurement problems or simply too little resources going into data and model quality control. Together, mis-specification and structural breaks, open the possibility that models with less causal content may turn out as the winner in a forecasting contest.

To illustrate the empirical relevance of these claims, we considered a model that has been used for forecasting the Norwegian economy. Forecasts for the period 1992(1)–1994(4) were calculated both for the incumbent EqCM version of the RIMINI model and the dVAR version of that model. Although the large-scale model holds its ground in this experiment, several of the theoretical points that have been made about the dVAR approach seem to have considerable practical relevance. We have seen demonstrated the automatic intercept correction of the dVAR forecasts (parameter change prior to forecast), and there were instances when the lower causal-content of the dVAR insulated forecast errors in one part of that system from contaminating the forecasts of other variables. Similarly, the large-scale EqCMs and its dVAR counterparts offer less protection against wrong inputs (of the exogenous variables) provided by the forecaster than the more 'naive' models. The overall impression is that the automatic intercept correction of the dVAR systems is most helpful for short forecast horizons. For longer horizons, the bias in the dVAR forecasts that are due to mis-specification tends to dominate, and the EqCM model performs relatively better.

Given that operational EqCMs are multi-purpose models that are used both for policy analysis and forecasting, while the dVAR is only suitable for forecasting, one would perhaps be reluctant to give up the EqCM, even in a situation where its forecasts are consistently less accurate than dVAR forecast. We do not find evidence of such dominance, overall the EqCM forecasts stand up well compared to the dVAR forecasts in this 'one-off' experiment. Moreover, in an actual forecasting situation, intercept corrections are used to correct EqCM forecast for parameter changes occurring before the start of the forecast. From the viewpoint of practical forecast preparation, one interesting development would be to automatise intercept correction based on simple dVAR forecast, or through differencing the EqCM term in order to insulate against a shift in the mean.

The strong linkage between forecasting and policy analysis makes the role of econometric models more important than ever. Policy makers face a menu of different models and an explicit inflation target implies that the central bank's conditional forecast 1–2 years ahead becomes the operational target of monetary policy. The presence of non-stationary data and frequent structural breaks makes inevitable a tradeoff between the gain and importance of correct structural modelling and their cost in terms of forecasting robustness. We have explored the importance of this tradeoff for inflation forecasting.

Specifically, we considered the two popular inflation models, namely Phillips curves and wage curve specifications. We establish that Phillips curve forecasts are robust to types of structural breaks that harm the wage-curve forecasts, but exaggerate forecast uncertainty in periods with no breaks. Moreover, omitted relevant equilibrium correction terms induce omitted variables bias in the usual way. Conversely, for the wage curve model, the potential biases in after-break forecast errors can be remedied by intercept corrections. As a conclusion, using a well-specified model of wage-price dynamics offers the best prospect of successful inflation forecasting.

Appendix

A.1   The Lucas critique

This appendix gives a proof of (4.29):

$$\operatorname*{plim}_{T\to\infty} \hat{\beta}_{\mathrm{OLS}} = \alpha_1^2 \beta,$$

in Chapter 4, Section 4.5.

Since $\operatorname{plim}_{T\to\infty}\hat{\beta}_{\mathrm{OLS}}$ is equal to the true regression coefficient between y_t and x_t, we express the regression coefficient in terms of the parameters of the expectations model. To simplify, we assume that $\{y_t, x_t\}$ are independently normally distributed:

$$\begin{bmatrix} y_t \\ x_t \end{bmatrix} \Big| \mathcal{I}_{t-1} \sim \mathsf{N}\begin{pmatrix} 0 & \alpha_1\beta \\ 0 & \alpha_1 \end{pmatrix}\begin{pmatrix} y_{t-1} \\ x_{t-1} \end{pmatrix}, \quad \begin{pmatrix} \sigma_{\epsilon_y}^2 & 0 \\ 0 & \sigma_{\epsilon_x}^2 \end{pmatrix}, \quad |\alpha_1| < 1. \tag{A.1}$$

From (4.27), the conditional expectation of y_t is:

$$\mathsf{E}[y_t \mid x_t] = x_t\beta + \mathsf{E}[\eta_t \mid x_t], \tag{A.2}$$

and, from (4.28):

$$\mathsf{E}[\eta_t \mid x_t] = \mathsf{E}[\epsilon_{y,t} \mid x_t] - \beta\mathsf{E}[\epsilon_{x,t} \mid x_t] = -\beta\mathsf{E}[\epsilon_{x,t} \mid x_t]. \tag{A.3}$$

Due to normality, $\mathsf{E}[\epsilon_{x,t} \mid x_t]$ is given by the linear regression

$$\mathsf{E}[\epsilon_{x,t} \mid x_t] = \delta_0 + \delta_1 x_t, \tag{A.4}$$

implying

$$\delta_1 = \frac{\mathsf{E}[\epsilon_{x,t} x_t]}{\mathsf{Var}[x_t]} = \frac{\mathsf{E}[\epsilon_{x,t}(\alpha_1 x_{t-1} + \epsilon_{x,t})]}{\mathsf{Var}[x_t]} = \frac{\sigma_{\epsilon_x}^2}{\mathsf{Var}[x_t]}. \tag{A.5}$$

Since $\mathsf{Var}[z_t] = \sigma_{\epsilon_x}^2/(1 - \alpha_1^2)$, we obtain

$$\delta_1 = (1 - \alpha_1^2), \tag{A.6}$$

which gives:

$$\mathsf{E}[\eta_t \mid x_t] = -\beta(1 - \alpha_1^2)x_t, \tag{A.7}$$

since $\delta_0 = 0$. Finally, using (A.7) in (A.2) yields the regression

$$E[y_t \mid x_t] = \alpha_1^2 \beta x_t, \tag{A.8}$$

and hence the true regression coefficient which is estimated consistently by $\hat{\beta}_{OLS}$ is $\alpha_1^2 \beta$ (not β).

A.2 Solving and estimating rational expectations models

To make the exposition self-contained, this appendix illustrates solution and estimation of simple models with forward looking variables—the illustration being the hybrid 'New Keynesian Phillips curve'. Finally, we comment on a problem with observational equivalence, or lack of identification within this class of models.

A sufficiently rich data generating process (DGP) to illustrate the techniques are

$$\Delta p_t = b_{p1}^f E_t \Delta p_{t+1} + b_{p1}^b \Delta p_{t-1} + b_{p2} x_t + \varepsilon_{pt}, \tag{A.9}$$

$$x_t = b_x x_{t-1} + \varepsilon_{xt}, \tag{A.10}$$

where all coefficients are assumed to be between zero and one. All of the techniques rely on the law of iterated expectations,

$$E_t E_{t+k} x_{t+j} = E_t x_{t+j}, \qquad k < j,$$

saying that your average revision of expectations, given more information, will be zero.

A.2.1 Repeated substitution

This method is the brute force solution, and therefore cumbersome. But since it is also instructive to see exactly what goes on, we begin with this method.

We start by using a trick to get rid of the lagged dependent variable, following Pesaran (1987, pp. 108–109), by implicitly defining π_t as

$$\Delta p_t = \pi_t + \alpha \Delta p_{t-1}, \tag{A.11}$$

where α will turn out to be the backward stable root of the process of Δp_t.

We take expectations one period ahead

$$E_t \Delta p_{t+1} = E_t \pi_{t+1} + \alpha E_t \Delta p_t,$$

$$E_t \Delta p_{t+1} = E_t \pi_{t+1} + \alpha \pi_t + \alpha^2 \Delta p_{t-1}.$$

Appendix

Next, we substitute for $E_t\Delta p_{t+1}$ into original model:

$$\pi_t + \alpha\Delta p_{t-1} = b_{p1}^f\left(E_t\pi_{t+1} + \alpha\pi_t + \alpha_{t-1}^2\Delta p_{t-1}\right) + b_{p1}^b\Delta p_{t-1} + b_{p2}x_t + \varepsilon_{pt}$$

$$\pi_t = \left(\frac{b_{p1}^f}{1 - b_{p1}^f\alpha}\right)E_t\pi_{t+1} + \left(\frac{b_{p1}^f\alpha^2 - \alpha + b_{p1}^b}{1 - b_{p1}^f\alpha}\right)\Delta p_{t-1}$$

$$+ \left(\frac{b_{p2}}{1 - b_{p1}^f\alpha}\right)x_t + \left(\frac{1}{1 - b_{p1}^f\alpha}\right)\varepsilon_t.$$

The parameter α is defined by

$$b_{p1}^f\alpha^2 - \alpha + b_{p1}^b = 0$$

or

$$\alpha^2 - \frac{1}{b_{p1}^f}\alpha + \frac{b_{p1}^b}{b_{p1}^f} = 0 \tag{A.12}$$

with the solutions

$$\left.\begin{array}{c}\alpha_1\\\alpha_2\end{array}\right\} = \frac{1 \pm \sqrt{1 - 4b_{p1}^f b_{p1}^b}}{2b_{p1}^f}. \tag{A.13}$$

The model will typically have a saddle-point behaviour with one root bigger than one and one smaller than one in absolute value. In the following we will use the backward stable solution, defined by:

$$\left|\alpha_1 = \frac{1 - \sqrt{1 - 4b_{p1}^f b_{p1}^b}}{2b_{p1}^f}\right| < 1.$$

In passing it might be noted that the restriction $b_{p1}^b = 1 - b_{p1}^f$ often imposed in the literature implies the roots

$$\alpha_1 = \frac{1 - b_{p1}^f}{b_{p1}^f} \leq 1,$$

$$\alpha_2 = 1.$$

as given in (A.13) as before. We choose $|\alpha_1| < 1$ in the following.

So we now have a pure forward-looking model

$$\pi_t = \left(\frac{b_{p1}^f}{1 - b_{p1}^f\alpha_1}\right)E_t\pi_{t+1} + \left(\frac{b_{p2}}{1 - b_{p1}^f\alpha_1}\right)x_t + \left(\frac{1}{1 - b_{p1}^f\alpha_1}\right)\varepsilon_{pt}.$$

Finally, using the relationship

$$\alpha_1 + \alpha_2 = \frac{1}{b_{p1}^f}$$

between the roots,[1] so:
$$1 - b^f_{p1}\alpha_1 = b^f_{p1}\alpha_2, \tag{A.14}$$
the model becomes
$$\pi_t = \left(\frac{1}{\alpha_2}\right)E_t\pi_{t+1} + \left(\frac{b_{p2}}{b^f_{p1}\alpha_2}\right)x_t + \left(\frac{1}{b^f_{p1}\alpha_2}\right)\varepsilon_{pt} \tag{A.15}$$
$$\pi_t = \gamma E_t\pi_{t+1} + \delta x_t + v_{pt}. \tag{A.16}$$

Following Davidson (2000, pp. 109–110), we now derive the solution in two steps:

1. Find $E_t\pi_{t+1}$.
2. Solve for π_t.

Find $E_t\pi_{t+1}$ Define the expectations errors as:
$$\eta_{t+1} = \pi_{t+1} - E_t\pi_{t+1}. \tag{A.17}$$
We start by reducing the model to a single equation:
$$\pi_t = \gamma\pi_{t+1} + \delta b_x x_{t-1} + \delta\varepsilon_{xt} + v_{pt} - \gamma\eta_{t+1}.$$
Solving forwards then produces:
$$\begin{aligned}
\pi_t &= \gamma(\gamma\pi_{t+2} + \delta b_x x_t + \delta\varepsilon_{xt+1} + v_{pt+1} - \gamma\eta_{t+2}) \\
&\quad + \delta b_x x_{t-1} + \delta\varepsilon_{xt} + v_{pt} - \gamma\eta_{t+1} \\
&= (\delta b_x x_{t-1} + \delta\varepsilon_{xt} + v_{pt} - \gamma\eta_{t+1}) \\
&\quad + \gamma(\delta b_x x_t + \delta\varepsilon_{xt+1} + v_{pt+1} - \gamma\eta_{t+2}) + (\gamma)^2\pi_{t+2} \\
&= \sum_{j=0}^{n}(\gamma)^j(\delta b_x x_{t+j-1} + \delta\varepsilon_{xt+j} + v_{pt+j} - \gamma\eta_{t+j+1}) + (\gamma)^{n+1}\pi_{t+n+1}.
\end{aligned}$$
By imposing the transversality condition:
$$\lim_{n\to\infty}(\gamma)^{n+1}\pi_{t+n+1} = 0$$
and then taking expectations conditional at time t, we get the 'discounted solution':
$$\begin{aligned}
E_t\pi_{t+1} &= \sum_{j=0}^{\infty}(\gamma)^j(\delta b_x E_t x_{t+j} + \delta E_t\varepsilon_{xt+j+1} + E_t v_{pt+j+1} - \gamma E_t\eta_{t+j+2}) \\
&= \sum_{j=0}^{\infty}(\gamma)^j(\delta b_x E_t x_{t+j}).
\end{aligned}$$

[1] See, for example, Chiang (1984, p. 506).

However, we know the process for the forcing variable, so:

$$E_{t-1}x_t = b_x x_{t-1},$$
$$E_t x_t = x_t,$$
$$E_t x_{t+1} = b_x x_t,$$
$$E_t x_{t+2} = E_t(E_{t+1}x_{t+2}) = E_t b_x x_{t+1} = b_x^2 x_t,$$
$$E_t x_{t+j} = b_x^j x_t.$$

We can therefore substitute in:

$$E_t \pi_{t+1} = \sum_{j=0}^{\infty} (\gamma)^j (\delta b_x b_x^j x_t)$$

$$= \delta b_x \sum_{j=0}^{\infty} (\gamma b_x)^j x_t$$

$$= \left(\frac{\delta b_x}{1 - \gamma b_x} \right) x_t$$

and substitute back the expectation into the original equation:

$$\pi_t = \gamma E_t \pi_{t+1} + \delta x_t + v_{pt}$$

$$= \gamma \left(\frac{\delta b_x}{1 - \gamma b_x} \right) x_t + \delta x_t + v_{pt}.$$

Solve for π_t Finally, using (A.11) and (A.16) we get the complete solution:

$$\Delta p_t - \alpha_1 \Delta p_{t-1} = \left(\frac{b_{p1}^f}{b_{p1}^f \alpha_2} \right) \left(\frac{(b_{p2}/b_{p1}^f \alpha_2) b_x}{1 - (b_{p1}^f/b_{p1}^f \alpha_2) b_x} \right) x_t$$

$$+ \left(\frac{b_{p2}}{b_{p1}^f \alpha_2} \right) x_t + \left(\frac{1}{b_{p1}^f \alpha_2} \right) \varepsilon_{pt}$$

$$= \left(\frac{1}{\alpha_2} \right) \left(\frac{b_{p2} b_x}{b_{p1}^f (\alpha_2 - b_x)} \right) x_t + \left(\frac{b_{p2}}{b_{p1}^f \alpha_2} \right) x_t + \left(\frac{1}{b_{p1}^f \alpha_2} \right) \varepsilon_{pt},$$

$$\Delta p_t = \alpha_1 \Delta p_{t-1} + \left(\frac{b_{p2}}{b_{p1}^f (\alpha_2 - b_x)} \right) x_t + \left(\frac{1}{b_{p1}^f \alpha_2} \right) \varepsilon_{pt}. \qquad (A.18)$$

A.2.2 Undetermined coefficients

This method is more practical. It consists of the following steps:

1. Make a guess at the solution.
2. Derive the expectations variable.

3. Substitute back into the guessing solution.
4. Match coefficients.

We will first use the technique, following the excellent exposition of Blanchard and Fisher (1989: ch. 5), to derive the solution conditional upon the expected path of the forcing variable, as in Galí *et al.* (2001), so we will ignore any information about the process of the forcing variable.

In the following we will define

$$z_t = b_{p2}x_t + \varepsilon_{pt}.$$

Since the solution must depend on the future, a guess would be that the solution will consist of the lagged dependent variable and the expected values of the forcing value:

$$\Delta p_t = \alpha \Delta p_{t-1} + \sum_{i=0}^{\infty} \beta_i E_t z_{t+i}. \tag{A.19}$$

We now take the expectation of the solution of the next period, using the law of iterated expectations, to find the expected outcome

$$E_t \Delta p_{t+1} = \alpha \Delta p_t + \sum_{i=0}^{\infty} \beta_i E_t z_{t+1+i},$$

which we substitute in the guessing solution

$$\Delta p_t = b_{p1}^f \left(\alpha \Delta p_t + \sum_{i=0}^{\infty} \beta_i E_t z_{t+1+i} \right) + b_{p1}^b \Delta p_{t-1} + z_t,$$

$$\Delta p_t = \left(\frac{b_{p1}^b}{1 - \alpha b_{p1}^f} \right) \Delta p_{t-1} + \left(\frac{1}{1 - \alpha b_{p1}^f} \right) z_t + \sum_{i=0}^{\infty} \left(\frac{\beta_i b_{p1}^f}{1 - \alpha b_{p1}^f} \right) E_t z_{t+1+i}. \tag{A.20}$$

Finally, the undetermined coefficients are now found by matching the coefficients of the variables between (A.19) and (A.20).

We start by matching the coefficients of Δp_{t-1}:

$$\alpha = \frac{b_{p1}^b}{1 - \alpha b_{p1}^f}.$$

This gives, as above, the second-order polynomial in α:

$$\alpha^2 - \frac{1}{b_{p1}^f} \alpha + \frac{b_{p1}^b}{b_{p1}^f} = 0$$

with the solutions given in (A.13).

Using α_1, we may now match the remaining undetermined coefficients of $E_t z_{t+i}$, giving

$$z_t: \qquad \beta_0 = \frac{1}{1 - b_{p1}^f \alpha_1},$$

$$E_t z_{t+1}: \quad \beta_1 = \frac{b_{p1}^f}{1 - b_{p1}^f \alpha_1} \beta_0,$$

$$E_t z_{t+i}: \quad \beta_i = \frac{b_{p1}^f}{1 - b_{p1}^f \alpha_1} \beta_{i-1},$$

so, using (A.14), the coefficients can therefore be written as

$$z_t: \qquad \beta_0 = \frac{1}{b_{p1}^f \alpha_2},$$

$$E_t z_{t+1}: \quad \beta_1 = \left(\frac{1}{b_{p1}^f \alpha_2} \right) \left(\frac{1}{\alpha_2} \right),$$

$$E_t z_{t+i}: \quad \beta_i = \left(\frac{1}{b_{p1}^f \alpha_2} \right) \left(\frac{1}{\alpha_2} \right)^i,$$

declining as time move forwards.

Substituting back for

$$z_t = b_{p2} x_t + \varepsilon_{pt},$$

the solution can therefore be written

$$\Delta p_t = \alpha_1 \Delta p_{t-1} + \left(\frac{b_{p2}}{b_{p1}^f \alpha_2} \right) \sum_{i=0}^{\infty} \left(\frac{1}{\alpha_2} \right)^i E_t x_{t+i} + \left(\frac{1}{b_{p1}^f \alpha_2} \right) \varepsilon_{pt}, \qquad (A.21)$$

which is the same as in Galí *et al.* (2001), except the error term which they ignore.

To derive the complete solution, we need to substitute in for the forcing process x_t. We can either do this already in the guessing solution, or by substituting in for the expected terms $E_t x_{t+i}$. Here we choose the latter solution. The expectations, conditional on information at time t, are:

$$E_t x_t = x_t,$$
$$E_t x_{t+1} = b_x x_t,$$
$$E_t x_{t+2} = E_t(E_{t+1} x_{t+2}) = E_t b_x x_{t+1} = b_x^2 x_t,$$
$$E_t x_{t+j} = b_x^j x_t,$$

where we again have used the law of iterated expectations. So the solution becomes

$$\Delta p_t = \alpha_1 \Delta p_{t-1} + \left(\frac{b_{p2}}{b_{p1}^f \alpha_2}\right) \sum_{i=0}^{\infty} \left(\frac{b_x}{\alpha_2}\right)^i x_t + \left(\frac{1}{b_{p1}^f \alpha_2}\right) \varepsilon_{pt}$$

$$\Delta p_t = \alpha_1 \Delta p_{t-1} + \left(\frac{b_{p2}}{b_{p1}^f \alpha_2}\right) \left(\frac{1}{1-(b_x/\alpha_2)}\right) x_t + \left(\frac{1}{b_{p1}^f \alpha_2}\right) \varepsilon_{pt}$$

$$\Delta p_t = \alpha_1 \Delta p_{t-1} + \left(\frac{b_{p2}}{b_{p1}^f (\alpha_2 - b_x)}\right) x_t + \left(\frac{1}{b_{p1}^f \alpha_2}\right) \varepsilon_{pt},$$

as in (A.18).

A.2.3 Factorization

Finally, we shall take a look at this very elegant method introduced by Sargent. It consists of the following steps:

1. Write the model in terms of lead- and lag-polynomials in expectations.
2. Factor the polynomials, into one-order polynomials, deriving the roots.
3. Invert the factored one-order polynomials into the directions of converging forward polynomials of expectations.

Again, we use the simplifying definition

$$z_t = b_{p2} x_t + \varepsilon_{pt},$$

so the model is again

$$\Delta p_t = b_{p1}^f E_t \Delta p_{t+1} + b_{p1}^b \Delta p_{t-1} + z_t.$$

Note that the forward, or lead, operator, F, and lag operator, L, only work on the variables and not expectations, so:

$$L E_t z_t = E_t z_{t-1}$$
$$F E_t z_t = E_t z_{t+1}$$
$$L^{-1} = F.$$

The model can then be written in terms of expectations as:

$$-b_{p1}^f E_t \Delta p_{t+1} + E_t \Delta p_t - b_{p1}^b E_t \Delta p_{t-1} = E_t z_t,$$

and using the lead- and lag-operators:

$$(-b_{p1}^f F + 1 - b_{p1}^b L) E_t \Delta p_t = E_t z_t,$$

or, as a second-order polynomial in the lead operator:

$$\left[F^2 - \left(\frac{1}{b_{p1}^f} \right) F + \frac{b_{p1}^b}{b_{p1}^f} \right] LE_t \Delta p_t = - \left(\frac{1}{b_{p1}^f} \right) E_t z_t.$$

The polynomial in brackets is exactly the same as the one in (A.12), so we know it can be factored into the roots (A.13):

$$\left[(F - \alpha_1)(F - \alpha_2) \right] LE_t \Delta p_t = - \left(\frac{1}{b_{p1}^f} \right) E_t z_t$$

$$(F - \alpha_1) LE_t \Delta p_t = - \left(\frac{1}{b_{p1}^f (F - \alpha_2)} \right) E_t z_t$$

$$(1 - \alpha_1 L) \Delta p_t = \left(\frac{1}{b_{p1}^f (\alpha_2 - F)} \right) E_t z_t$$

$$(1 - \alpha_1 L) \Delta p_t = \left(\frac{1}{b_{p1}^f \alpha_2} \right) \left(\frac{1}{1 - (1/\alpha_2)F} \right) E_t z_t.$$

However, we know that $(1/1 - (1/\alpha_2)F) = \sum_{i=0}^{\infty} (1/\alpha_2)^i F^i$, since $|1/\alpha_2| < 1$, so we can write down the solution immediately:

$$\Delta p_t = \alpha_1 \Delta p_{t-1} + \left(\frac{b_{p2}}{b_{p1}^f \alpha_2} \right) \sum_{i=0}^{\infty} \left(\frac{1}{\alpha_2} \right)^i E_t x_{t+i} + \left(\frac{1}{b_{p1}^f \alpha_2} \right) \varepsilon_{pt},$$

where we have also substituted back for z_t.

To derive the complete solution, we have to solve for

$$\sum_{i=0}^{\infty} \left(\frac{1}{\alpha_2} \right)^i E_t x_{t+i}$$

given

$$(1 - b_x L) x_t = \varepsilon_{xt}.$$

We can now appeal to the results of Sargent (1987, p. 304) that work as follows. If the model can be written in the form

$$y_t = \lambda E_t y_{t+1} + x_t a(L) x_t + e_t,$$

$$a(L) = 1 - \sum_{j=1}^{r} a_j L^j$$

with the partial solution

$$y_t = \zeta \sum_{i=0}^{\infty} (\lambda)^i E_t x_{t+i},$$

then the complete solution

$$y_t = \zeta g(L) x_t$$

is determined by

$$g(L) = \frac{1 - \lambda a(\lambda)^{-1} a(L) L^{-1}}{1 - \lambda L^{-1}}$$

$$= a(\lambda)^{-1} \left[1 + \sum_{j=1}^{r-1} \left(\sum_{k=j+1}^{r} \lambda^{k-j} a_k \right) L^j \right].$$

In our case

$$\zeta = \frac{b_{p2}}{b_{p1}^f \alpha_2},$$

$$\lambda = \frac{1}{\alpha_2},$$

$$a(L) = 1 - b_x L,$$

so $g(L)$ will have the form

$$g(L) = (1 - a_1 \lambda)^{-1}$$

$$= \frac{1}{1 - b_x (1/\alpha_2)}.$$

The solution therefore becomes

$$\Delta p_t - \alpha_1 \Delta p_{t-1} = \left(\frac{1}{b_{p1}^f \alpha_2} \right) \left(\frac{1}{1 - b_x (1/\alpha_2)} \right) x_t + \left(\frac{1}{b_{p1}^f \alpha_2} \right) \varepsilon_{pt}$$

$$\Delta p_t = \alpha_1 \Delta p_{t-1} + \left(\frac{b_{p2}}{b_{p1}^f (\alpha_2 - b_x)} \right) x_t + \left(\frac{1}{b_{p1}^f \alpha_2} \right) \varepsilon_{pt},$$

as before.

A.2.4   Estimation

Remember that the model is

$$\Delta p_t = b_{p1}^f E_t \Delta p_{t+1} + b_{p1}^b \Delta p_{t-1} + b_{p2} x_t + \varepsilon_{pt},$$

which can be rewritten as

$$\pi = \gamma E_t \pi_{t+1} + \delta x_t + v_{pt}.$$

The model is usually estimated by means of instrumental variables, using the 'errors in variables' method (evm)—where expected values are replaced by actual values and the expectational errors:

$$\pi_t = \gamma \pi_{t+1} + \delta x_t + v_{pt} - \gamma \eta_{t+1}. \qquad (A.22)$$

The implications of estimating the model by means of the 'errors in variables' method is to induce moving average errors. Following Blake (1991), this can be readily seen using the expectational errors as follows.

1. Lead (A.15) one period and subtract the expectation to find the RE error:

$$\eta_{t+1} = \gamma E_t \pi_{t+2} + \delta x_{t+1} + v_{pt+1} - E_t \pi_{t+1}$$

$$= \gamma \left(\frac{\delta b_x}{1 - \gamma b_x} \right) x_{t+1} + \delta x_{t+1} + v_{pt+1} - \left(\frac{\delta b_x}{1 - \gamma b_x} \right) x_t$$

$$= \left(\frac{\delta}{1 - \gamma b_x} \right) (x_{t+1} - b_x x_t) + v_{pt+1}$$

$$= \left(\frac{\delta}{1 - \gamma b_x} \right) \varepsilon_{xt+1} + v_{pt+1}$$

2. Substitute into (A.22):

$$\pi_t = \gamma \pi_{t+1} + \delta x_t + v_{pt} - \gamma v_{pt+1} - \left(\frac{\gamma \delta}{1 - \gamma b_x} \right) \varepsilon_{xt+1}.$$

3. Finally, re-express in terms of original variables, again using $\Delta p_t = \pi_t + \alpha \Delta p_{t-1}$:

$$\Delta p_t - \alpha_1 \Delta p_{t-1} = \left(\frac{1}{\alpha_2} \right) (\Delta p_{t+1} - \alpha_1 \Delta p_t) + \left(\frac{b_{p2}}{b_{p1}^f \alpha_2} \right) x_t + \left(\frac{1}{b_{p1}^f \alpha_2} \right) \varepsilon_{pt}$$

$$- \left(\frac{1}{\alpha_2} \right) \left(\frac{1}{b_{p1}^f \alpha_2} \right) \varepsilon_{pt+1} - \left(\frac{(1/\alpha_2)(b_{p2}/b_{p1}^f \alpha_2)}{1 - (1/\alpha_2) b_x} \right) \varepsilon_{xt+1},$$

$$\Delta p_t \left(\frac{1}{b_{p1}^f \alpha_2} \right) = \left(\frac{1}{b_{p1}^f \alpha_2} \right) b_{p1}^f \Delta p_{t+1} + \left(\frac{1}{\alpha_2 b_{p1}^f} \right) b_{p1}^b \Delta p_{t-1} + \left(\frac{1}{b_{p1}^f \alpha_2} \right) b_{p2} x_t$$

$$+ \left(\frac{1}{b_{p1}^f \alpha_2} \right) \varepsilon_{pt} - \left(\frac{1}{b_{p1}^f \alpha_2} \right) \left(\frac{1}{\alpha_2} \right) \varepsilon_{pt+1}$$

$$- \left(\frac{1}{b_{p1}^f \alpha_2} \right) \left(\frac{b_{p2}/\alpha_2}{1 - (1/\alpha_2) b_x} \right) \varepsilon_{xt+1},$$

$$\Delta p_t = b_{p1}^f \Delta p_{t+1} + b_{p1}^b \Delta p_{t-1} + b_{p2} x_t + \varepsilon_{pt}$$

$$- \left(\frac{1}{\alpha_2} \right) \varepsilon_{pt+1} - \left(\frac{b_{p2}}{\alpha_2 - b_x} \right) \varepsilon_{xt+1},$$

where we have exploited the two well-known relationships between the roots:

$$\alpha_1 + \alpha_2 = \frac{1}{b_{p1}^f},$$

$$\alpha_1 \alpha_2 = \frac{b_{p1}^b}{b_{p1}^f}.$$

So even though the original model has white noise errors, the estimated model will have first-order moving average errors.

A.2.5   Does the MA(1) process prove that the forward solution applies?

Assume that the true model is

$$\Delta p_t = b_{p1} \Delta p_{t-1} + \varepsilon_{pt}, \qquad |b_{p1}| < 1$$

and the the following model is estimated by means of instrumental variables

$$\Delta p_t = b_{p1}^f \Delta p_{t+1} + \varepsilon_{pt}^f.$$

What are the properties of ε_{pt}^f?

$$\varepsilon_{pt}^f = \Delta p_t - b_{p1}^f \Delta p_{t+1}.$$

Assume, as is common in the literature, that we find that $b_{p1}^f \approx 1$. Then

$$\varepsilon_{pt}^f \approx \Delta p_t - \Delta p_{t+1} = -\Delta^2 p_{t+1}$$
$$= -[\varepsilon_{pt+1} + (b_{p1} - 1)\varepsilon_{pt} + \cdots].$$

So we get a model with a moving average residual, but this time the reason is not forward-looking behaviour but mis-specification.

A.3   Calculation of interim multipliers in a linear dynamic model: a general exposition

Interim multipliers provide a simple yet powerful way to describe the dynamic properties of a dynamic model. We follow Lütkepohl (1991) and derive the dynamic multipliers in a simultaneous system of n linear dynamic equations with n endogenous variables y_t and m exogenous variables x_t. The *structural form* of the model is given by:

$$\boldsymbol{\Gamma}_0 \mathbf{y}_t = \sum_{i=1}^{q} \boldsymbol{\Gamma}_i \mathbf{y}_{t-i} + \sum_{i=0}^{q} \mathbf{D}_i \mathbf{x}_{t-i} + \boldsymbol{\varepsilon}_t. \tag{A.23}$$

To investigate the dynamic properties of the model it will be more convenient to work with the *reduced form* of the model:

$$\mathbf{y}_t = \sum_{i=1}^{q} \mathbf{A}_i \mathbf{y}_{t-i} + \sum_{i=0}^{q} \mathbf{B}_i \mathbf{x}_{t-i} + \mathbf{u}_t \tag{A.24}$$

defining the $n \times n$ matrices $\mathbf{A}_i = \boldsymbol{\Gamma}_0^{-1} \boldsymbol{\Gamma}_i$, $i = 1, \dots, q$, and the $n \times m$ matrices $\mathbf{B}_i = \boldsymbol{\Gamma}_0^{-1} \mathbf{D}_i$, $i = 0, \dots, q$. The reduced form residuals are given by $\mathbf{u}_t = \boldsymbol{\Gamma}_0^{-1} \boldsymbol{\varepsilon}_t$.

It is also useful to define the autoregressive *final form* of the model as:

$$\mathbf{y}_t = \mathbf{A}(L)^{-1} \mathbf{B}(L) \mathbf{x}_t + \mathbf{A}(L)^{-1} \mathbf{u}_t \tag{A.25}$$
$$= \mathbf{D}(L) \mathbf{x}_t + \mathbf{v}_t,$$

where the polynomials are

$$\mathbf{A}(L) = \mathbf{I} - \mathbf{A}_1 L - \cdots - \mathbf{A}_q L^q,$$
$$\mathbf{B}(L) = \mathbf{B}_0 + \mathbf{B}_1 L + \cdots + \mathbf{B}_q L^q,$$

and the final form coefficients are given by the (infinite) rational lag polynomial

$$\mathbf{D}(L) = \mathbf{A}(L)^{-1} \mathbf{B}(L)$$
$$= \mathbf{D}_0 + \mathbf{D}_1 L + \cdots + \mathbf{D}_j L^j + \cdots.$$

To obtain a simple expression for the interim multipliers it is useful to rewrite the reduced form representation of the model in its *companion* form as:

$$\mathbf{Z}_t = \boldsymbol{\Phi} \mathbf{Z}_{t-1} + \boldsymbol{\Psi} \mathbf{x}_t + \mathbf{U}_t \tag{A.26}$$

forming stacked $(n + m)q \times 1$ vectors with new variables

$$\mathbf{Z}_t = (\mathbf{y}_t', \dots, \mathbf{y}_{t-q+1}', \mathbf{x}_t', \dots, \mathbf{x}_{t-q+1}')'$$

and

$$\mathbf{U}_t = (\mathbf{u}_t', \mathbf{0}, \dots, \mathbf{0})'$$

and defining a selection matrix

$$\mathbf{J}_{n \times (n+m)q} = (\mathbf{I}_n, \mathbf{0}_n, \dots, \mathbf{0}_n | \mathbf{0}_{n,m}, \dots, \mathbf{0}_{n,m}).$$

The matrices $\Phi_{(n+m)q\times(n+m)q}$ and $\Psi_{(n+m)q\times m}$ are formed by stacking the (reduced form) coefficient matrices A_i, B_i for $\forall i$ in the following way:

$$
\Phi = \left[
\begin{array}{ccccc|cccc}
\mathbf{A}_1 & \cdots & \mathbf{A}_{q-1} & \mathbf{A}_q & \mathbf{B}_1 & \cdots & \mathbf{B}_{q-1} & \mathbf{B}_q \\
\mathbf{I}_n & & \mathbf{0}_n & \mathbf{0}_n & \mathbf{0}_n & & \mathbf{0}_n & \mathbf{0}_n \\
& \ddots & \vdots & \vdots & \vdots & \ddots & \vdots & \vdots \\
\mathbf{0}_n & \cdots & \mathbf{I}_n & \mathbf{0}_n & \mathbf{0}_n & \cdots & \mathbf{0}_n & \mathbf{0}_n \\
\hline
& & & & \mathbf{0}_m & \cdots & \mathbf{0}_m & \mathbf{0}_m \\
& \multicolumn{3}{c}{\mathbf{0}_{mq\times nq}} & \mathbf{I}_m & & \mathbf{0}_m & \mathbf{0}_m \\
& & & & & \ddots & \vdots & \vdots \\
& & & & \mathbf{0}_m & \cdots & \mathbf{I}_m & \mathbf{0}_m
\end{array}
\right], \quad
\Psi = \left[
\begin{array}{c}
\mathbf{B}_0 \\
\mathbf{0}_n \\
\vdots \\
\mathbf{0}_n \\
\hline
\mathbf{I}_m \\
\mathbf{0}_m \\
\vdots \\
\mathbf{0}_m
\end{array}
\right]
$$

$$(\text{A.27})$$

The eigenvalues (characteristic roots) of the system matrix Φ are useful to summarise the characteristics of the dynamic behaviour of the complete system, like whether it will generate 'oscillations' as in the case when there is (at least) one pair of complex conjugate roots, or 'exploding' behaviour when (at least) one root has modulus greater than 1.

A different way to address the dynamic properties is to calculate the 'interim multipliers' of the model, which has the additional advantage that they can be easily graphed.

Successive substitution of $\mathbf{Z}_t$ in equation (A.26) yields:

$$
\mathbf{Z}_t = \Phi \mathbf{Z}_{t-1} + \Psi \mathbf{x}_t + \mathbf{U}_t \tag{A.28}
$$

$$
= \Phi^i \mathbf{Z}_{t-i} + \sum_{j=0}^{i-1} \Phi^j \Psi \mathbf{x}_{t-j} + \sum_{j=0}^{i-1} \mathbf{U}_{t-j}
$$

$$\Downarrow$$

$$
\mathbf{y}_t = \sum_{j=0}^{\infty} \mathbf{J}\Phi^j \Psi \mathbf{x}_{t-j} + \sum_{j=0}^{\infty} \mathbf{J}\Phi^j \mathbf{J}' \mathbf{u}_{t-j}, \tag{A.29}
$$

since Φ^i is assumed to disappear as i grows sufficiently large. The dynamic multipliers D_j and the interim multipliers M_i can be obtained from (A.29) as the partial derivatives $\mathbf{D}_j = \partial \mathbf{y}_t/\partial \mathbf{x}_{t-j}$ and their cumulated sums

$$
\mathbf{M}_i = \sum_{j=0}^{i} \mathbf{D}_j = \sum_{j=0}^{i} \frac{\partial \mathbf{y}_t}{\partial \mathbf{x}_{t-j}},
$$

respectively. We obtain estimates of the multipliers $\hat{\mathbf{D}}_i$ and $\widehat{\mathbf{M}}_i$ by inserting estimates of the parameters in (A.24) into the companion form matrices $\hat{\Phi}$ and $\hat{\Psi}$.

$$
\hat{\mathbf{D}}_i = \mathbf{J}\hat{\Phi}^i \hat{\Psi}, \quad i = 0, \ldots \tag{A.30}
$$

and the interim multipliers are defined in terms of their cumulated sums $\widehat{M}_i$:

$$\widehat{M}_i = \sum_{j=0}^{i} \hat{D}_j \tag{A.31}$$

$$= J\sum_{j=0}^{i} \hat{\Phi}^j \hat{\Psi}$$

$$= J(I + \hat{\Phi} + \hat{\Phi}^2 + \cdots + \hat{\Phi}^i)\hat{\Psi}.$$

The long-run multipliers are given by

$$\widehat{M}_\infty = \sum_{j=0}^{\infty} \hat{D}_j = J(I - \hat{\Phi})^{-1}\hat{\Psi} \tag{A.32}$$

$$= \widehat{A(1)}^{-1}\widehat{B(1)}.$$

A.3.1 An example

As an example, and in the process of illustrating different techniques, we will work out the dynamic properties of the wage–price model of Section 9.2.2. This involves evaluating the stability of the model, and the long-run and dynamic multipliers. Disregarding taxes and short-run effects, the systematic part of the model is on matrix form:

$$\begin{bmatrix} 1 & -0.81 \\ -0.14 & 1 \end{bmatrix}\begin{bmatrix} \Delta w \\ \Delta p \end{bmatrix}_t = \begin{bmatrix} 0 & 0 \\ 0.1 & 0 \end{bmatrix}\begin{bmatrix} \Delta w \\ \Delta p \end{bmatrix}_{t-1} \begin{bmatrix} 0 & 0 \\ 0 & 0.16 \end{bmatrix}\begin{bmatrix} \Delta w \\ \Delta p \end{bmatrix}_{t-2}$$

$$+ \begin{bmatrix} 0.082 & 0 & 0 \\ -0.015 & 0 & 0.026 \end{bmatrix}\begin{bmatrix} \Delta a \\ \Delta u \\ \Delta pi \end{bmatrix}_t + \begin{bmatrix} -0.16 & 0 \\ 0 & -0.055 \end{bmatrix}$$

$$\times \begin{bmatrix} 1 & -1 & -1 & 0.1L & 0 \\ -0.7L & 1L^2 & 0.7 & 0 & -0.3 \end{bmatrix}\begin{bmatrix} w \\ p \\ a \\ u \\ pi \end{bmatrix}_{t-1}$$

Steady-state properties from cointegration   The long-run elasticities of the model are, from the cointegration analysis:

$$w = p + a - 0.1u$$
$$p = 0.7(w - a) + 0.3pi,$$

so the long-run multipliers of the system should be easily obtained by solving for wages and prices. For wages:

$$w = 0.7(w - a) + 0.3p + a - 0.1u$$
$$w(1 - 0.7) = -0.7a + 0.3pi + a - 0.1u$$
$$w = \frac{0.3}{0.3}a - \frac{0.1}{0.3}u + pi$$
$$w = a - 0.33u + pi.$$

Then for prices:

$$p = 0.7(w - a) + 0.3pi$$
$$= 0.7(-0.33u + pi) + 0.3pi$$
$$p = -0.23u + pi.$$

So the reduced form long-run multipliers of wages and prices with respect to the exogenous variables are

$$w = a - 0.33u + pi$$
$$p = -0.23u + pi.$$

Note that the long-run multipliers of the real wage are given from the wage curve alone

$$w - p = a - 0.1u.$$

Imposing long-run properties of exogenous variables

- $\Delta a = g_a$
- $\Delta u = 0$
- $\Delta pi = g_{pi}$

gives the long-run multipliers for inflation

$$\pi = g_p = \Delta p = g_{pi}.$$

Finally, the steady-state growth path of the nominal system is

$$g_w = g_a + g_{pi}$$
$$g_p = g_{pi}.$$

Dynamic properties from difference equations   Now, let us try to see if this holds for the dynamic system. Intuitively, the same steady state—and therefore the same multipliers—should be obtained if no invalid restrictions are imposed.

For the dynamic analysis of the system below, following Wallis (1977), it will be more convenient to work with the model in lag-polynomial form

$\tilde{\mathbf{A}}(L)\mathbf{y}_t = \tilde{\mathbf{B}}(L)\mathbf{x}_t$. This is easily achieved with the steps:

$$\begin{bmatrix} 1 & -0.81 \\ -0.14 - 0.1L & 1 - 0.16L^2 \end{bmatrix} \begin{bmatrix} \Delta w \\ \Delta p \end{bmatrix}_t =$$

$$\begin{bmatrix} 0.082 & 0 & 0 \\ -0.015 & 0 & 0.026 \end{bmatrix} \begin{bmatrix} \Delta a \\ \Delta u \\ \Delta pi \end{bmatrix}_t + \begin{bmatrix} -0.16 & 0 \\ 0 & -0.055 \end{bmatrix}$$

$$\times \begin{bmatrix} L & -L & -L & 0.1L^2 & 0 \\ -0.7L^2 & L^3 & 0.7L & 0 & -0.3L \end{bmatrix} \begin{bmatrix} w \\ p \\ a \\ u \\ pi \end{bmatrix}_t$$

or:

$$\begin{bmatrix} 1 - 1L & -0.81 + 0.81L \\ -0.14 - 0.1L - (-0.14 - 0.1L)L & 1 - 0.16L^2 - (1 - 0.16L^2)L \end{bmatrix} \begin{bmatrix} w \\ p \end{bmatrix}_t =$$

$$\begin{bmatrix} 0.082 - 0.082L & 0 & 0 \\ -0.015 + 0.015L & 0 & 0.026 - 0.026L \end{bmatrix} \begin{bmatrix} a \\ u \\ pi \end{bmatrix}_t$$

$$+ \begin{bmatrix} -0.16L & 0.16L & 0.16L & -0.016L^2 & 0 \\ 0.0385L^2 & -0.055L^3 & -0.0385L & 0 & 0.0165L \end{bmatrix} \begin{bmatrix} w \\ p \\ a \\ u \\ pi \end{bmatrix}_t$$

and collecting terms:

$$\underbrace{\begin{bmatrix} 1 - 0.84L & -0.81 + 0.65L \\ -0.14 + 0.04L + 0.0615L^2 & 1 - 0.16L^2 - 1L + 0.215L^3 \end{bmatrix}}_{\tilde{\mathbf{A}}(L)} \underbrace{\begin{bmatrix} w \\ p \end{bmatrix}_t}_{\mathbf{y}_t} =$$

$$\underbrace{\begin{bmatrix} 0.082 + 0.078L & -0.016L^2 & 0 \\ -0.015 - 0.0235L & 0 & 0.026 - 0.0095L \end{bmatrix}}_{\tilde{\mathbf{B}}(L)} \underbrace{\begin{bmatrix} a \\ u \\ pi \end{bmatrix}_t}_{\mathbf{x}_t}$$

Checking stability For the system to be stable, the autoregressive part needs to have all roots outside the unit circle.

The autoregressive polynomial is

$$\tilde{\mathbf{A}}(L) = \begin{bmatrix} 1 - 0.84L & -0.81 + 0.65L \\ -0.14 + 0.04L + 0.0615L^2 & 1 - 0.16L^2 - 1L + 0.215L^3 \end{bmatrix},$$

with determinant:

$$|\tilde{\mathbf{A}}(L)| = 0.8866 - 1.7166L + 0.703815L^2 + 0.309425L^3 - 0.1806L^4.$$

The model is stable if all the roots of

$$0.8866 - 1.7166z + 0.703815z^2 + 0.309425z^3 - 0.1806z^4 = 0$$

are outside the unit circle. Here the polynomial can be factored (approximately) as

$$-0.1806(z + 2.26942781)(z - 1.03041478)(z - 1.19380201)(z - 1.75852774) = 0$$

so the roots are

$$\begin{Bmatrix} -2.26942781 \\ 1.03041478 \\ 1.19380201 \\ 1.75852774 \end{Bmatrix}.$$

So all roots of $|\tilde{\mathbf{A}}(z)| = 0$ are outside the unit circle. Also, in this case, the roots are real, so the adjustment from a shock back towards steady state will be monotonic and non-cyclical.

Deriving the long-run multipliers—the hard way Next the long-run multipliers are $\tilde{\mathbf{A}}^{-1}(1)\tilde{\mathbf{B}}(1)$. Here $\tilde{\mathbf{A}}(1)$ is given as:

$$\begin{aligned} \tilde{\mathbf{A}}(1) &= \begin{bmatrix} 1 - 0.84 & -0.81 + 0.65 \\ -0.14 + 0.04 + 0.0615 & 1 - 0.16 - 1 + 0.215 \end{bmatrix} \\ &= \begin{bmatrix} 0.16 & -0.16 \\ -0.0385 & 0.055 \end{bmatrix}, \end{aligned}$$

while

$$\begin{aligned} \tilde{\mathbf{B}}(1) &= \begin{bmatrix} 0.082 + 0.078 & -0.016 & 0 \\ -0.015 - 0.0235 & 0 & 0.026 - 0.0095 \end{bmatrix} \\ &= \begin{bmatrix} 0.16 & -0.016 & 0 \\ -0.0385 & 0 & 0.0165 \end{bmatrix} \end{aligned}$$

giving the long-run multipliers

$$\begin{aligned} \tilde{\mathbf{A}}^{-1}(1)\tilde{\mathbf{B}}(1) &= \begin{bmatrix} 0.16 & -0.16 \\ -0.0385 & 0.055 \end{bmatrix}^{-1} \begin{bmatrix} 0.16 & -0.016 & 0 \\ -0.0385 & 0 & 0.0165 \end{bmatrix} \\ &= \begin{bmatrix} 1.0 & -0.33 & 1.0 \\ 0 & -0.23 & 1.0 \end{bmatrix} \end{aligned}$$

or

$$\begin{bmatrix} w \\ p \end{bmatrix} = \begin{bmatrix} 1.0 & -0.33 & 1.0 \\ 0 & -0.23 & 1.0 \end{bmatrix} \begin{bmatrix} a \\ u \\ pi \end{bmatrix},$$

which corresponds to the long-run multipliers derived directly from the cointegration analysis.

So the cointegration relationships is therefore the steady-state of the dynamic system; it ties down the long-run solution of the dynamic system, and the comparative static properties—the long-run multipliers. In fact, this is nothing else than Samuelson's correspondence principle in disguise.

Deriving the long-run multipliers—the easy way   To show that cointegration is nothing but steady-state with growing variables is just finding the long-run multipliers as in Bårdsen (1989), but now for systems. The reduced form of the model is:

$$\begin{bmatrix} \Delta w \\ \Delta p \end{bmatrix}_t = \begin{bmatrix} 0.09 & 0 \\ 0.113 & 0 \end{bmatrix}\begin{bmatrix} \Delta w \\ \Delta p \end{bmatrix}_{t-1} + \begin{bmatrix} 0 & 0.146 \\ 0 & 0.18 \end{bmatrix}\begin{bmatrix} \Delta w \\ \Delta p \end{bmatrix}_{t-2}$$

$$+ \begin{bmatrix} 0.079 & 0 & 0.024 \\ -0.004 & 0 & 0.029 \end{bmatrix}\begin{bmatrix} \Delta a \\ \Delta u \\ \Delta pi \end{bmatrix}_t + \mathbf{\Pi}(L)\begin{bmatrix} w \\ p \\ a \\ u \\ pi \end{bmatrix}_{t-1}$$

with the cointegration part alone:

$$\underbrace{\begin{bmatrix} -0.18+0.035L & 0.18-0.05L^2 & 0.145 & -0.018L & 0.015 \\ -0.025+0.042L & 0.025-0.06L^2 & -0.017 & -0.0025L & 0.018 \end{bmatrix}}_{\mathbf{\Pi}(L)}\begin{bmatrix} w \\ p \\ a \\ u \\ pi \end{bmatrix}_{t-1},$$

or when evaluated at the same date, so in steady-state:

$$\begin{bmatrix} -0.145 & 0.13 & 0.145 & -0.018 & 0.015 \\ 0.017 & -0.035 & -0.017 & -0.0025 & 0.018 \end{bmatrix}\begin{bmatrix} w \\ p \\ a \\ u \\ pi \end{bmatrix}_t$$

The long-run multipliers are therefore simply:

$$\begin{bmatrix} w \\ p \end{bmatrix} = \begin{bmatrix} -0.145 & 0.13 \\ 0.017 & -0.035 \end{bmatrix}^{-1}\begin{bmatrix} 0.145 & -0.018 & 0.015 \\ -0.017 & -0.0025 & 0.018 \end{bmatrix}\begin{bmatrix} a \\ u \\ pi \end{bmatrix}$$

$$\begin{bmatrix} w \\ p \end{bmatrix} = \begin{bmatrix} -1 & 0.33 & -1 \\ -0 & 0.23 & -1 \end{bmatrix}\begin{bmatrix} a \\ u \\ pi \end{bmatrix},$$

as before.

Dynamic multipliers   The dynamic multipliers of the model are given as

$$\tilde{\mathbf{A}}^{-1}(L)\tilde{\mathbf{B}}(L) = \mathbf{D}(L) = \begin{bmatrix} \delta_{11}(L) & \delta_{12}(L) & \delta_{13}(L) \\ \delta_{21}(L) & \delta_{22}(L) & \delta_{23}(L) \end{bmatrix},$$

while the interim multipliers are the sums of the dynamic multipliers.

The simplest solution is to match coefficients of $\tilde{\mathbf{B}}(L) = \tilde{\mathbf{A}}(L)\mathbf{D}(L)$ for powers of L and solve for $\delta(L)$.

Let us assume we are only interested in the first three dynamic and interim multipliers of productivity on wages:

$$\delta_{11}(L) = \delta_{11,0} + \delta_{11,1}L + \delta_{11,2}L^2. \tag{A.33}$$

The inverse autoregressive matrix polynomials are of course the product of the inverse of the determinant and the adjoint

$$\tilde{\mathbf{A}}^{-1}(L) = \begin{bmatrix} 1 - 0.84L & -0.81 + 0.65L \\ -0.14 + 0.04L + 0.0615L^2 & 1 - 0.16L^2 - 1L + 0.215L^3 \end{bmatrix}^{-1}$$

$$= \frac{1}{0.89 - 1.72L + 0.7L^2 + 0.31L^3 - 0.18L^4}$$
$$\times \begin{bmatrix} 1 - 0.16L^2 - L + 0.215L^3 & 0.81 - 0.65L \\ 0.14 - 0.04L - 0.0615L^2 & 1 - 0.84L \end{bmatrix}.$$

The matrix of distributed lag-polynomials was

$$\tilde{\mathbf{B}}(L) = \begin{bmatrix} 0.082 + 0.078L & -0.016L^2 & 0 \\ -0.015 - 0.0235L & 0 & 0.026 - 0.0095L \end{bmatrix}.$$

Therefore

$$\mathbf{D}(L) = \begin{bmatrix} \delta_{11}(L) & \delta_{12}(L) & \delta_{13}(L) \\ \delta_{21}(L) & \delta_{22}(L) & \delta_{23}(L) \end{bmatrix}$$

$$= \frac{1}{0.89 - 1.72L + 0.7L^2 + 0.31L^3 - 0.18L^4}$$
$$\times \begin{bmatrix} 1 - L - 0.16L^2 + 0.215L^3 & 0.81 - 0.65L \\ 0.14 - 0.04L - 0.0615L^2 & 1 - 0.84L \end{bmatrix}$$
$$\times \begin{bmatrix} 0.082 + 0.078L & -0.016L^2 & 0 \\ -0.015 - 0.0235L & 0 & 0.026 - 0.0095L \end{bmatrix}$$

$$= \frac{1}{0.89 - 1.72L + 0.7L^2 + 0.31L^3 - 0.18L^4}$$
$$\times \begin{bmatrix} (0.07 - 0.01L - 0.08L^2 & (-0.02L^2 + 0.02L^3 & (0.02 - 0.02L \\ +0.01L^3 + 0.02L^4) & +0.003L^4 - 0.003L^5) & +0.006L^2) \\ (-0.004 - 0.003L & (-0.002L^2 + 0.0006L^3 & (0.03 - 0.03L \\ +0.01L^2 - 0.005L^3) & +0.001L^4) & +0.008L^2) \end{bmatrix}.$$

So to find the dynamic multipliers of wages with respect to productivity $\delta_{11,i}$, for period $i = 0, 1, 2$, we have to solve

$$0.07 - 0.013L - 0.076L^2 + 0.005L^3 + 0.02L^4$$
$$= (0.89 - 1.72L + 0.7L^2 + 0.31L^3 - 0.18L^4)(\delta_{11,0} + \delta_{11,1}L + \delta_{11,2}L^2)$$
$$= 0.89\delta_{11,0} + (0.89\delta_{11,1} - 1.72\delta_{11,0})L + (0.89\delta_{11,2} - 1.72\delta_{11,1} + 0.70\delta_{11,0})L^2$$
$$+ (-1.72\delta_{11,2} + 0.70\delta_{11,1} + 0.31\delta_{11,0})L^3$$
$$+ (0.70\delta_{11,2} + 0.31\delta_{11,1} - 0.18\delta_{11,0})L^4$$
$$+ (0.31\delta_{11,2} - 0.18\delta_{11,1})L^5 - 0.18\delta_{11,2}L^6$$

for the δ's by evaluating the polynomials for powers of L:

$$L = 0: \quad \delta_{11,0} = \frac{0.07}{0.89} = 0.079,$$

$$L = 1: \quad \delta_{11,1} = \frac{1.72\delta_{11,0} - 0.013}{0.89} = 0.138,$$

$$L = 2: \quad \delta_{11,2} = \frac{1.72\delta_{11,1} - 0.70\delta_{11,0} - 0.076}{0.89} = 0.119.$$

Bibliography

Akerlof, G. A. (1979). Irving Fisher on his head: The consequences of constant target-threshold monitoring of money holdings. *Quarterly Journal of Economics*, 93, 169–188.

Akerlof, G. A. (2002). Behavioural macroeconomics and macroeconomic behaviour. *American Economic Review*, 92(3), 411–433.

Akram, Q. F. (2004). Oil prices and exchange rates: Norwegian evidence. *Econometrics Journal*, 7, 476–504.

Akram, Q. F., G. Bårdsen, Ø. Eitrheim, and E. S. Jansen (2003). Interest rate rules in a macroeconometric of a small open economy. Unpublished paper, presented at ESEM03 in Stockholm 20–24 August 2003, Norges Bank.

Aldrich, J. (1989). Autonomy. *Oxford Economic Papers*, 41, 15–34.

Alogoskoufis, G. S. and A. Manning (1988). On the persistence of unemployment. *Economic Policy*.

Andersen, T. M. (1994). *Price Rigidity. Causes and Macroeconomic Implications*. Clarendon Press, Oxford.

Anderson, T. W. (1951). Estimating linear restrictions on regression coefficients for multivariate normal distributions. *Annals of Mathematical Statistics*, 22, 327–351.

Anderson, T. W. and H. Rubin (1949). Estimation of the parameters of a single equation in a complete system of stochastic equations. *Annals of Mathematical Statistics*, 20, 46–63.

Anderson, T. W. and H. Rubin (1950). The asymptotic properties of estimates of the parameters of a single equation in a complete system of stochastic equations. *Annals of Mathematical Statistics*, 21, 570–582.

Apel, M. and P. Jansson (1999). System estimates of potential output and the NAIRU. *Empirical Economics*, 24, 373–388.

Aukrust, O. (1977). Inflation in the open economy. A Norwegian model. In L. B. Klein and W. S. Sâlant (eds.), *World Wide Inflation. Theory and Recent Experience*, pp. 107–153. Brookings, Washington DC.

Backhouse, R. E. (1995). *Interpreting Macroeconomics. Explorations in the History of Macroeconomic Thought*. Routledge, London and New York.

Backhouse, R. E. (2000). Theory, evidence and the labour market. In R. E. Backhouse and A. Salanti (eds.), *Macroeconomics and the Real World. Volume 2: Keynesian Economics, Unemployment and Policy*, chap. 7A, pp. 145–155. Oxford University Press, Oxford.

Ball, L. (1999). Policy rules for open economies. In J. B. Taylor (ed.), *Monetary Policy Rules*, A National Bureau of Economic Research Conference Report, chap. 3, pp. 127–144. University of Chicago Press, Chicago.

Bårdsen, G. (1989). Estimation of long-run coefficients in error-correction models. *Oxford Bulletin of Economics and Statistics*, 51, 345–350.

Bårdsen, G. (1992). Dynamic modelling of the demand for narrow money in Norway. *Journal of Policy Modelling*, 14, 363–393.

Bårdsen, G. and P. G. Fisher (1999). Economic theory and econometric dynamics in modelling wages and prices in the United Kingdom. *Empirical Economics*, 24(3), 483–507.

Bårdsen, G., P. G. Fisher, and R. Nymoen (1998). Business cycles: Real facts or fallacies? In S. Strøm (ed.), *Econometrics and Economic Theory in the 20th Century: The Ragnar Frisch Centennial Symposium*, no. 32 in Econometric Society Monograph Series, chap. 16, pp. 499–527. Cambridge University Press, Cambridge.

Bårdsen, G., E. S. Jansen, and R. Nymoen (2002a). Model specification and inflation forecast uncertainty. *Annales d'Économie et de Statistique*, 67/68, 495–517.

Bårdsen, G., E. S. Jansen, and R. Nymoen (2002b). Testing the New Keynesian Phillips curve. Working paper ano 2002/5. Research Department, Norges Bank (Central Bank of Norway).

Bårdsen, G., E. S. Jansen, and R. Nymoen (2003). Econometric inflation targeting. *Econometrics Journal*, 6, 429–460.

Bårdsen, G., E. S. Jansen, and R. Nymoen (2004). Econometric evaluation of the New Keynesian Phillips curve. *Oxford Bulletin of Economics and Statistics*, 66, 611–686 (supplement).

Bårdsen, G. and J. T. Klovland (2000). Shaken or stirred? Financial deregulation and the monetary transmission mechanism in Norway. *Scandinavian Journal of Economics*, 102(4), 563–583.

Bårdsen, G. and R. Nymoen (2003). Testing steady-state implications for the NAIRU. *Review of Economics and Statistics*, 85, 1070–1075.

Barkbu, B. B., R. Nymoen, and K. Røed (2003). Wage coordination and unemployment dynamics in Norway and Sweden. *Journal of Socio-Economics*, 32, 37–58.

Bates, J. M. and C. W. J. Granger (1969). The combination of forecasts. *Operations Research Quarterly*, 20, 451–468.

Batini, N. and A. G. Haldane (1999). Forward-looking rules for monetary policy. In J. B. Taylor (ed.), *Monetary Policy Rules*, A National Bureau of Economic Research Conference Report, chap. 4, pp. 157–192. University of Chicago Press, Chicago.

Batini, N., R. Harrison, and S. P. Millard (2001). Monetary policy rules for an open economy. Working paper 149, London: Bank of England.

Batini, N., B. Jackson, and S. Nickell (2000). Inflation dynamics and the labour share in the UK. Discussion Paper 2, External MPC Unit, Bank of England.

Baumol, W. J. (1952). The transaction demand for cash: An inventory theoretic approach. *Quarterly Journal of Economics*, 66, 545–556.

Bean, C. R. (1994). European unemployment: A survey. *Journal of Economic Literature*, XXXII, 573–619.

Bernanke, B. S., T. Laubach, F. S. Mishkin, and A. S. Posen (1999). *Inflation Targeting: Lessons from International Experience*. Princeton University Press, Princeton, NJ.

Beyer, A., J. A. Doornik, and D. F. Hendry (2000). Reconstructing aggregate Euro-zone data. *Journal of Common Market Studies*, 38, 613–624.

Beyer, A., J. A. Doornik, and D. F. Hendry (2001). Constructing historical Euro-zone data. *Economic Journal*, 111, F102–F121.

Bjerkholt, O. (1998). Interaction between model builders and policy makers in the Norwegian tradition. *Economic Modelling*, 15, 317–339.

Blake, D. (1991). The estimation of rational expectations models: A survey. *Journal of Economic Studies*, 18(3), 31–70.

Blanchard, O. J. (1987). The wage price spiral. *Quarterly Journal of Economics*, 101, 543–565.

Blanchard, O. J. and S. Fisher (1989). *Lectures on Macroeconomics*. The MIT Press, Cambridge, MA.

Blanchard, O. J. and L. Katz (1997). What we know and do not know about the natural rate of unemployment. *Journal of Economic Perspectives*, 11, 51–72.

Blanchard, O. J. and L. Katz (1999). Wage dynamics: Reconciling theory and evidence. *American Economic Review*, 89, 69–74.

Blanchard, O. J. and N. Kiyotaki (1987). Monopolistic competition and the effects of aggregate demand. *American Economic Review*, 77, 647–666.

Blanchard, O. J. and L. H. Summers (1986). Hysteresis and the European unemployment problem. *NBER Macroeconomics Manual*, 1, 15–78.

Blanchflower, D. G. and A. J. Oswald (1994). *The Wage Curve*. The MIT Press, Cambridge, MA.

Bodkin, R. G., L. R. Klein, and K. Marwah (1991). *A History of Macroeconometric Model-Building*. Edward Elgar, Aldershot.

Bofinger, P. (2000). Inflation targeting: Much ado about nothing new. Working paper, University of Würzburg.

Bomhoff, E. J. (1991). Stability of velocity in the major industrial countries. *IMF Staff Papers*, 38, 626–642.

Bontemps, C. and G. E. Mizon (2003). Congruence and encompassing. In B. P. Stigum (ed.), *Econometrics and the Philosophy of Economics*, chap. 15, pp. 354–378. Princeton University Press, Princeton, NJ.

Bordo, M. D. and L. Jonung (1990). The long-run behaviour of velocity: The institutional approach revisited. *Journal of Policy Modelling*, 12, 165–197.

Bowitz, E. and Å. Cappelen (2001). Modelling incomes policies: some Norwegian experiences 1973–1993. *Economic Modelling*, 18, 349–379.

Box, G. E. P. and G. M. Jenkins (1970). *Time Series Analysis, Forecasting and Control*. Holden Day, San Francisco, CA.

Brockwell, P. J. and R. A. Davies (1991). *Time Series: Theory and Methods*, 2nd edn. Springer, New York.

Brodin, P. A. and R. Nymoen (1989). The consumption function in Norway. Breakdown and reconstruction. Working paper 1989/7, Oslo: Norges Bank.

Brodin, P. A. and R. Nymoen (1992). Wealth effects and exogeneity: The Norwegian consumption function 1966(1)–1989(4). *Oxford Bulletin of Economics and Statistics*, 54, 431–454.

Bruno, M. (1979). Price and output adjustment: Micro foundations and macro theory. *Journal of Monetary Economics*, 5, 187–212.

Bruno, M. and J. Sachs (1984). *Economics of World Wide Stagflation.* Blackwell, Oxford.

Calmfors, L. (1977). Inflation in Sweden. In L. B. Klein and W. S. Sâlant (eds.), *World Wide Inflation. Theory and Recent Experience.* Brookings, Washington DC.

Calmfors, L. and A. Forslund (1991). Real-wage determination and labour market policies: The Swedish experience. *The Economic Journal*, 101, 1130–1148.

Calmfors, L. and R. Nymoen (1990). Nordic employment. *Economic Policy*, 5(11), 397–448.

Calvo, G. A. (1983). Staggered prices in a utility maximizing framework. *Journal of Monetary Economics*, 12, 383–398.

Carlin, W. and D. Soskice (1990). *Macroeconomics and the Wage Bargain.* Oxford University Press, Oxford.

Carruth, A. A. and A. J. Oswald (1989). *Pay Determination and Industrial Prosperity.* Oxford University Press, Oxford.

Cassino, V. and R. Thornton (2002). Do changes in structural factors explain movements in the equilibrium rate of unemployment? Working paper 153, Bank of England.

Chadha, B., P. R. Masson, and G. Meredith (1992). Models of inflation and the costs of disinflation. *IMF Staff Papers*, 39, 395–431.

Chatterjee, S. (2002). The Taylor curve and the unemployment–inflation tradeoff. Business Review Q3 2002, Federal Reserve Bank of Philadelphia.

Chiang, A. C. (1984). *Fundamental Methods of Mathematical Economics*, 3rd edn. McGraw-Hill.

Chow, G. C. (1960). Tests of equality between sets of coefficients in two linear regressions. *Econometrica*, 28, 591–605.

Christ, C. F. (1966). *Econometric Models and Methods.* John Wiley, New York.

Clarida, R., J. Galí, and M. Gertler (1999). The science of monetary policy: A New Keynesian perspective. *Journal of Economic Literature*, 37(4), 1661–1707.

Clements, M. P. and D. F. Hendry (1993). On the limitations of comparing mean squared forecasts errors. *Journal of Forecasting*, 12, 669–676 (with discussion).

Clements, M. P. and D. F. Hendry (1995a). Forecasting in cointegrating systems. *Journal of Applied Econometrics*, 10, 127–147.

Clements, M. P. and D. F. Hendry (1995b). Macro-economic forecasting and modelling. *The Economic Journal*, 105, 1001–1013.

Clements, M. P. and D. F. Hendry (1996). Intercept corrections and structural breaks. *Journal of Applied Econometrics*, 11, 475–494.

Clements, M. P. and D. F. Hendry (1998). *Forecasting Economic Time Series*. Cambridge University Press, Cambridge.

Clements, M. P. and D. F. Hendry (1999a). *Forecasting Non-stationary Economic Time Series*. The MIT Press, Cambridge, MA.

Clements, M. P. and D. F. Hendry (1999b). Some methodological implications of forecast failure. Working paper, Institute of Economics and Statistics, University of Oxford.

Clements, M. P. and D. F. Hendry (2002). Modelling methodology and forecast failure. *Econometrics Journal*, 5, 319–344.

Coenen, G. and J. L. Vega (2001). The demand for M3 in the Euro area. *Journal of Applied Econometrics*, 16, 727–748.

Coenen, G. and V. Wieland (2002). Inflation dynamics and international linkages: A model of the United States, the Euro area and Japan. Working paper 181, European Central Bank.

Courbis, R. (1974). Liason internationale des prix et inflation importé. *Economie Appliquée*, 27, 205–220.

Cromb, R. (1993). A survey of recent econometric work on the NAIRU. *Journal of Economic Studies*, 20(1/2), 27–51.

Cross, R. (ed.) (1988). *Unemployment, Hysteresis and the Natural Rate Hypothesis*. Basil Blackwell, Oxford.

Cross, R. (ed.) (1995). *The Natural Rate of Unemployment. Reflections on 25 Years of the Hypothesis*. Cambridge University Press, Cambridge.

Dasgupta, A. K. (1985). *Epochs of Economic Theory*. Blackwell, Oxford.

Davidson, J. (2000). *Econometric Theory*. Blackwell, Oxford.

Davies, E. P. and L. Schøtt-Jensen (1994). Wage and price dynamics in EU-Countries: Preliminary empirical estimates. European Monetary Institute.

de Grauwe, P. and M. Polan (2001). Is inflation always and everywhere a monetary phenomenon. CEPR Discussion paper 2841, Centre for Economic Policy Research.

Desai, M. (1984). Wages, prices and unemployment a quarter of a century after the Phillips curve. In D. F. Hendry and K. F. Wallis (eds.), *Econometrics and Quantitative Economics*, chap. 9, pp. 253–273. Blackwell, Oxford.

Desai, M. (1995). *The Natural Rate of Unemployment: A Fundamentalist Keynesian View*, chap. 16, pp. 346–361. Cambridge University Press.

Doornik, J. A. (1996). Testing vector autocorrelation and heteroscedasticity in dynamic models. Working paper, Nuffield College, University of Oxford.

Doornik, J. A. and H. Hansen (1994). A practical test of multivariate normality. Unpublished paper, Nuffield College, University of Oxford.

Doornik, J. A. and D. F. Hendry (1997a). The implications for econometric modelling of forecast failure. *Scottish Journal of Political Economy*, 44, 437–461.

Doornik, J. A. and D. F. Hendry (1997b). *Modelling Dynamic Systems Using PcFiml 9 for Windows*. International Thomson Publishing, London.

Dornbusch, R. (1976). Expectations and exchange rate dynamics. *Journal of Political Economy*, 84, 1161–1176.

Drèze, J. and C. R. Bean (eds.) (1990). *Europe's Unemployment Problem; Introduction and Synthesis*. MIT Press, Cambridge.

Driehuis, W. and P. de Wolf (1976). A sectoral wage and price model for the Netherlands' economy. In H. Frisch (ed.), *Inflation in Small Countries*, pp. 283–339. Springer-Verlag, New York.

Dunlop, J. T. (1944). *Wage Determination under Trade Unions*. Reprints of Economic Classic, 1966. Augustus M. Kelley Publishers, New York.

Edgren, G., K.-O. Faxén, and C.-E. Odhner (1969). Wages, growth and distribution of income. *Swedish Journal of Economics*, 71, 133–160.

Eika, K. H., N. R. Ericsson, and R. Nymoen (1996). Hazards in implementing a monetary conditions index. *Oxford Bulletin of Economics and Statistics*, 58(4), 765–790.

Eitrheim, Ø. (1998). The demand for broad money in Norway, 1969–1993. *Empirical Economics*, 23, 339–354.

Eitrheim, Ø. (2003). Testing the role of money in the inflation process. Unpublished paper, presented at EEA03 in Stockholm 20–24 August 2003, Norges Bank.

Eitrheim, Ø., T. A. Husebø, and R. Nymoen (1999). Equilibrium-correction versus differencing in macroeconomic forecasting. *Economic Modelling*, 16, 515–544.

Eitrheim, Ø., T. A. Husebø, and R. Nymoen (2002a). Empirical comparisons of models' forecast accuracy. In M. P. Clements and D. F. Hendry (eds.), *A Companion to Economic Forecasting*, chap. 16, pp. 354–385. Blackwell, Oxford.

Eitrheim, Ø., E. S. Jansen, and R. Nymoen (2002*b*). Progress from forecast failure: The Norwegian consumption function. *Econometrics Journal*, 5, 40 64.

Eitrheim, Ø. and R. Nymoen (1991). Real wages in a multisectoral model of the Norwegian economy. *Economic Modelling*, 8(1), 63–82.

Elmeskov, J. (1994). Nordic unemployment in a European perspective. *Swedish Economic Policy Review*, 1(1–2 Autumn 1994), 27–70.

Elmeskov, J. and M. MacFarland (1993). Unemployment persistence. *OECD Economic Studies*, no. 21, 59–88.

Engle, R. F. and C. W. J. Granger (1987). Co-integration and error correction: Representation, estimation and testing. *Econometrica*, 55, 251–276.

Engle, R. F. and C. W. J. Granger (eds.) (1991). *Long-Run Economic Relationships. Readings in Cointegration.* Oxford University Press, Oxford.

Engle, R. F. and D. F. Hendry (1993). Testing super exogeneity and invariance in regression models. *Journal of Econometrics*, 56, 119–139.

Engle, R. F., D. F. Hendry, and J.-F. Richard (1983). Exogeneity. *Econometrica*, 51, 277–304.

Ericsson, N. R. (1992). Parameter constancy, mean square forecast errors, and measuring forecast performance: An exposition, extensions and illustration. *Journal of Policy Modelling*, 14, 465–495.

Ericsson, N. R. (2005). *Empirical Modeling of Economic Time Series.* Oxford University Press, Oxford (forthcoming).

Ericsson, N. R. and D. Hendry (1999). Encompassing and rational expectations: How sequential corroboration can imply refutation. *Empirical Economics*, 24(1), 1–21.

Ericsson, N. R. and D. F. Hendry (eds.) (2001). *Understanding Economic Forecasts.* MIT Press, Cambridge, MA and London, England.

Ericsson, N. R. and J. S. Irons (eds.) (1994). *Testing Exogeneity.* Oxford University Press, Oxford.

Ericsson, N. R. and J. S. Irons (1995). The Lucas critique in practice: Theory without measurement. In K. D. Hoover (ed.), *Macroeconometrics: Developments, Tensions and Prospects*, chap. 8, pp. 263–312. Kluwer Academic Publishers, Dortrecht.

Ericsson, N. R., E. S. Jansen, N. A. Kerbeshian, and R. Nymoen (1997). Understanding a Monetary Conditions Index. Paper presented at the European Meeting of the Econometric Society in Toulouse, 1997.

Estrella, A. and F. S. Mishkin (1997). Is there a role for monetary aggregates in the conduct of monetary policy? *Journal of Monetary Economics*, 40, 279–304.

Fagan, G. and J. Henry (1998). Long run money demand in the EU: Evidence for area-wide aggregates. *Empirical Economics*, 23, 483–506.

Fagan, G., J. Henry, and R. Mestre (2001). An area-wide model (AWM) for the Euro area. Working paper 42, European Central Bank.

Fair, R. C. (1984). *Specification, Estimation, and Analysis of Macroeconometric Models*. Harvard University Press, Cambridge, MA.

Fair, R. C. (1994). *Testing Macroeconometric Models*. Harvard University Press, Cambrigde, MA.

Fair, R. C. (2000). Testing the NAIRU model for the United States. *Review of Economics and Statistics*, LXXXII, 64–71.

Faust, J. and C. H. Whiteman (1995). Commentary (on Grayham E. Mizon's Progressive modelling of macroeconomic time series: The LSE methodology). In K. D. Hoover (ed.), *Macroeconometrics: Developments, Tensions and Prospects*, pp. 171–180. Kluwer Academic Publishers, Dortrecht.

Faust, J. and C. H. Whiteman (1997). General-to-specific procedures for fitting a data-admissible, theory-inspired, congruent, parsimonious, weakly-exogenous, identified, structural model to the DGP. *Carnegie-Rochester Conference Series on Economic Policy*, 47, 121–162.

Favero, C. A. (2001). *Applied Macroeconometrics*. Oxford University Press, Oxford.

Favero, C. A. and D. F. Hendry (1992). Testing the Lucas critique: A review. *Econometric Reviews*, 11, 265–306.

Ferri, P. (2000). Wage dynamics and the Phillips curve. In R. E. Backhouse and A. Salanti (eds.), *Macroeconomics and the Real World. Volume 2: Keynesian Economics, Unemployment and Policy*, chap. 5, pp. 97–111. Oxford University Press, Oxford.

Fieller, E. C. (1954). Some problems in interval estimation. *Journal of the Royal Statistical Society*, Series B, 16(2), 175–185.

Fisher, I. (1911). *The Purchasing Power of Money*. MacMillan, New York.

Forslund, A. and O. Risager (1994). Wages in Sweden. New and Old Results. Memo 1994–22, University of Aarhus, Institute of Economics.

Friedman, M. (1963). *Inflation, Causes and Consequences*. Asia Publishing House, Bombay.

Friedman, M. (1968). The role of monetary policy. *American Economic Review*, 58(1), 1–17.

Frisch, H. (1977). The Scandinavian model of inflation. A generalization and empirical evidence. *Atlantic Economic Journal*, 5, 1–14.

Frisch, R. (1938). Statistical versus theoretical relationships in economic macro-dynamics. Memorandum, League of Nations. Reprinted in *Autonomy of Economic Relationships*, Memorandum 6. November 1948, Universitetets Socialøkonomiske Institutt, Oslo.

Fuhrer, J. C. (1995). The Phillips curve is alive and well. *New England Economic Review*, March/April, 41–56.

Fuhrer, J. C. (1997). The (un)importance of forward-looking behavior in price specifications. *Journal of Money, Credit and Banking*, 29, 338–350.

Fuhrer, J. C. and G. A. Moore (1995). Inflation persistence. *Quarterly Journal of Economics*, 110, 127–159.

Galí, J. (2003). New perspectives on monetary policy, inflation, and the business cycle. In M. Dewatripont, L. P. Hansen, and S. J. Turnovsky (eds.), *Advances in Economics and Econometrics. Theory and Applications. Eight World Congress, Volume III*, chap. 5, pp. 151–197. Cambridge University Press, Cambridge.

Galí, J. and M. Gertler (1999). Inflation dynamics: A structural econometric analysis. *Journal of Monetary Economics*, 44(2), 233–258.

Galí, J., M. Gertler, and J. D. López-Salido (2001). European inflation dynamics. *European Economic Review*, 45, 1237–1270.

Gerlach, S. and L. E. O. Svensson (2003). Money and inflation in the Euro area: A case for monetary indicators? *Journal of Monetary Economics*, 50, 1649–1672.

Godfrey, L. G. (1978). Testing for higher order serial correlation when the regressors include lagged dependent variables. *Econometrica*, 46, 1303–1313.

Gordon, R. J. (1983). 'Credibility' vs. 'mainstream': Two views of the inflation process. In W. D. Nordhaus (ed.), *Inflation: Prospects and Remedies, Alternatives for the 1980s*, pp. 25–39. Center for National Policy, Washington.

Gordon, R. J. (1997). The time-varying NAIRU and its implications for economic policy. *Journal of Economic Perspectives*, 11(1), 11–32.

Gourieroux, C. and A. Monfort (1997). *Time Series and Dynamic Models*. Cambridge University Press, Cambridge, New York.

Granger, C. W. J. (1981). Some properties of time series data and their use in econometric model specification. *Journal of Econometrics*, 16, 121–130.

Granger, C. W. J. (1990). General introduction: Where are the controversies in econometric methodology? In C. W. J. Granger (ed.), *Modelling Economic Series. Readings in Econometric Methodology*, pp. 1–23. Oxford University Press, Oxford.

Granger, C. W. J. (1992). Fellow's opinion: Evaluating economic theory. *Journal of Econometrics*, 51, 3–5.

Granger, C. W. J. (1999). *Empirical Modeling in Economics. Specification and Evaluation*. Cambridge University Press, Cambridge.

Granger, C. W. J. and N. Haldrup (1997). Separation in cointegrated systems and persistent-transitory decompositions. *Oxford Bulletin of Economics and Statistics*, 59, 449–463.

Granger, C. W. J. and P. Newbold (1974). Spurious regressions in econometrics. *Journal of Econometrics*, 2, 111–120.

Granger, C. W. J. and P. Newbold (1986). *Forecasting Economic Time Series*. Academic Press, San Diego.

Gregory, A. M. and M. R. Veall (1985). Formulating Wald tests of nonlinear restrictions. *Econometrica*, 53(6), 1465–1468.

Grubb, D. (1986). Topics in the OECD Phillips curve. *Economic Journal*, 96, 55–79.

Gruen, D., A. Pagan, and C. Thompson (1999). The Phillips curve in Australia. *Journal of Monetary Economics*, 44, 223–258.

Haavelmo, T. (1944). The probability approach in econometrics. *Econometrica*, 12, 1–118 (Supplement).

Haldane, A. G. and C. K. Salmon (1995). Three issues on inflation targets. In A. G. Haldane (ed.), *Targeting Inflation*, pp. 170–201. Bank of England, London.

Hallman, J. J., R. D. Porter, and D. H. Small (1991). Is the price level tied to the M2 monetary aggregate in the long run? *American Economic Review*, 81, 841–858.

Hansen, B. E. (1992). Testing for parameter instability in linear models. *Journal of Policy Modelling*, 14, 517–533.

Hansen, B. E. (1996). Methodology: Alchemy or science. *Economic Journal*, 106, 1398–1431.

Hansen, L. P. (1982). Large sample properties of generalized method of moments estimators. *Econometrica*, 50, 1029–1054.

Heckman, J. J. (1992). Haavelmo and the birth of modern econometrics: A review of the history of econometric ideas by Mary Morgan. *Journal of Economic Literature*, 30, 876–886.

Hecq, A., F. C. Palm, and J.-P. Urbain (2002). Separation, weak exogeneity, and P-T decomposition in cointegrated VAR systems with common features. *Econometric Reviews*, 21, 273–307.

Hendry, D. F. (1988). The encompassing implications of feedback versus feedforward mechanisms in econometrics. *Oxford Economic Papers*, 40, 132–149.

Hendry, D. F. (1993*a*). *Econometrics. Alchemy or Science?* Blackwell, Oxford.

Hendry, D. F. (1993*b*). The Roles of Economic Theory and Econometrics in Time Series Economics. Paper presented at the European Meeting of the Econometric Society in Uppsala, 1993.

Hendry, D. F. (1995*a*). *Dynamic Econometrics*. Oxford University Press, Oxford.

Hendry, D. F. (1995*b*). Econometrics and business cycle empirics. *The Economic Journal*, 105, 1622–1636.

Hendry, D. F. (1997*a*). The econometrics of macro-economic forecasting. *Economic Journal*, 107, 1330–1357.

Hendry, D. F. (1997*b*): On congruent econometric relations: A comment. *Carnegie-Rochester Conference Series on Public Policy*, 47, 163–190.

Hendry, D. F. (1998). Structural breaks in economic forecasting. Mimeo Nuffield College, Oxford.

Hendry, D. F. (2001*a*). How economists forecast. In N. R. Ericsson and D. F. Hendry (eds.), *Understanding Economic Forecasts*, chap. 2, pp. 15–41. The MIT Press, Cambridge, MA.

Hendry, D. F. (2001*b*). Modelling UK inflation, 1875–1991. *Journal of Applied Econometrics*, 16, 255–275.

Hendry, D. F. (2002). Applied econometrics without sinning. *Journal of Economic Surveys*, 16, 591–614.

Hendry, D. F. and N. R. Ericsson (1991). Modeling the demand for narrow money in the United Kingdom and the United States. *European Economic Review*, 35, 833–886.

Hendry, D. F. and H.-M. Krolzig (1999). Improving on 'Data Mining Reconsidered' by K. D. Hoover and S. J. Perez. *Econometrics Journal*, 2, 41–58.

Hendry, D. F. and H.-M. Krolzig (2001). *Automatic Econometric Model Selection Using PcGets*. Timberlake Consultants Ltd, London.

Hendry, D. F. and G. E. Mizon (1993). Evaluating dynamic econometric models by encompassing the VAR. In P. C. B. Phillips (ed.), *Models, Methods and Applications of Econometrics*, chap. 18, pp. 272–300. Blackwell, Oxford.

Hendry, D. F. and G. E. Mizon (2000). Reformulating empirical macroeconometric modelling. *Oxford Review of Economic Policy*, 16(4), 138–157.

Hendry, D. F. and M. S. Morgan (1995). *The Foundations of Econometric Analysis*. Cambridge University Press, Cambridge.

Hendry, D. F. and A. Neale (1988). Interpreting long-run equilibrium solutions in conventional macro models: A comment. *Economic Journal*, 98, 809–817.

Hendry, D. F. and J. F. Richard (1982). On the formulation of empirical models in dynamic econometrics. *Journal of Econometrics*, 20, 3–33.

Hendry, D. F. and J. F. Richard (1983). The econometric analysis of economic time series. *International Statistical Review*, 51, 111–163.

Hendry, D. F. and J. F. Richard (1989). Recent developments in the theory of encompassing. In B. Cornet and H. Tulkens (eds.), *Contributions to Operations Research and Econometrics. The XXth Anniversary of CORE*, chap. 12, pp. 393–440. MIT Press, Cambridge, MA.

Hendry, D. F., A. Spanos, and N. R. Ericsson (1989). The contributions to econometrics in Trygve Haavelmo's The probability approach in econometrics. *Sosialøkonomen, 43*(11), 12–17.

Hodrick, R. J. and E. C. Prescott (1997). Postwar U.S. business cycles: An empirical investigation. *Journal of Money, Credit and Banking*, 29, 1–16.

Hoel, M. and R. Nymoen (1988). Wage formation in Norwegian manufacturing. An empirical application of a theoretical bargaining model. *European Economic Review*, 32, 977–997.

Holden, S. (1990). Wage drift in Norway: A bargaining approach. In L. Calmfors (ed.), *Wage Formation and Macroeconomic Policy in the Nordic Countries*, chap. 7. Oxford University Press, Oxford. (With comment by T. Eriksson and P. Skedinger.)

Holden, S. (2003). Wage setting under different monetary regimes. *Economica*, 70, 251–266.

Holden, S. and R. Nymoen (2002). Measuring structural unemployment: NAWRU estimates in the Nordic countries. *The Scandinavian Journal of Economics*, 104(1), 87–104.

Hoover, K. D. (1991). The causal directions between money and prices. *Journal of Monetary Economics*, 27, 381–423.

Hoover, K. D. and S. J. Perez (1999). Data mining reconsidered: Encompassing and the general-to-specific approach to specification search. *Econometrics Journal*, 2, 1–25.

Hunt, B., D. Rose, and A. Scott (2000). The core model of the Reserve Bank of New Zealand's Forecasting and Policy System. *Economic Modelling*, 17, 247–274.

Jacobson, T., P. Jansson, A. Vredin, and A. Warne (2001). Monetary policy analysis and inflation targeting in a small open economy: A VAR approach. *Journal of Applied Econometrics*, 16, 487–520.

Jansen, E. S. (2002). Statistical issues in macroeconomic modelling (with discussion). *Scandinavian Journal of Statistics*, 29, 193–217.

Jansen, E. S. (2004). Modelling inflation in the Euro area. Working paper 322, European Central Bank.

Jansen, E. S. and T. Teräsvirta (1996). Testing parameter constancy and super exogeneity in econometric equations. *Oxford Bulletin of Economics and Statistics*, 58, 735–763.

Johansen, K. (1995a). Norwegian wage curves. *Oxford Bulletin of Economics and Statistics*, 57, 229–247.

Johansen, L. (1977). *Lectures on Macroeconomic Planning. Volume 1. General Aspects*. North-Holland, Amsterdam.

Johansen, L. (1982). Econometric models and economic planning and policy. Some trends and developments. In M. Hazewinkel and A. H. G. Rinnooy Kan (eds.), *Current Developments in the Interface: Economics, Econometrics, Mathematics*, chap. 5, pp. 91–122. D. Reidel Publishing Company, Dordrecht.

Johansen, S. (1988). Statistical analysis of cointegration vectors. *Journal of Economic Dynamics and Control*, 12, 231–254.

Johansen, S. (1991). Estimation and testing of cointegrating vectors in Gaussian vector autoregressive models. *Econometrica*, 59, 1551–1580.

Johansen, S. (1992). Cointegration in partial systems and the efficiency of single-equation analysis. *Journal of Econometrics*, 52, 389–402.

Johansen, S. (1995*b*). *Likelihood-Based Inference in Cointegrated Vector Autoregressive Models*. Oxford University Press, Oxford.

Johansen, S. (2002). Discussion (of E. S. Jansen: Statistical issues in macroeconomic modeling). *Scandinavian Journal of Statistics*, 29, 213–216.

Juselius, K. (1992). Domestic and foreign effects on prices in an open economy: The case of Denmark. *Journal of Policy Modeling*, 14, 401–428.

Kendall, M. G. and A. Stuart (1973). *The Advanced Theory of Statistics: Volume 2, Inference and Relationship*, 3rd edn. Charles Griffin and Company, London.

Keuzenkamp, H. A. and J. R. Magnus (1995). On tests and significance in econometrics. *Journal of Econometrics*, 67, 5–24.

King, M. (1998). The Employment Policy Institute's Fourth Annual Lecture: Lessons from the UK labour market. BIS Review 103, Bank for International Settlement.

King, R. G., C. I. Plosser, J. H. Stock, and M. W. Watson (1991). Stochastic trends and economic fluctuations. *American Economic Review*, 81, 819–840.

Klein, L. R. (1950). *Economic Fluctuations in the United States 1921–1941*. Cowles Commission Monograph 11. Wiley, New York.

Klein, L. R. (1953). *A Textbook of Econometrics*. Row, Peterson & Co, Evanston, IL.

Klein, L. R. (1983). *Lectures in Econometrics*. North-Holland, Amsterdam.

Klein, L. R. (1988). The statistical approach to economics. *Journal of Econometrics*, 37, 7–26.

Klein, L. R., A. Welfe, and W. Welfe (1999). *Principles of Macroeconometric Modeling*. North-Holland, Amsterdam.

Klovland, J. T. (1983). The demand for money in the secular perspective: The case of Norway, 1867–1980. *European Economic Review*, 22, 193–218.

Kolsrud, D. and R. Nymoen (1998). Unemployment and the open economy wage–price spiral. *Journal of Economic Studies*, 25, 450–467.

Koopmans, T. C. and W. C. Hood (1953). The estimation of simultaneous linear economic relationships. In W. C. Hood and T. C. Koopmans (eds.), *Studies in Econometric Method*. Cowles Commission Monograph 14, chap. 6, pp. 112–199. Wiley, New York.

Koopmans, T. C., H. Rubin, and R. B. Leibnik (1950). Measuring the equation systems of dynamic economics. In T. C. Koopmans (ed.), *Statistical Inference in Dynamic Economic Models*. Cowles Commission Monograph 10, chap. 2, pp. 53–237. Wiley, New York.

Kremers, J. J. M., N. R. Ericsson, and J. J. Dolado (1992). The power of cointegration tests. *Oxford Bulletin of Economics and Statistics*, 54, 325–348.

Kydland, F. E. and E. C. Prescott (1991). The econometrics of the general equilibrium approach to business cycles. *Scandinavian Journal of Economics*, 93, 161–178.

Layard, R. and S. Nickell (1986). Unemployment in Britain. *Economica*, 53, 121–166 (Special issue).

Layard, R., S. Nickell, and R. Jackman (1991). *Unemployment*. Oxford University Press, Oxford.

Layard, R., S. Nickell, and R. Jackman (1994). *The Unemployment Crises*. Oxford University Press, Oxford.

Leamer, E. E. (1983). Lets take the 'Con' out of econometrics. *American Economic Review*, 73, 31–43.

Levin, A., V. Wieland, and J. C. Williams (2003). The performance of forecast-based monetary policy rules under model uncertainty. *American Economic Review*, 93(3), 622–645.

Lindbeck, A. (1993). *Unemployment and Macroeconomics*. The MIT Press, Cambridge, MA.

Lipsey, R. G. (1960). The relationship between unemployment and the rate of change in money wages in the United Kingdom 1862–1957: A further analysis. *Economica*, 27, 1–31.

Lucas, R. E., Jr. (1972). Expectations and the neutrality of money. *Journal of Economic Theory*, 4, 103–124.

Lucas, R. E., Jr. (1976). Econometric policy evaluation: A critique. *Carnegie-Rochester Conference Series on Public Policy*, 1, 19–46.

Lucas, R. E., Jr. and L. A. Rapping (1969). Real wages, employment and inflation. *Journal of Political Economy*, 77, 721–754.

Lucas, R. E., Jr. and L. A. Rapping (1970). Price expectations and the Phillips curve. *American Economic Review*, 59, 342–349.

Lütkepohl, H., Berlin. (1991). *Introduction to Multiple Time Series Analysis*. Springer-Verlag, Berlin.

MacKinnon, J. G. (1991). Critical values for cointegration tests. In R. F. Engle and C. W. J. Granger (eds.), *Long-Run Economic Relationships: Readings in Cointegration*, chap. 13, pp. 267–276. Oxford University Press, Oxford.

Manning, A. (1993). Wage bargaining and the Phillips curve: The identification and specification of aggregate wage equations. *The Economic Journal*, 103, 98–117.

Mavroeidis, S. (2002). *Econometric Issues in Forward-Looking Monetary Models*. Ph.D. thesis, Nuffield College, Oxford University.

Miller, M. H. and D. Orr (1966). A model of the demand for money by firms. *Quarterly Journal of Economics*, 80, 413–434.

Mizon, G. M. (1995). Progressive modelling of macroeconomic time series: The LSE methodology. In K. D. Hoover (ed.), *Macroeconometrics: Developments, Tensions and Prospects*, chap. 4, pp. 107–170. Kluwer Academic Publishers, Dortrecht.

Mizon, G. M. and J. F. Richard (1986). The encompassing principle and its application to testing non-nested hypotheses. *Econometrica*, 54, 657–678.

Morgan, M. S. (1990). *The History of Econometric Ideas*. Cambridge University Press, Cambridge.

Nickell, S. (1987). Why is wage inflation in Britain so high? *Oxford Bulletin of Economics and Statistics*, 49, 103–128.

Nickell, S. (1993). Unemployment: A survey. *Economic Journal*, 100(401), 391–439.

Nickell, S. J. and M. J. Andrews (1983). Unions, real-wages and employment in Britain 1951–79. *Oxford Economic Papers (Supplement)*, 35, 183–206.

Nicoletti Altimari, S. (2001). Does money lead inflation in the Euro area? Working paper 63, European Central Bank.

Nilsson, C. (2002). Rixmod—The Riksbank's macroeconomic model for monetary policy analysis. *Sveriges Riksbank Economic Review*, 2002(2), 46–71.

Nordhaus, W. D. (1972). The world wide wage explosion. *Brookings Papers on Economic Activity*, 2, 431–464.

Nymoen, R. (1989*a*). Modelling wages in the small open economy: An error-correction model of Norwegian manufacturing wages. *Oxford Bulletin of Economics and Statistics*, 51, 239–258.

Nymoen, R. (1989*b*). Wages and the length of the working day. An empirical test based on Norwegian quarterly manufacturing data. *Scandinavian Journal of Economics*, 91, 599–612.

Nymoen, R. (1990). *Empirical Modelling of Wage–Price Inflation and Employment Using Norwegian Quarterly Data*. Ph.D. thesis, University of Oslo.

Nymoen, R. (1991). A small linear model of wage- and price-inflation in the Norwegian economy. *Journal of Applied Econometrics*, 6, 255–269.

Nymoen, R. (1992). Finnish manufacturing wages 1960–1987: Real-wage flexibility and hysteresis. *Journal of Policy Modelling*, 14, 429–451.

Nymoen, R. (2002). Faulty watch towers—'Structural' models in Norwegian monetary policy analysis. Unpublished manuscript, University of Oslo.

Nymoen, R. and A. Rødseth (2003). Explaining unemployment: Some lessons from Nordic wage formation. *Labor Economics*, 10, 1–29.

OECD (1997*a*). *Economic Survey for Norway*. OECD, Paris.

OECD (1997*b*). *Employment Outlook*. July 1997. OECD, Paris.

Olsen, K. and F. Wulfsberg (2001). The role of assessments and judgement in the macroeconomic model RIMINI. *Economic Bulletin (Norges Bank)*, 72, 55–64.

Orphanides, A. (2001). Monetary policy rules based on real-time data. *American Economic Review*, 91, 964–985.

Orphanides, A. (2003). The quest for prosperity without inflation. *Journal of Monetary Economics*, 50, 663–693.

Pagan, A. (2003). Report on modelling and forecasting at the Bank of England. *Bank of England Quarterly Bulletin* (Spring), 1–29.

Pesaran, M. H. (1987). *The Limits to Rational Expectations*. Blackwell, Oxford.

Pesaran, M. H. and R. P. Smith (1998). Structural analysis of cointegrating VARs. *Journal of Economic Surveys*, 12, 471–505.

Phelps, E. S. (1967). Phillips curves, expectations and inflation, and optimal unemployment over time. *Economica*, 34, 254–281.

Phelps, E. S. (1968). Money–wage dynamics and labour market equilibrium. *Journal of Political Economy*, 76, 678–711.

Phelps, E. S. (1978). Disinflation without recession: Adaptive guideposts and monetary policy. *Weltwirtschaftliches Archiv*, 100, 239–265.

Phelps, E. S. (1995). The origins and further developments of the natural rate of unemployment. In R. Cross (ed.), *The Natural Rate of Unemployment. Reflections on 25 Years of the Hypothesis*, chap. 2, pp. 15–31. Cambridge University Press, Cambridge.

Phillips, A. W. (1958). The Relation between Unemployment and the Rate of Change of Money Wage Rates in the United Kingdom, 1861–1957. *Economica*, 25, 283–299.

Poloz, S., D. Rose, and R. Tetlow (1994). The Bank of Canada's New Quarterly Projection Model (QPM): An introduction—Le Nouveau Modèle Trimesteriel de Prévision (MTP) de la Banque Du Canada: Un Aperçu. *Bank of Canada Review—Revue de la Banque du Canada, 1994* (Autumn), 23–28.

Qvigstad, J. F. (1975). Noen emner fra inflasjonsteorien (Topics from the theory of inflation). Memorandum 13.2.1975, Institute of Economics University of Oslo. Notes from Professor Haavelmo's lectures spring 1974.

Rasche, R. H. (1987). M1-velocity and money-demand functions: Do stable relationships exist? *Carnegie-Rochester Conference Series on Public Policy*, 27, 9–88.

Richardson, P., L. Boone, L. Giorno, C. Meacci, M. Rae, and D. Turner (2000). The concept, policy use and measurement of structural unemployment: Estimating a time varying NAIRU across 21 OECD countries. Economics Department Working paper 250, OECD.

Roberts, J. M. (1995). New Keynesian economics and the Phillips curve. *Journal of Money, Credit and Banking*, 27, 975–984.

Roberts, J. M. (2001). How well does the New Keynesian sticky-price model fit the data? Finance and Economics Discussion Series 13, Federal Reserve Board of Governors.

Rødseth, A. (2000). *Open Economy Macroeconomics*. Cambridge University Press, Cambridge.

Rødseth, A. and S. Holden (1990). Wage formation in Norway. In L. Calmfors (ed.), *Wage Formation and Macroeconomic Policy in the Nordic Countries*, chap. 5, pp. 237–280. Oxford University Press, Oxford.

Røed, K. (1994). Hysteresis versus persistence: Does it make any difference? Unpublished paper, Economics Department, University of Oslo.

Røisland, Ø. and R. Torvik (2004). Exchange Rate versus Inflation Targeting: A Theory of Output Fluctuations in Traded and Non-Traded Sectors. *Journal of International Trade and Economic Development*, 13, 265–285.

Romer, D. (1996). *Advanced Macroeconomics*. McGraw-Hill, New York.

Rorty, R. (1984). The histography of philosophy: Four genres. In R. Rorty, J. B. Schneewind, and Q. Skinner (eds.), *Philosophy in History*, chap. 3, pp. 49–75. Cambridge University Press, Cambridge.

Rotemberg, J. J. (1982). Sticky prices in the United States. *Journal of Political Economy*, 90, 1187–1211.

Rowlatt, P. A. (1992). *Inflation*. Chapman and Hall, London.

Rowthorn, R. E. (1977). Conflict, inflation and money. *Cambridge Journal of Economics*, 1, 215–239.

Royal Swedish Academy of Science (1990). The Nobel Memorial Prize in Economics. Press release from the Royal Swedish Academy of Sciences. *Scandinavian Journal of Economics*, 92, 11–15.

Rudd, J. and K. Whelan (2004). New tests of the New Keynesian Phillips curve. *Journal of Monetary Economics* (forthcoming).

Samuelson, P. A. (1941). The stability of equilibrium: Comparative statics and dynamics. *Econometrica*, 9, 97–120.

Samuelson, P. A. and R. M. Solow (1960). Analytical aspects of anti inflation policy. *American Economic Review*, 50, 177–194.

Sargan, J. D. (1964). Wages and prices in the United Kingdom: A study of econometric methodology. In P. E. Hart, G. Mills, and J. K. Whitaker (eds.), *Econometric Analysis for National Economic Planning*, pp. 25–63. Butterworth Co., London.

Sargan, J. D. (1980). A model of wage–price inflation. *Review of Economic Studies*, 47, 113–135.

Sargan, J. D. (1988). *Lectures on Advanced Econometric Theory*. Blackwell, Oxford.

Sargent, T. J. (1987). *Macroeconomic Theory*, 2nd edn. Academic Press, Orlando.

Scarpetta, S. (1996). Assessing the role of labour markets policies and institutional settings on unemployment: A cross-country study. *OECD Economic Studies*, no. 26, 43–98.

Sgherri, S. and K. F. Wallis (1999). Policy Analysis with Macroeconometric Models: Inflation Targetry in a Small Structural Model. Unpublished manuscript, Department of Economics, University of Warwick.

Siklos, P. L. (1993). Income velocity and institutional change: Some new time series evidence, 1870–1986. *Journal of Money, Credit and Banking*, 25, 377–392.

Silvey, S. D. (1975). *Statistical Inference*. Chapman and Hall, London.

Sims, C. A. (1980). Macroeconomics and reality. *Econometrica*, 48, 1–48.

Smets, F. and R. Wouters (2003). An estimated dynamic stochastic general equilibrium model of the Euro area. *Journal of the European Economic Association*, 1, 1123–1175.

Smith, A. A. (1993). Estimating nonlinear time-series models using simulated vector autoregressions. *Journal of Applied Econometrics*, 8, S63–S84. (Special issue: Econometric inference using simulation techniques.)

Solow, R. M. (1986). Unemployment: Getting the questions right. *Economica*, 53, S23–S34 (Special Issue).

Spanos, A. (1989). On rereading Haavelmo: A retrospective view of econometric modeling. *Econometric Theory*, 5, 405–429.

Staiger, D., J. H. Stock, and M. W. Watson (1997). The NAIRU, unemployment and monetary policy. *Journal of Economic Perspectives*, 11, 33–49.

Staiger, D., J. H. Stock, and M. W. Watson (2001). Prices, wages and the U.S. NAIRU in the 1990s. In A. B. Kruger and R. M. Solow (eds.), *The Roaring Nineties*, chap. 1, pp. 3–60. Russell Sage Foundation, New York.

Stanley, T. D. (2000). An empirical critique of the Lucas critique. *Journal of Socio-Economics*, 29, 91–107.

Stock, J. H., J. H. Wright, and M. Yogo (2002). A survey of weak instruments and weak identification in generalized method of moments. *Journal of Business and Economic Statistics*, 20(4), 518–529.

Stølen, N. M. (1990). Is there a Nairu in Norway. Working paper 56, Central Bureau of Statistics.

Stølen, N. M. (1993). *Wage Formation and the Macroeconometric Functioning of the Norwegian Labour Market*. Ph.D. thesis, University of Oslo.

Summers, L. H. (1991). The scientific illusions in empirical macroeconomics. *Scandinavian Journal of Economics*, 93, 129–148.

Svensson, L. E. O. (2000). Open economy inflation targeting. *Journal of International Economics*, 50, 155–183.

Taylor, J. B. (1979a). Estimation and control of a macroeconomic model with rational expectations. *Econometrica*, 47, 1267–1286.

Taylor, J. B. (1979b). Staggered wage setting in a macro model. *American Economic Review*, 69, 108–113.

Taylor, J. B. (1980). Aggregate dynamics and staggered contracts. *Journal of Political Economy*, 88, 1–23.

Taylor, J. B. (1999). Introduction. In J. B. Taylor (ed.), *Monetary Policy Rules, A National Bureau of Economic Research Conference Report*, pp. 1–14. University of Chicago Press, Chicago.

Teräsvirta, T. (1998). Modelling economic relationships with smooth transition regressions. In A. Ullah and D. E. Giles (eds.), *Handbook of Applied Economic Statistics*, chap. 15, pp. 507–552. Marcel Dekker, Inc., New York.

Tinbergen, J. (1937). *An Econometric Approach to Business Cycle Problems*. Hermann & Cie, Paris.

Tintner, G. (1952). *Econometrics*. Wiley, New York.

Tobin, J. (1956). The interest-elasticity of transaction demand for cash. *Review of Economics and Statistics*, 38, 241–247.

Tödter, K.-H. and H.-E. Reimers (1994). P-Star as a link between money and prices in Germany. *Weltwirtschaftliches Archiv*, 130, 273–289.

Trecroci, C. and J. L. Vega (2002). The information content of M3 for future inflation. *Weltwirtschaftliches Archiv*, 138, 22–53.

Wald, A. (1943). Tests of statistical hypotheses concerning several parameters when the number of observations is large. *Transactions of the American Mathematical Society*, 54, 426–482.

Wallis, K. F. (1977). Multiple time series analysis and the final form of econometric models. *Econometrica*, 45, 1481–1497.

Wallis, K. F. (1989). Macroeconomic forecasting. A survey. *Economic Journal*, 99, 28–61.

Wallis, K. F. (1993). On macroeconomic policy and macroeconometric models. *The Economic Record*, 69, 113–130.

Wallis, K. F. (1994). Introduction. In K. F. Wallis (ed.), *Macroeconometric Modelling, Volume 2*. Edward Elgar, Aldershot.

Wallis, K. F. (1995). Large-scale macroeconometric modeling. In M. H. Pesaran and M. R. Wickens (eds.), *Handbook of Applied Econometrics. Volume I: Macroeconomics*, chap. 6, pp. 312–355. Blackwell, Oxford.

Wallis, K. F., M. J. Andrews, D. N. F. Bell, P. G. Fisher, and J. D. Whitley (1984). *Models of the UK Economy. A Review by the ESRC Macroeconomic Modelling Bureau*. Oxford University Press, Oxford.

Walsh, C. E. (1999). Monetary policy trade-offs in the open economy. Unpublished.

Walsh, C. E. (2003). Speed limit policies: The output gap and optimal monetary policy. *American Economic Review*, 93(1), 265–278.

White, H. (1980). A heteroskedasticity-consistent covariance matrix estimator and a direct test of heteroskedasticity. *Econometrica*, 48, 817–838.

Willman, A., M. Kortelainen, H.-L. Männistö, and M. Tujula (2000). The BOF5 macroeconomic model of Finland, structure and dynamic microfoundations. *Economic Modelling*, 17, 275–303.

Woodford, M. (2000). Pitfalls of forward-looking monetary policy. *American Economic Review*, 90, 100–104.

Woodford, M. (2003). *Interest and Prices. Foundations of a Theory of Monetary Policy*. Princeton University Press, Princeton, NJ.

Wren-Lewis, S. and R. Moghadam (1994). Are wages forward looking? *Oxford Economic Papers*, 46, 403–424.

Wright, S. (1992). Equilibrium real-exchange rates. *The Manchester School*, LX, 63–84.

Author Index

Subscripts

a - the reference is abbreviated (e.g. GG for Galí and Gertler (1999) in
    Chapter 7)

n - the reference is in a footnote

c - the reference is to a co-author of a paper with more than two authors
    (e.g. Wallis et. al. (1984))

c,n - the reference is to a co-author of a paper with more than two authors
    occuring in a footnote

Akerlof, G. A. 5, 149, 150, 303
Akram, Q. F. 207, 208, 226n, 303
Aldrich, J. 28, 303
Alogoskoufis, G. S. 77, 303
Andersen, T. M. 92, 303
Anderson, T. W. 22, 32, 69n, 205n, 303
Andrews, M. J. 73, 142, 319, 324
Apel, M. 109, 303
Aukrust, O. 35, 36, 36n, 37, 38, 39, 40,
       41, 42, 43, 45, 45n, 47, 48, 73, 84,
       87, 95, 100, 137, 304

Backhouse, R. E. 43, 53, 109, 304, 311
Ball, L. 199, 228, 304
Bårdsen, G. 3, 10, 63, 64, 86, 89, 107,
       113, 128n, 134n, 136, 137, 143,
       143n, 144, 144n, 145, 182, 189,
       201n, 203, 210, 214, 215, 219, 223,
       226c,n, 268, 299, 303, 304, 305
Barkbu, B. B. 66, 305
Bates, J. M. 246, 305
Batini, N. 128, 129, 141, 142, 143, 143n,
       199, 228, 305

Baumol, W. J. 149, 305
Bean, C. R. 79, 105, 106, 106n, 109, 268,
       305, 309
Bell, D. N. F. 72c,n, 324
Bernanke, B. S. 199, 305
Beyer, A. 161n, 305
Bjerkholt, O. 11n, 36n, 305
Blake, D. 135, 291, 305
Blanchard, O. J. 8, 54, 73, 80, 90, 106,
       137, 142, 268, 286, 305, 306
Blanchflower, D. G. 8, 85, 306
Bodkin, R. G. 18, 21, 306
Bofinger, P. 165, 306
Bomhoff, E. J. 149, 306
Bontemps, C. 28, 306
Boone, L. 109c, 321
Bordo, M. D. 149, 306
Bowitz, E. 273, 306
Box, G. E. P. 21, 306
Brockwell, P. J. 10, 37, 59n, 130, 306
Brodin, P. A. 29, 29a, 30a, 31a, 32a,
       259n, 306
Bruno, M. 73, 306, 307

327

Subject Index